AF794788

Hybrid Materials for Piezoelectric Energy Harvesting and Conversion

Edited by

S. Wazed Ali
Indian Institute of Technology Delhi
New Delhi, India

Satyaranjan Bairagi
University of Glasgow
Glasgow, UK

Shahid Ul Islam
Jamia Millia Islamia
New Delhi, India

WILEY

Copyright © 2024 by John Wiley & Sons, Inc. All rights reserved.

Published by John Wiley & Sons, Inc., Hoboken, New Jersey.
Published simultaneously in Canada.

No part of this publication may be reproduced, stored in a retrieval system, or transmitted in any form or by any means, electronic, mechanical, photocopying, recording, scanning, or otherwise, except as permitted under Section 107 or 108 of the 1976 United States Copyright Act, without either the prior written permission of the Publisher, or authorization through payment of the appropriate per-copy fee to the Copyright Clearance Center, Inc., 222 Rosewood Drive, Danvers, MA 01923, (978) 750-8400, fax (978) 750-4470, or on the web at www.copyright.com. Requests to the Publisher for permission should be addressed to the Permissions Department, John Wiley & Sons, Inc., 111 River Street, Hoboken, NJ 07030, (201) 748-6011, fax (201) 748-6008, or online at http://www.wiley.com/go/permission.

Trademarks: Wiley and the Wiley logo are trademarks or registered trademarks of John Wiley & Sons, Inc. and/or its affiliates in the United States and other countries and may not be used without written permission. All other trademarks are the property of their respective owners. John Wiley & Sons, Inc. is not associated with any product or vendor mentioned in this book.

Limit of Liability/Disclaimer of Warranty: While the publisher and author have used their best efforts in preparing this book, they make no representations or warranties with respect to the accuracy or completeness of the contents of this book and specifically disclaim any implied warranties of merchantability or fitness for a particular purpose. No warranty may be created or extended by sales representatives or written sales materials. The advice and strategies contained herein may not be suitable for your situation. You should consult with a professional where appropriate. Further, readers should be aware that websites listed in this work may have changed or disappeared between when this work was written and when it is read. Neither the publisher nor authors shall be liable for any loss of profit or any other commercial damages, including but not limited to special, incidental, consequential, or other damages.

For general information on our other products and services or for technical support, please contact our Customer Care Department within the United States at (800) 762-2974, outside the United States at (317) 572-3993 or fax (317) 572-4002.

Wiley also publishes its books in a variety of electronic formats. Some content that appears in print may not be available in electronic formats. For more information about Wiley products, visit our web site at www.wiley.com.

Library of Congress Cataloging-in-Publication Data Applied for:

Hardback ISBN: 9781394150342

Cover Design: Wiley
Cover Image: © filo/Getty Images

Set in 9.5/12.5pt STIXTwoText by Straive, Chennai, India

Contents

List of Contributors

Ramasamy Alagirusamy
Department of Textile and Fibre Engineering
Indian Institute of Technology Delhi
Hauz Khas, New Delhi
India

S. Wazed Ali
Department of Textile and Fibre Engineering
Indian Institute of Technology Delhi
Hauz Khas, New Delhi
India

and

School of Interdisciplinary Research (SIRe)
Indian Institute of Technology Delhi
Hauz Khas, New Delhi
India

Akshaya Aliyana
Research Institute for Flexible Materials, School of Textiles and Design
Heriot-Watt University
Edinburgh
UK

Arunachalakasi Arockiarajan
Department of Applied Mechanics
Indian Institute of Technology Madras (IIT Madras)
Chennai, Tamil Nadu
India

and

Ceramic Technology Group-Center of Excellence in Materials and Manufacturing Futuristic Mobility
Indian Institute of Technology Madras (IIT Madras)
Chennai, Tamil Nadu
India

Aiswarya Baburaj
Department of Electronics
Mangalore University
Mangaluru
India

Satyaranjan Bairagi
Materials and Manufacturing Research Group
James Watt School of Engineering, University of Glasgow
Scotland
UK

Sourav Banerjee
School of Interdisciplinary Research (SIRe)
Indian Institute of Technology Delhi
Hauz Khas, New Delhi
India

Swagata Banerjee
Department of Textile and Fibre Engineering
Indian Institute of Technology Delhi
Hauz Khas, New Delhi
India

Dipankar Chattopadhyay
Department of Polymer Science and Technology
University of Calcutta
Kolkata, West Bengal
India

and

Center for Research in Nanoscience and Nanotechnology
Acharya Prafulla Chandra Roy Sikhsha Prangan
University of Calcutta
Kolkata
India

Ipsita Chinya
Department of Chemical Engineering
Dr. V R Godhania College of Engineering and Technology
Porbandar, Gujarat
India

and

Academy of Scientific and Innovative Research
CSIR-CGCRI
Kolkata, West Bengal
India

and

Functional Material and Device Division
CSIR-Central Glass & Ceramic Research Institute
Kolkata, West Bengal
India

Anupam Chowdhury
Department of Textile and Fibre Engineering
Indian Institute of Technology Delhi
Hauz Khas, New Delhi
India

Srijan Das
Department of Textile and Fibre Engineering
Indian Institute of Technology Delhi
Hauz Khas, New Delhi
India

Sheer Khanyisile Dhlamini
Department of Chemical Sciences
University of Johannesburg
Johannesburg
South Africa

and

DST-CSIR National Centre for Nanostructured Materials
Council for Scientific and Industrial Research
Pretoria
South Africa

Sahal Ebrahim
Department of Nanoscience and Technology
University of Calicut
Kerala
India

Adrija Ghosh
Department of Polymer Science and Technology
University of Calcutta
Kolkata, West Bengal
India

Priyanka Gupta
Department of Textile and Fibre Engineering
Indian Institute of Technology Delhi
Hauz Khas, New Delhi
India

Mukta Nitin Mirlekar
Department of Textile and Fibre Engineering
Indian Institute of Technology Delhi
Hauz Khas, New Delhi
India

Satya Narayan Naik
Centre for Rural Development and Technology
Indian Institute of Technology Delhi
Hauz Khas, New Delhi
India

and

School of Interdisciplinary Research (SIRe)
Indian Institute of Technology Delhi
Hauz Khas, New Delhi
India

Naveen Kumar Sindhughatta Kalaiah
Department of Electronics
Mangalore University
Mangaluru
India

Jonathan Tersur Orasugh
Department of Chemical Sciences
University of Johannesburg
Johannesburg
South Africa

and

DST-CSIR National Centre for Nanostructured Materials
Council for Scientific and Industrial Research
Pretoria
South Africa

Deepalekshmi Ponnamma
Materials Science and Technology Program
Department of Mathematics, Statistics and Physics
College of Arts and Sciences
Qatar University
Doha
Qatar

Suprakash Sinha Ray
Department of Chemical Sciences
University of Johannesburg
Johannesburg
South Africa

and

DST-CSIR National Centre for Nanostructured Materials
Council for Scientific and Industrial Research
Pretoria
South Africa

Abhishek Sasmal
Functional Materials and Devices Division
CSIR-Central Glass & Ceramic Research Institute
Kolkata, West Bengal
India

and

Department of Applied Mechanics
Indian Institute of Technology Madras (IIT Madras)
Chennai, Tamil Nadu
India

Shrabanee Sen
Functional Materials and Devices Division
CSIR-Central Glass & Ceramic Research Institute
Kolkata, West Bengal
India

Chirantan Shee
Department of Textile and Fibre Engineering
Indian Institute of Technology Delhi
Huaz Khas, New Delhi
India

Ankur Shukla
Indian Institute of Technology Delhi
Hauz Khas, New Delhi
India

Mayuri Srivastava
School of Interdisciplinary Research
Department of Textile and Fibre Engineering
Indian Institute of Technology Delhi
New Delhi
India

Preface

The field of materials science has continually evolved to meet the ever-growing demands of modern technology. As we stand on the precipice of a new era characterized by the fusion of diverse disciplines and the relentless pursuit of sustainability, it is increasingly evident that the future of materials lies in their ability to be both versatile and environmentally friendly. In this context, the realm of piezoelectric materials has emerged as a cornerstone of innovation, offering a unique combination of properties that make them exceptionally promising for a wide range of applications.

Hybrid Materials for Piezoelectric Energy Harvesting and Conversion represents a comprehensive exploration of the cutting-edge developments and transformative potential of piezoelectric materials in the context of hybrid materials and sustainable technologies. Piezoelectric materials, known for their ability to convert mechanical energy into electrical energy and vice versa, have found their way into an array of applications, from sensors and energy harvesters to actuators and medical devices. Their versatility is undeniable, but it is their integration into hybrid systems and the ingenious combinations with other materials that hold the key to unlocking new horizons in science and engineering.

As the world faces unprecedented challenges related to energy generation, environmental sustainability, and the quest for multifunctional materials, the fusion of piezoelectric materials with other novel materials becomes imperative. The amalgamation of piezoelectricity with materials such as polymers, composites, nanoparticles, and even biological substances has paved the way for revolutionary breakthroughs. The resulting hybrid materials not only inherit the piezoelectric properties but also acquire additional functionalities, such as enhanced mechanical properties, biocompatibility, and tailored electrical conductivities.

In this book, we embark on a captivating journey through the diverse landscape of hybrid piezoelectric materials. We delve into the fundamental principles governing piezoelectricity, exploring the physics and chemistry that underpin these exceptional materials. We also examine the synthesis and fabrication techniques that enable the creation of hybrid structures, shedding light on the delicate balance required to harness the synergies between different materials.

The book then takes readers on a tour of the myriad applications where hybrid piezoelectric materials have begun to make their mark. From their use in sustainable energy generation to the development of smart materials for healthcare and aerospace, these materials are driving innovation across multiple domains.

Throughout these pages, we have endeavored to bring together contributions from experts in the field, each offering their unique perspective and insights. The chapters encompass a wide spectrum of topics, from theoretical considerations to practical implementations, providing a holistic view of the field's current state and prospects.

As editors of this book, our goal is to inspire and inform researchers, students, and practitioners in the fields of materials science, engineering, and beyond. We hope that the content presented here will ignite new ideas and contribute to the continued evolution of hybrid piezoelectric materials as a driving force for technological advancement.

We extend our heartfelt gratitude to all the authors who have contributed their expertise to this endeavor as well as to the readers who embark on this exciting exploration of hybrid piezoelectric materials. The journey ahead promises a world of possibilities, and we invite you to join us on this captivating expedition.

Editors

1

Introduction to Hybrid Piezoelectric Materials

Sheer Khanyisile Dhlamini[1,2]*, Jonathan Tersur Orasugh*[1,2]*,*
Suprakash Sinha Ray[1,2]*, and Dipankar Chattopadhyay*[3,4]

[1] *Department of Chemical Sciences, University of Johannesburg, Johannesburg, South Africa*
[2] *DST-CSIR National Centre for Nanostructured Materials, Council for Scientific and Industrial Research, Pretoria, South Africa*
[3] *Department of Polymer Science and Technology, University of Calcutta, Kolkata, West Bengal, India*
[4] *Center for Research in Nanoscience and Nanotechnology, Acharya Prafulla Chandra Roy Sikhsha Prangan, University of Calcutta, Kolkata, India*

1.1 Introduction

The exhaustion of conventional fossil fuels and pollution resulting from nonrenewable energy sources have created a global energy crisis. Another problem resulting from the energy crisis is that long-lasting batteries have a limited lifespan, requiring replacement every few years; replacing these becomes expensive when hundreds of sensors are in secluded areas [1]. This has spurred significant interest in energy harvesting as an alternative. The development of green energy harvesters becomes more and more important as the world's demand for alternative energy increases. Energy garnering is a favorable technology that could help solve the global energy crisis with no depletion of natural resources. Energy garnering technologies allow low-power equipment to operate for an unlimited period, eliminating the need to replace batteries where it would be impractical, hazardous, or too costly [2]. Using energy-harvesting technologies would also enable running self-sustained remote sensors, powering portable devices, and wireless electronics [3]. Piezoelectric energy-harvesting technology offers several advantages over other energy-harvesting methods, comprising smaller sizes, higher output voltage, a straightforward mechanism, and high sensitivity to applied strain [4]. In the recent decade, piezoelectric hybrid materials have received considerable attention due to their advantages

Hybrid Materials for Piezoelectric Energy Harvesting and Conversion, First Edition.
Edited by S. Wazed Ali, Satyaranjan Bairagi, and Shahid Ul Islam.
© 2024 John Wiley & Sons, Inc. Published 2024 by John Wiley & Sons, Inc.

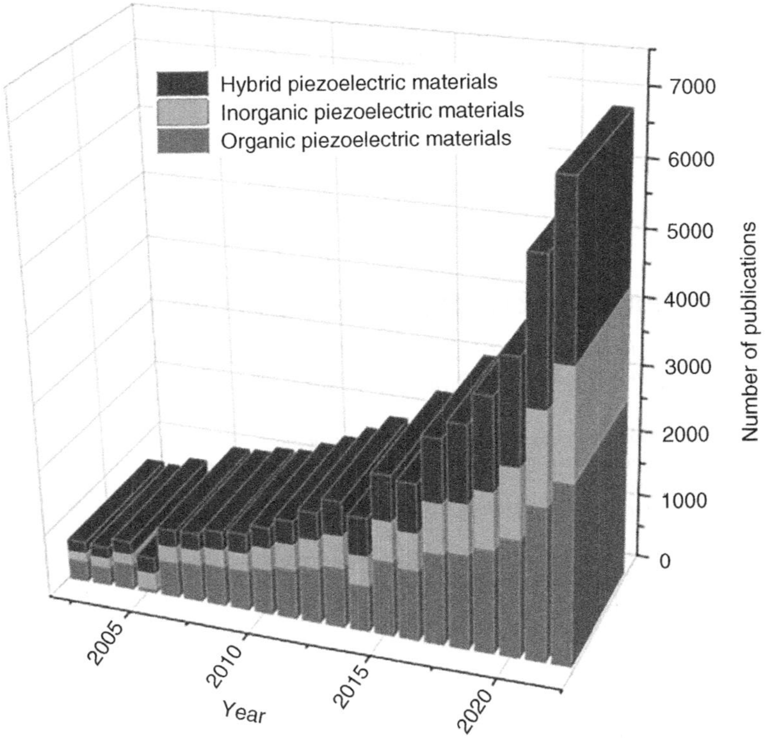

Figure 1.1 The number of articles on hybrid, organic, and inorganic piezoelectric materials published from 2002 to 2023 in ScienceDirect. (Search keywords used: Hybrid piezoelectric materials; Organic piezoelectric materials; and Inorganic piezoelectric materials. The search was performed on 12 June 2023).

over traditional organic and inorganic piezoelectric materials, as shown in Figure 1.1. They are one of the most emerging material classes at the forefront of technological advancement.

In many high-performance applications, traditional piezoelectric materials such as quartz and lead zirconate titanate (PZT) ceramics still exhibit excellent piezoelectric properties and are advantageous [5]. Both synthetic and naturally occurring materials can produce piezoelectric effects. Cane sugar, quartz, topaz, tendon, Rochelle salt, tourmaline, DNA, and bone are examples of naturally occurring materials that exhibit piezoelectricity [6]. Piezoelectric materials play an essential role in numerous applications in our everyday life, as shown in Figure 1.2. At home, the piezo buzzer in our alarms wakes us up in the morning, and the piezoelectric speakers in our headphones allow us to listen to music, and on our mobile phones, we use

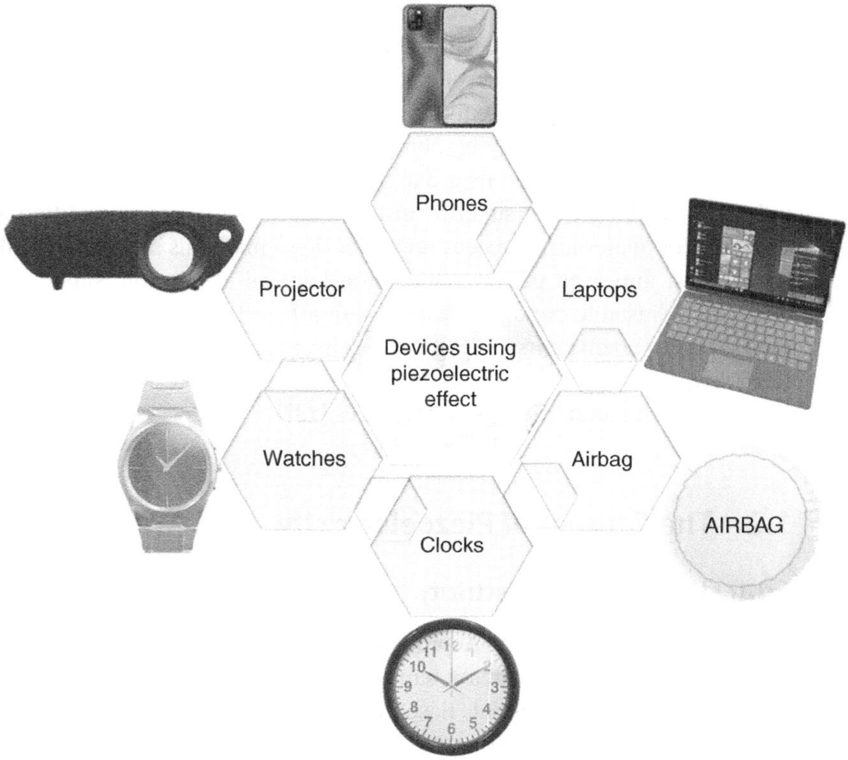

Figure 1.2 Selected devices that make use of the piezoelectric effect (self-drawn).

the piezo motor to adjust the focus of the camera. When driving to work, the car uses piezoelectric fuel injection to enhance fuel efficiency. In case of an accident/collision, the airbags used to protect us use a piezoelectric acceleration sensor [7]. By combining piezoelectric materials with two or more other materials leads to the development of hybrid materials that have improved properties and can be applied to novel applications [8]. Hybrid piezoelectric materials offer additional flexibility and tailored functionality, allowing them to offer unique solutions in certain situations. With hybrid piezoelectric materials, existing technologies can be enhanced, and new devices and systems can be developed, ranging from healthcare, ultrasonic transducers, energy storage, smart fabrics, sensors and actuators, energy-harvesting systems, and robotics [9–11].

In hybrid piezoelectric materials, two or more piezoelectric materials are combined to enhance the piezoelectric properties of the resulting material [8]. The combination of these materials results in enhanced piezoelectric performance and can be tailored to have specific properties. The choice of the inorganic components, along with the synthesis routes, conditions, and connectivity, may allow one to tailor the properties of the hybrid materials to meet specific requirements [12]. Due to the special combination of qualities that may be obtained from the inorganic and organic components, hybrid materials are a great choice for usage in a variety of industries, making them superior in the majority of facets to traditional one-function materials. A recent review of these materials showed several advantages: improved performance, strong optical absorption, superconductivity, adjustable bandgaps, easy solution-based processing, increased piezoelectric coefficients, high photoconversion efficiency, enhanced flexibility, tailored mechanical properties, high charge mobility, and light-harvesting properties for solar devices [13].

1.2 The Concept of Piezoelectricity

1.2.1 History of Piezoelectricity

Carolus Linnaeus and Franz Aepinus studied the pyroelectric effect in the mid-eighteenth century, when a temperature change caused a material to generate an electric potential. Based on this knowledge, both René Just Haüy and Antoine César Becquerel proposed that mechanical stress and electric charge were related; however, their experiments were inconclusive [14]. These scientists observed the piezoelectric effect, but it was not fully understood until two French physicists, Jacques and Pierre Curie, demonstrated it in 1880 [15]. While studying quartz, tourmaline, and Rochelle salt crystals, they discovered that applying pressure on these materials generates electrical charges on their surface [15]. Upon mechanical pressure, all of these materials exhibited characteristic crystal structures composed of a lattice of molecules having asymmetric dipole moments; as a result, they are commonly referred to as piezoelectric crystals. In 1881, Wilhelm G. Henkel named this phenomenon "piezoelectricity" after the Greek words piezein, which means to press or squeeze, and elektron, meaning amber, which refers to substances that attract other substances when rubbed like amber [16]. However, the Curie brothers did not predict the converse piezoelectric effect. In 1881, Gabriel Lippmann mathematically demonstrated the converse effect from fundamental thermodynamic principles; he deduced that if crystals exhibit the direct piezoelectric effect (electricity produced by applied stress), they would also exhibit the converse piezoelectric effect (stress caused by applied electric field) [17]. Immediately after discovering the converse effect, the Curie brothers obtained quantitative proof of the complete reversibility of electro-elasto-mechanical deformations in piezoelectric crystals. With the publication of Woldemar Voigt's Lehrbuch der Kristallphysik (Textbook on Crystal Physics), Voigt was able to describe the 20 natural crystal classes that exhibited piezoelectricity and define the piezoelectric constants using tensor analysis [18]. In 1893, Lord Kelvin developed an atomic model to describe the piezoelectric effect [19].

In 1917, during World War I, Paul Langevin and his French colleagues developed sonar, which was the first application of piezoelectricity. They built an ultrasonic submarine detector using a quartz crystal transducer between two steel plates and a hydrophone [20]. In 1935, the quest for understanding new synthetic piezoelectric materials led to the discovery and development of potassium dihydrogen phosphate (KDP) and ammonium dihydrogen phosphate (ADP) crystals [21]. A breakthrough in ferroelectrics with a perovskite structure was achieved in the early 1940s. During World War II, researchers from three countries (United States, Japan, and Russia) independently discovered the ferroelectric ceramic material barium titanate ($BaTiO_3$) in 1946. Since then, piezoelectric materials have made rapid progress [22]. In 1952, physicists at the Tokyo Institute of Technology developed the first piezoelectric material PZT, which exhibits better piezoelectricity than $BaTiO_3$ and operates at a higher temperature [23]. In the search for ferroelectric materials suitable for high temperatures, Aurivillius discovered bismuth layer-structured ferroelectrics (BLSFs) in 1949. Piezoelectric activity in BLSFs can be further improved by suitable doping. The BLSFs are promising candidates for applications in high-temperature sensors. Around 1950, the first evidence of piezoelectricity in wood was reported, followed shortly afterward by evidence of it in collagenous tissues such as tendons and bones. This research was carried out by Eiichi Fukada. In 1969, Heiji Kawai discovered piezoelectric effects in the synthetic polymer poly(vinylidene fluoride) (PVDF) polymers, which can be used for new sensors that cannot be realized with piezoceramics or single crystals [24]. For the purpose of simplistic understanding, the whole history is presented in Figure 1.3.

1.2.2 The Piezoelectric Effect

The piezoelectric effect transpires when a material experiences mechanical stress that generates a small electrical potential across its sides. When a piezoelectric material is compressed, electricity is produced, and there is a conversion of kinetic or mechanical energy due to crystal deformation into electrical energy, as shown in Figure 1.4a. An electric field is generated

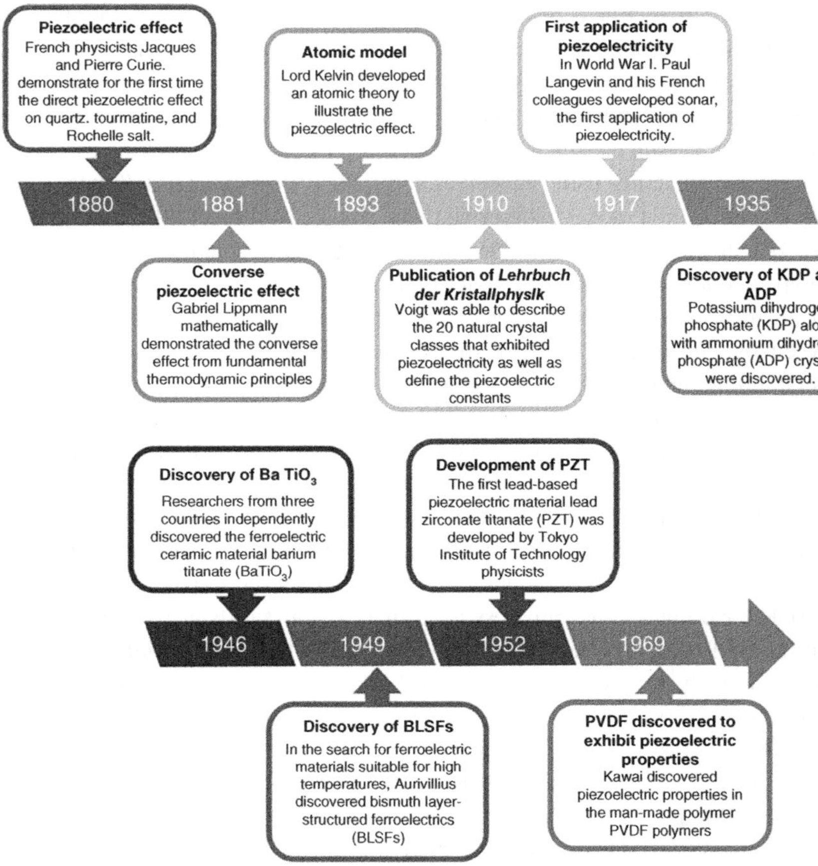

Figure 1.3 The history of piezoelectricity (self-drawn).

when the positive and negative charge centers of piezoelectric materials shift under mechanical stress; this is referred to as direct piezoelectric effect. Whenever a voltage is applied to opposite sides of a material, it will undergo the opposite effect, known as the converse piezoelectric effect. On the other hand, converse piezoelectric effect takes place whenever an electric voltage is applied across a material, as shown in Figure 1.4b, causing its balanced electric charges to become unbalanced. In order to rebalance the electric charges, the material's atoms must be rearranged, and this causes the material to deform slightly.

Piezoelectric materials behave and perform differently depending on how many electric dipoles they contain. The piezoelectric property is greatly affected by crystal symmetry. A crystal is any solid with constituents

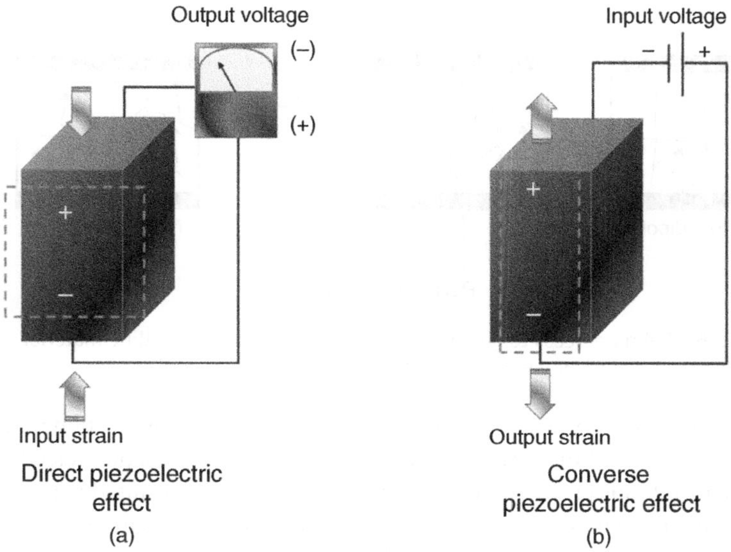

Figure 1.4 The two varieties of piezoelectric phenomenon: (a) direct, (b) converse piezoelectric effect (self-drawn).

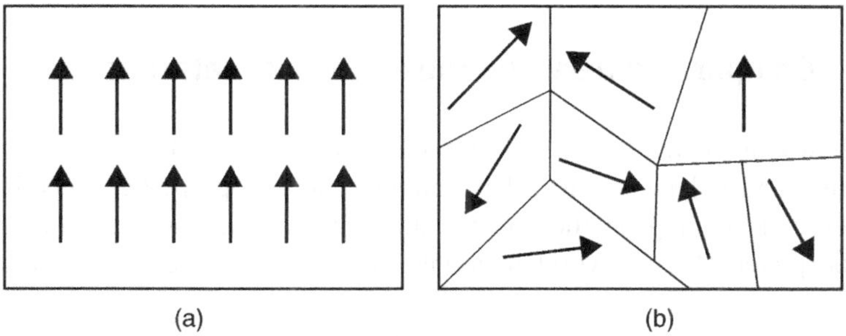

Figure 1.5 Dipole arrangement in (a) monocrystalline and (b) polycrystalline materials (self-drawn).

(atoms or molecules) arranged orderly based on repetitions of the same basic atomic building block (the unit cell). In most crystals, such as your monocrystals, the polar axes of all the dipole moments lie in one direction, as shown in Figure 1.5a, and they are symmetrical. In contrast, in piezoelectric crystals, which are polycrystalline, as shown in Figure 1.5b, there are different regions with different polar axes causing irregular alignment of dipoles and asymmetry. Cutting the polycrystal results in the two remaining pieces having different polar axes. The material, at this stage, does not exhibit piezoelectricity, as all domains need to be polarized in one direction

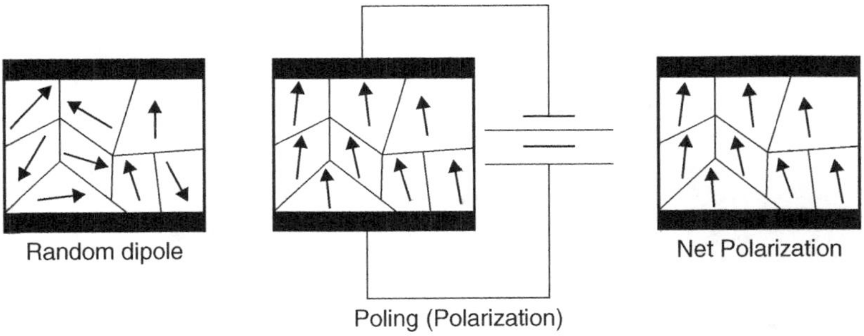

Figure 1.6 Poling process to generate polarization of a polycrystalline material transduction principle of piezoelectronics (self-drawn).

through a process known as poling in order to obtain a net polarization [25]. Poling is the procedure by which an electric field is applied to a piezoelectric material, and thus creates an instinctive polarization; as a result, due to the electric field, the molecules are freer to move, and the dipoles in the crystal are forced to align and face in nearly the same direction (the electric field's applied path), as shown in Figure 1.6. Changing the poling conditions can easily tailor the degree of polarization [26].

1.3 Comparison between Piezoelectric Materials

As shown in Figure 1.7, hybrid materials have several advantages over ceramics and polymers, since they combine the best properties of both. However, depending on the targeted application, ceramics or polymers can be preferred over hybrid materials. A typical ceramic piezoelectric

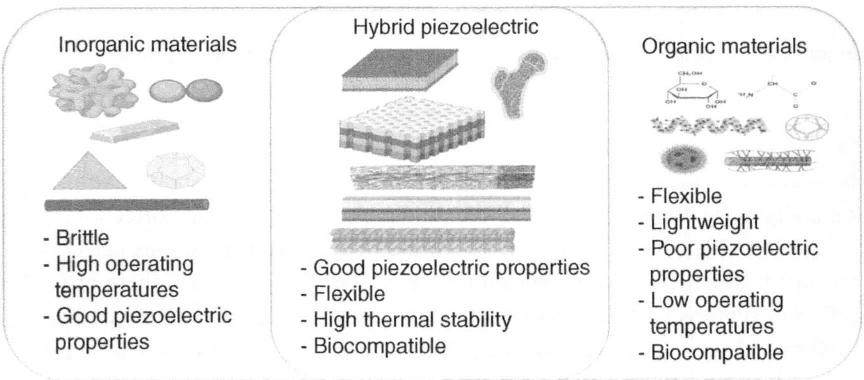

Figure 1.7 The different classes of piezoelectric materials (self-drawn).

material is used in microelectromechanical system (MEMS) devices and transducers, such as sonar, which require piezoelectric response without requiring flexibility [8]. The majority of piezoelectric ceramic materials have excellent piezoelectric characteristics, including strong electromechanical coupling coefficients and a high piezoelectric constant. However, because of their tremendous stiffness, these materials have high resonant frequencies. Due to the brittleness of ceramic materials, repetitive cycles of deformation can result in early failure [27]. Piezoelectric polymers/polymer materials are especially suitable for biomedical applications due to their compatibility with the human body; they are also flexible and lightweight. However, in general, their piezoelectric coefficients are not very high. They are also incapable of withstanding high temperatures [28]. Due to their weak mechanical properties, piezoelectric polymers are not recommended for applications in which they will be subjected to heavy loads [29].

Piezoelectric hybrid materials offer several advantages over both ceramics and polymers as they combine the properties of these two materials, with the possibility of tailoring properties by modifying their structure and composition [30]. A hybrid material made of both ceramic and polymer piezoelectric components can provide a product with a good balance of piezoelectric performance and acoustic match that can be tailored to meet the requirements of the intended application. By choosing the right combinations and amounts of constituent materials, the adoption of composite materials toward the design and fabrication of their hybrid counterparts enables the tuning of specific features. The mechanical, thermal, electrical, and other properties of the composite can be tailored for a particular application by carefully selecting the kind of orientation and volume fraction of the reinforcing components [31]. Comparing composites to ceramics and polymers alone, they also provide more design flexibility. They can be created to have precise geometries and optimal performance by being developed to have complex shapes and structures, resulting in lightweight and high-performance systems [8]. Because hybrid materials contain diverse materials with complementing properties, they can display greater strength and stiffness than ceramics and polymers alone. They can be strengthened using a polymer matrix that contains ceramic fibers or particles, which can considerably improve the composite's mechanical characteristics and make it stronger and stiffer [28]. Compared to polymers alone, composites and/or hybrids can provide improved thermal stability. Hybrid materials can resist greater temperatures without significant degradation by adding thermally stable reinforcing elements, increasing their suitability in high-temperature environments [8]. When used in a hybrid structure, the piezoceramics' acoustic match to tissue or water is substantially improved, since the stiff and dense ceramic is swapped out for a less rigid and more pliable polymer [8]. The advantages of hybrids can vary depending on the exact application, the choice of constituent materials, and the production process. To choose the best material, it is important to carefully analyze both the application's needs and the desired features.

1.4 Piezoelectric Material Types

Due to their ability to directly convert mechanical energy into electrical energy, piezoelectric materials are an essential part of energy-harvesting systems, sensors, and actuators. The investigations on piezo-materials have resulted in a wide variety of piezoelectric materials available today. Approximately 200 distinct categories of piezoelectric materials are employed in energy-harvesting applications [32]. The main categories of piezoelectric materials are natural and synthetic, which are also further classified into ceramics, single crystals, polymers, and nanocomposites, as depicted in Figure 1.8. In this section, we will focus more on hybrid materials. The development of piezoelectric nanostructures has allowed the exploitation of these materials, which have been shown to have enhanced properties [33, 34].

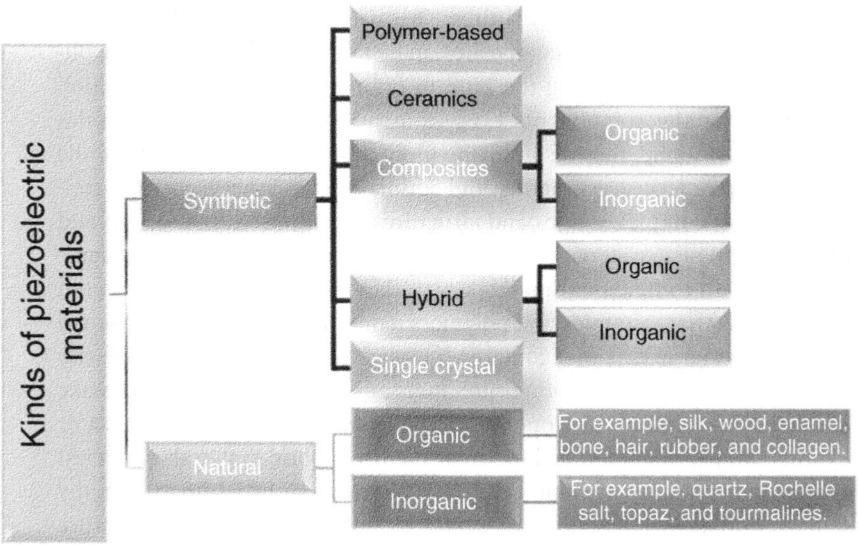

Figure 1.8 Categorization of piezoelectric materials (self-drawn).

1.4.1 Inorganic Piezoelectric Materials

1.4.1.1 Single Crystal-Based Piezoelectric Materials

Piezoelectric crystals are single crystals that grow in a long-range order based on the crystal matrix dot space. The single crystal structure's regular ordering of positive and negative ions causes the dipoles to be properly aligned, which is essentially what causes their piezoelectric action [35, 36]. Due to their greater piezoelectric coefficients, these materials are used in a variety of electromechanical fields, including sonar transducers, medical ultrasonic devices, actuators, hydrophones, accelerometers, and sensors [37]. Rochelle salt and quartz are naturally occurring piezoelectric single crystals; however, to enhance their piezoelectric properties, they must be aligned and cut in a particular crystallographic direction. Lithium niobate ($LiNbO_3$), lead magnesium niobate–lead titanate solid solution (PMN–PT), as well as lead zinc niobate–lead titanate (PZN–PT), are the most used single-crystal piezoelectric materials.

1.4.1.1.1 Lithium Niobate ($LiNbO_3$) $LiNbO_3$ is a single-crystal oxide material with an ilmenite crystal structure, a high electromechanical coupling coefficient, spontaneous polarization coefficient, low-temperature resistance, electro-optic coefficient, corrosion resistance, and piezoelectric properties [38]. $LiNbO_3$ is also nontoxic, in contrast to lead-containing materials, making it suitable for biomedical applications [39]. The crystal was first synthesized using the flux method by Remeika and Matthias in 1949, who also discovered that this crystal has ferroelectric properties [40]. $LiNbO_3$ single crystals have extensively been used in acoustics and electro-optics due to their strong and stable electro-mechanical coupling capabilities [41]. $LiNbO_3$ has a very high Curie temperature (1142 °C); this high Curie temperature makes it possible to be used as a high-temperature acoustic transducer, such as an accelerometer for jet aircraft. As a single crystal with a high Curie temperature, $LiNbO_3$ can sustain stable piezoelectricity over a long period of time and a wide range of temperatures without experiencing problems with grain size and porosity [42]. $LiNbO_3$ is an excellent material for surface acoustic wave (SAW) devices. About 70% of radio-frequency (RF) filters, based on SAW, are fabricated using $LiNbO_3$ single crystals [43]. $LiNbO_3$ films need to be highly crystalline, possess a pure $LiNbO_3$ phase and a smooth surface, be twin-free and domain-free, and have a good texture quality in order to be used in electrically active acoustic and optical devices and to ensure high performance [43]. There are a number of crystal cuts in $LiNbO_3$ that are now widely used in piezoelectric applications, including 36° and 163° rotated Y-cuts [5].

1.4.1.1.2 Lead Magnesium Niobate–Lead Titanate Solid Solution (PMN–PT) PMN–PT is a single crystal composed of a solid-solution material, made of lead magnesium niobate (PMN) and lead titanate (PT). Each processing method for PMN–PT material, including single crystal, polycrystalline ceramic, thick film, and thin film, has unique functional properties that are appropriate for particular applications. For instance, PMN–PT single crystals have the best functional properties for actuator applications [44]. The PMN–PT single crystal of the composition on the morphotropic phase boundary (MPB) can have piezoelectric coefficient as high as 2820 pC/N. PMN–PT-based materials exhibit high piezoelectricity, high electrostriction, and high dielectric permittivity and can be used to make multilayer capacitors, actuators, sensors, and electro-optical devices [45]. The strong piezoelectric capabilities of PMN–PT solid solutions are related to the polarization rotation between adjacent rhombohedral and tetragonal phases via one (or more) low-symmetry intermediate phases, that is, a monoclinic (orthorhombic or triclinic) phase [44]. Most PMN–PT crystals can be grown using the high-temperature flux technique or the Bridgeman method. The Bridgman technique has proven to be a promising growth technology for PMN–PT crystals [37].

1.4.1.2 Ceramic-Based Piezoelectric Materials

Piezoelectric ceramics refer to polycrystalline materials having a perovskite crystal structural architecture, made of large number of identically chemically constituted single crystal "grains." However, because each grain's orientation varies from the other, the spacing between the ions also varies slightly. The most commonly used piezoelectric ceramics are potassium niobate ($KNbO_3$), $BaTiO_3$, PZT, and potassium sodium niobate (KNN).

1.4.1.2.1 Pb(Zr,Ti)O_3 (PZT) PZT is a perovskite-structured ceramic material that possesses superior piezoelectric properties as it is a solid solution of lead zirconate ($PbZrO_3$) and lead titanate ($PbTiO_3$). PZT is regarded as the major piezoelectric ceramic that is broadly utilized in applications, such as energy harvesting, actuators, transducers, sensors, and sonar systems [8]. These materials are distinguished by their exceptional piezoelectric characteristics. PZT has a high piezoelectric constant (500–600 pC/N), a high electromechanical coupling coefficient (0.69), good mechanical stability, and low dielectric loss, which results in a higher conversion from mechanical to electrical energy [46, 47]. PZT can be employed in a variety of applications since it is mechanically robust, possesses excellent chemical stability, and does not degrade in humid conditions [8]. PZT thick and thin films have been utilized extensively in a range of piezoelectricity as well as piezo-resistivity-based sensors, including pressure sensors and biosensing cantilevers. This newly found material's acceptance as a piezoelectric material was substantially accelerated by Sir Bernard Jaffe's study on it. The material also has high Curie temperatures of 350 °C and retains its piezoelectric properties at high temperatures [8]. Because of its high density and high Young's modulus (50 GPa), PZT is too brittle and unable to undergo significant deformation; hence, it cannot be used in applications that require flexibility [48]. Bulk PZT has a high resonance frequency due to its stiffness, making it unsuitable for capturing energy from low-frequency ambient vibrations [8]. Another disadvantage of PZT is that it cannot be used in some applications, such as devices implanted into the body, because lead is dangerous to both human health and the environment [48]. Due to their cost-efficiency, good piezoelectric properties, and ease of integration into energy-harvesting devices, PZT has been extensively used for energy-harvesting as well as storage applications [49].

1.4.1.2.2 Barium titanate ($BaTiO_3$) Another significant ceramic-based piezoelectric material is $BaTiO_3$. During World War II, scientists found that $BaTiO_3$ was the first perovskite-structured ferroelectric substance. It was extensively used in phonograph needles and sonar detection [50]. $BaTiO_3$ is a ferroelectric substance, which means that when an electric field is applied, it can switch between several polarization states. Pure, unmodified $BaTiO_3$ has a low piezoelectric constant (190 pC/N) and a low Curie temperature (130 °C) [5]. However, $BaTiO_3$ can be modified to have good piezoelectric properties; for example, it has been shown using different processing techniques that result in dramatic improvements in piezoelectric properties. For instance, Karaki et al. [51] successfully fabricated high-density $BaTiO_3$ piezoelectric ceramics while maintaining the average grain size by a two-step sintering method from hydrothermally synthesized $BaTiO_3$ nanoparticles of 100 nm, which showed improved piezoelectric properties compared to unmodified $BaTiO_3$. This ceramic material showed excellent piezoelectric properties, with a piezoelectric coefficient of 460 pC/N, an electromechanical coupling factor of 42%, and a dielectric constant of 5000. It has also been shown that the introduction of fine domains into these ceramics leads to a high piezoelectric constant. Takahashi et al. [52] reported that the introduction of fine domains with a size of 50 nm into $BaTiO_3$ ceramics fabricated by microwave sintering results in a high piezoelectric coefficient of 350 pC/N. Wada et al. [53] also showed that all piezoelectric-related constants of single-crystal $BaTiO_3$ increased with the decrease in ferroelectric domain size. They prepared $BaTiO_3$ grain-oriented ceramics by the templated grain growth (TGG) method and obtained a high piezoelectric coefficient of 788 pC/N. $BaTiO_3$ can be used in several applications, such as sonar systems, sensors, capacitors, and piezoelectric transformers [50].

1.4.1.2.3 Potassium Sodium Niobate (KNN) KNN is a lead-free ceramic piezoelectric material made up of $NaNbO_3$ and $KNbO_3$; it is a promising alternative to lead-based piezo ceramics [49]. Undoped KNN has a very high Curie temperature (above 400 °C). However, it has a low piezoelectric constant (80–160 pC/N) and is difficult to fabricate due to low sintered densification [54]. Modified KNN structures can result in higher piezoelectric properties closer to those of PZT materials; for instance, Saito et al. [55] created highly textured polycrystals with a high piezoelectric constant (416 pC/N) using the reactive template grain generation technique. Guo et al. [56] also obtained very high piezoelectric constant ranging from 200 to 235 pC/N in KNN ceramics by doping $LiNbO_3$ or $LiTaO_3$ using traditional solid-state methods. The biggest benefit of KNN is that it does not contain lead, making it suitable for use in energy harvesting and medical implants [54].

1.4.2 Organic Piezo Materials

Piezoelectric polymers are carbon-based substances with lengthy polymer chains that have mechanical flexibility; as a result, they can withstand high strains, making them more suitable for piezoelectric energy-harvesting applications. Piezoelectric polymers include materials such as PVDF, polylactic acid (PLA), polyamides (PA), copolymers, cellulose, and its derivatives. In order for polymers to exhibit considerable piezoelectricity, mechanical stretching and poling are typically required. The applied electrical field alters the orientation of positive and negative charges and dipoles, which in turn causes piezoelectricity [48, 57]. In comparison to inorganic piezoelectric materials, polymers have different properties, making them uniquely qualified to fill niches that inorganic materials cannot. The properties of polymers include high biocompatibility, excellent flexibility, environmental friendliness, and ease of processing. Since polymers have a lower piezoelectric strain constant (d_{31}) than inorganic materials, they tend to make better sensors. It is easier to make sensors from these materials because they are lightweight and can be cut and shaped into complex shapes. Due to their high strength, high impact resistance, low elastic stiffness, low density, and low dielectric constant, they have the advantage of having high voltage sensitivity, low acoustic impedance, and high mechanical impedance. Numerous energy-related applications, including medical transducers, vibrometers, audio transducers, displays, shock sensors, and pressure sensors, have found these materials to be acceptable. Nylons, polyimide (PI), PLA, PA, cellulose, as well as its derivatives, polyurea, and polyurethane (PU) are examples of other polymers that exhibit some degree of piezoelectric effect.

1.4.2.1 PVDF

The semicrystalline polymer PVDF is composed of chains aligned to form dipoles, and it was discovered by Kuwaii. In polycrystalline PVDF, polymer chains are arranged in zigzag patterns in the crystal grains, with an amorphous matrix surrounding the crystalline structure and it is usually 35–70% crystalline [48]. Due to its non-centrosymmetric crystal structure, PVDF exhibits piezoelectric activity, and it is preferred over other polymer materials [58]. Additionally, it has a low acoustic impedance, great electromechanical response, and good chemical stability. PVDF exhibits a negative piezoelectric constant, in contrast to other widely used piezoelectric materials such as PZT; this means that PVDF compresses rather than expands when exposed to an electric field [59]. The material is mechanically stretched to align the molecular chains and then poled under tension to give it its piezoelectric properties [59]. PVDF exists in three main polymorphic phases: alpha (α), beta (β), and gamma (γ). The β phase is the most piezoelectrically active phase of PVDF and has the strongest ferroelectric properties; however, a high amount of crystallinity must be attained initially in order for the phase to develop [60]. The β-phase content and degree of crystallinity are influenced by stretching ratio and temperature, which in turn will affect the electroactive properties of the polymer [61]. To induce and stabilize the desired phase, a variety of techniques can be used, such as mechanical stretching, electrical poling, or annealing. Moreover, PVDF is biocompatible, which makes it ideal for biomedical applications. Among its applications are biosensors, implants, and scaffolds for tissue engineering. One of the most promising PVDF copolymers investigated for its piezoelectric properties is trifluoroethylene (PVDF–TrFE). As PVDF–TrFE contains an extra fluorine, it increases the possibility of forming the β phase [48]. PVDF–TrFE has a higher piezoelectric power density (312.5 $\mu W/cm^3$) compared to PVDF with a piezoelectric power density of 81.3 $\mu W/cm^3$ [48].

1.4.2.2 Polylactic Acid (PLA)

PLA is a biodegradable and biocompatible polymer with very low piezoelectric activity; however, it can be modified to exhibit applicable piezoelectric activities. The alignment and orientation of Poly-L-lactic acid (PLLA)'s polymer chains are thought to be the cause of its piezoelectric characteristic. Stretching or poling are two examples of processing methods that can be used on PLLA to align the polymer chains in a desired orientation and produce a non-centrosymmetric structure that supports piezoelectricity. PLLA is suitable for applications that require flexible and lightweight piezoelectric materials. PLLA is biocompatible and it has potential to be used in biomedical applications, such as medical implants, biomedical sensors, and tissue engineering scaffolds.

1.4.3 Hybrid Piezoelectric Materials

Hybrid/nanocomposite materials have been developed to enhance the properties of different piezoelectric materials. Nanocomposites combine piezoelectric ceramic nanoparticles with a polymer matrix [28]. The ceramic nanoparticles contribute unique properties such as high mechanical strength, piezoelectricity, and thermal stability to the nanocomposite; the most used ceramic materials include PZT, ZnO, and $BaTiO_3$ nanoparticles. The polymer matrix provides flexibility, processability, and mechanical properties to the nanocomposite, with the most used polymer matrices used are PVDF, PU, and polystyrene (PS). Incorporating other components to construct hybrid systems can improve the piezoelectric performance of pure piezoelectric materials by compensating for the shortcomings of each component and exploiting synergistic effects [62]. In comparison to ceramics and polymers, nanocomposites have been shown to possess improved properties [63]. The most explored hybrid piezoelectric nanocomposite materials are PVDF–PZT, $BaTi0_3$/PVDF, and ZnO–PVDF, among others [28].

1.4.3.1 PVDF–PZT systems

One of the most researched polymer–ceramic piezoelectric hybrids/composites is PVDF–PZT [8]. The 0–3 types of polymer–ceramic piezoelectric composites were first reported by Kitayama and Sugawara [41]. A PVDF–PZT composite is a type of hybrid piezoelectric material that combines PVDF, which is the organic polymer material that acts as the organic matrix, with PZT, which is the inorganic ceramic material that acts as the ceramic filler [41]. PZT is too stiff to be employed in fluids such as water or blood, and PVDF has a comparatively low piezoelectric constant and low stiffness on its own [8]. Compared to pure PVDF or PZT materials alone, PVDF–PZT composites have improved piezoelectric properties. PZT ceramic fibers or particles can improve charge generation and mechanical-to-electrical energy conversion by increasing the piezoelectric response. PVDF–PZT composites benefit from the mechanical properties of PVDF and the high stiffness of PZT ceramics. The presence of PZT fillers improves the stiffness and mechanical strength of the composite, making it more robust and suitable for applications that require both piezoelectricity and mechanical integrity [28]. PVDF–PZT composites also exhibit enhanced mechanical properties and flexibility. The composite can be bent or shaped without losing its piezoelectric properties due to the flexibility of PDVF. This allows the composite to be used in flexible electronics and wearable technology, while PZT fillers increase the composite's stiffness and mechanical strength, making it more resilient [28].

PVDF–PZT composites can be used in various applications such as sensors, medical imaging, actuators, sonar systems, energy harvesting, and ultrasound transducers. It has been shown that the piezoelectric properties of the composite can be improved by making use of fibers through the electrospinning (ES) technique. Chang et al. [64] synthesized PZT–PDVF nanocomposite; the PZT nanofibers were prepared via ES, and the modified PZT nanofibers and PVDF were dispersed in dimethylformamide (DMF) to fabricate the PZT/PVDF nanocomposites. The results showed improved piezoelectric properties, with a piezoelectric coefficient of 87.4 pm/V with only 2.4 vol% of PZT nanofibers. Chamankar et al. [65] prepared flexible pressure sensors based on nanocomposite fibers with enhanced dielectric and piezoelectric properties. The flexible nanocomposite fibers were prepared from PZT ceramic particles and PVDF polymer. PZT ceramic particles were synthesized by the sol–gel process method, after which the prepared solution was electrospun to give PZN nanoparticles with a size of 85.85 nm. PVDF–PZT nanocomposite fibers were prepared through the ES method with a different volume fraction of PZT. The PVDF–PZT nanocomposite fibers were 338.9 nm. To fabricate the power-generating modules, the nanocomposite fibers were cut into a square (4 cm × 4 cm). A paper sheet was employed as a spacer between the top and bottom of the nanocomposite fiber web, and aluminum foils were positioned at the top and bottom of the web as the electrode. The results showed that increasing the PZT volume fraction increased the crystallinity and electroactive β phase, piezoelectric constant, dielectric constant, piezoelectric sensitivity, and output voltage of PVDF–PZT nanocomposite fibers.

1.4.3.2 $BaTiO_3$/PVDF

A $BaTiO_3$/PVDF composite is a type of hybrid piezoelectric material that combines $BaTiO_3$, which is the inorganic ceramic material that acts as the ceramic filler, and PVDF, which is the organic polymer material that acts as the organic matrix. Just like PZT–PVDF, $BaTiO_3$/PVDF composites also result in enhanced piezoelectric properties. The use of self-poled, highly crystalline PVDF and $BaTiO_3$ nanoparticles, both of which have positive piezoelectricity, gives the composite outstanding piezoelectric performance [50]. $BaTiO_3$, a ferroelectric material with a high dielectric permittivity, when combined with PVDF, creates a composite with increased dielectric characteristics, enhancing electrical performance and energy storage capacity. The mechanical characteristics of the composite are enhanced; the composite is made stiffer and stronger by the use of $BaTiO_3$ ceramics, which makes the composite more appropriate for applications that require both mechanical integrity and piezoelectricity [8].

The composites can be prepared using various methods, including ES, melt blending, and solution casting [66]. They can be developed into different materials, such as films, fibers, or coatings, depending on the targeted application. Dashtizad et al. [67] synthesized a $BaTiO_3$–PVDF nanocomposite with improved piezoelectric properties. PVDF–$BaTiO_3$ and PVDF–$BaTiO_3$–Ag nanocomposite fibers were prepared by the ES method. By using the sol–gel method, $BaTiO_3$ particles were synthesized, and $BaTiO_3$–Ag were subsequently irradiated with ultraviolet light for 2, 6, and 10 minutes in silver nitrate solution to precipitate silver on the $BaTiO_3$ particles. Using a specific force and a defined frequency, piezoelectric properties of each composite were measured. Based on the results, the nanocomposite showed better piezoelectric properties than PVDF fiber. PVDF–$BaTiO_3$–Ag composites showed a higher output voltage than PVDF–$BaTiO_3$ owing to the greater extent of β-phase formation. Kong et al. [68] fabricated low-cost, flexible piezoelectric pressure sensors using $BaTiO_3$–PVDF nanofibers by ES method. PVDF was dissolved in a mixture of first mix acetone and DMSO; the solution was stirred until a clear solution was obtained. $BaTiO_3$ was then added to the clear solution, and the solution was electrospun using near-field ES method. To obtain the hierarchical structured film, the barium titanate (BTO)/PVDF membrane was placed between molds and dried at a constant temperature as shown in Figure 1.9a. As a result, the fabricated flexible pressure sensor was able to detect human activity successfully; this shows it can be used in wearable bioelectronics, as shown in Figure 1.9c,d. The morphology of the membrane is depicted in Figure 1.9b. When a periodic load of 2 N at 2.5 Hz was applied to the 2 mm bulge high-entropy alloy (HHS) sensor, it produced a maximum output voltage of 2.32 V.

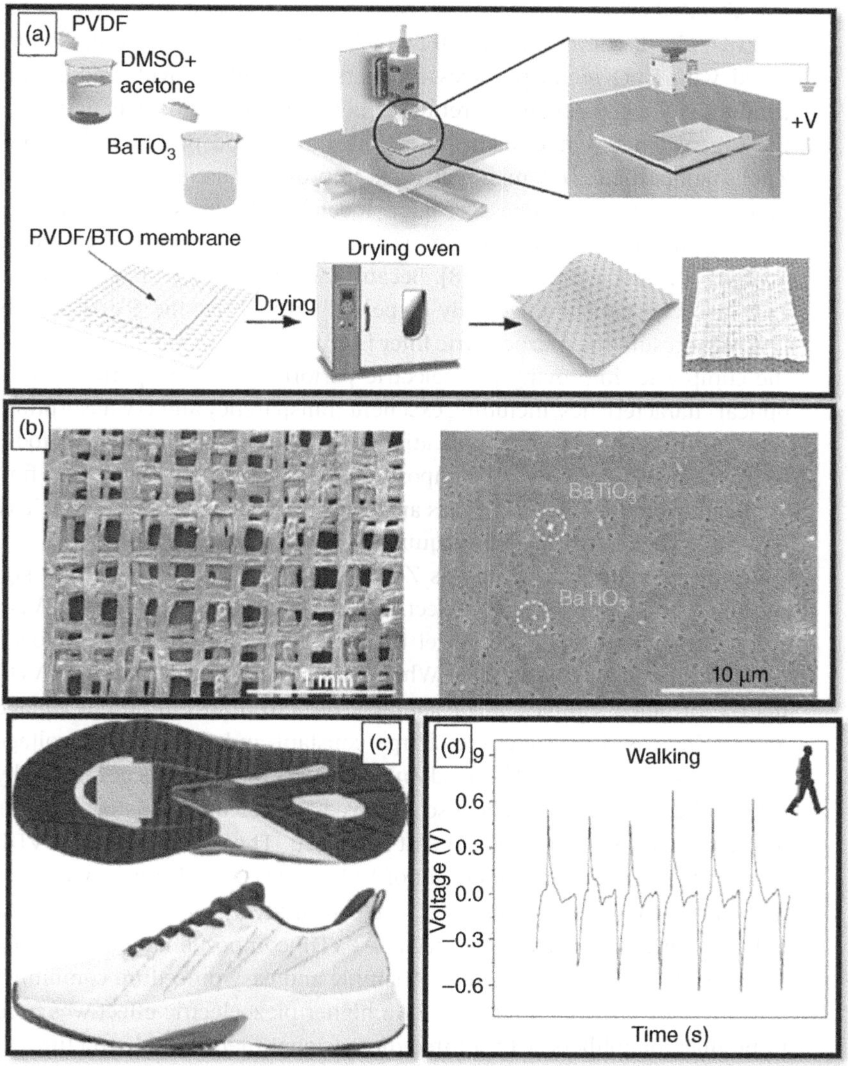

Figure 1.9 $BaTiO_3$–PVDF: (a) the schematic of the fabrication process, (b) the SEM image of the five-layer fiber membrane and a single fiber, (c) the photographs of the HHS pressure sensor integrated into the soles of shoes, and (d) the output voltage generated by walking. Source: Reproduced with permission from Kong et al. [68], © 2023/John Wiley & Sons, Inc.

1.4.3.3 ZnO–PVDF

In ZnO–PVDF composites, PVDF, which is the organic material, is combined with ZnO metal particles or fibers. The ZnO provides additional conductivity or mechanical reinforcement, while the PVDF acting as the organic matrix may contribute to flexibility and processability [69]. ZnO nanoparticles or microparticles improve the composite's overall piezoelectric responsiveness. Under mechanical stress, ZnO particles produce charges, and the PVDF matrix aids in charge separation, producing a stronger piezoelectric effect [8]. Because ZnO and PVDF are compatible, ZnO particles can be effectively dispersed throughout the PVDF matrix [70]. As a result, the piezoelectric filler is uniformly distributed throughout the composite, improving piezoelectric performance. ZnO particles have optical characteristics, including excellent transparency and UV absorption. As a result, optoelectronic applications such as UV sensors and photodetectors can use ZnO–PVDF composites [69]. These composites also find applications in flexible electronics and sensors, where both piezoelectricity and electrical conductivity are required [8].

Adding metallic fillers such as ZnO to PVDF-based materials can significantly increase their piezoelectric activity. In order to create PVDF composite films using the sol–gel method, Bhunia et al. [71] employed freestanding flexible nano-ZnO. When nano-ZnO was added to the PVDF matrix for poled samples, a considerable rise (10 times) in output voltage and an increased piezoelectric charge constant and piezoelectric voltage constant were observed. Al-doped ZnO–PVDF composite was developed by Ling and coworkers [72] who also observed an increase in the composite's dielectric constant as the Al content increased. The Al-doped ZnO–PVDF composite had a dielectric constant of 18.5 at 100 Hz, which was twice that of the pure ZnO/PVDF composites. A recent study by Fu et al. [73] showed enhanced piezoelectric activity in ZnO–PVDF composite nanofiber with a pine-needle-like bionic structure. A simple and easy procedure combining ES with hydrothermal treatment for a higher piezoelectric effect was used to produce nanofibers with a three-dimensional hierarchical structure, as shown in Figure 1.10. After 12 hours, the hydrothermal reaction's optimum outcome was obtained, showing consistent thickness and length, highly oriented organization along the PVDF nanofibers, and ZnO nanorods, as illustrated in Figure 1.10. ZnO–PVDF nanofiber's highest output voltage was 0.112, 0.208, 0.352, and 0.359 V for 1, 3, 6, and 12 hours, respectively, demonstrating improved properties.

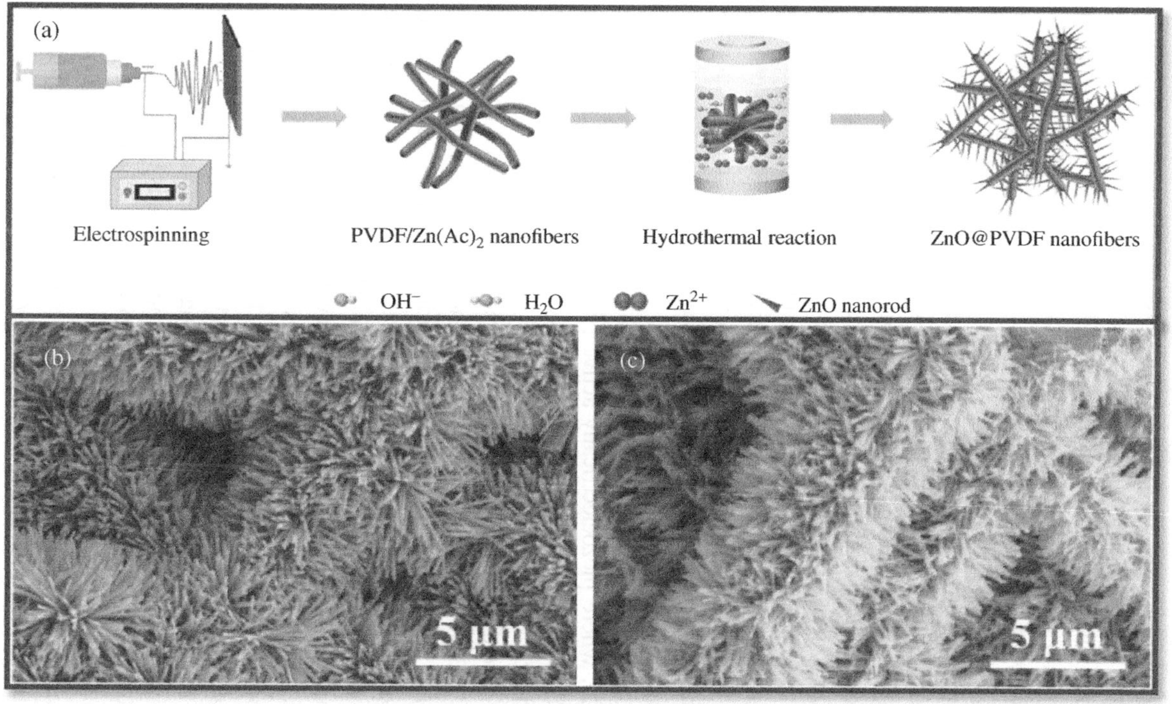

Figure 1.10 ZnO–PVDF nanofibers: (a) schematic preparation, (b) the effect of hydrothermal reaction on the morphology after 6 hours (c) the effect of hydrothermal reaction on the morphology after 12 hours. Source: Reproduced with permission from Fu et al. [73], Copyright 2022, Elsevier Science Ltd.

1.5 Connectivity of Composites Similar in Hybrid Systems

In composite materials, the properties are determined not only by the filler and matrix types but also by how they are connected. Depending on how the connections are constructed, the physical properties of the composite materials can vary by many orders of magnitude [74]. The concept of connectivity was first introduced by Newnham et al. [74] for classifying different PZT: polymer composites [75]. In composites, active materials such as piezoelectric ceramics are combined with passive materials, such as polymers, and they are attached to form different connectivities. The connectivity of the components of a composite strongly influences its properties, such as electric flux patterns and mechanical properties [75]. The arrangement of active as well as passive phases in a composite is referred to as connectivity. The connectivity of each phase in a two-phase composite is identified; if the phase is connected in all three directions, a phase is designated 3; if the phase is connected in two directions, a phase is designated 2; and if the phase is connected in one direction, a phase is designated 1 [41, 76]. There are several different connectivity types, ranging from (0–0), (0–1), (0–2), (0–3), (1–1), (1–2), (2–2), (1–3), (2–3), and (3–3), where piezoelectrically active phase (filler) refers to the first digit within the parenthesis and the inactive phase (matrix) refers to the second digit.

The 0–3 and 1–3 configurations are the most often studied composites, and the 0–3 family is the most popular since it is simple to prepare as a randomly distributed mixture at a low price [74]. The 1–3 connection maximizes the geometrical advantages of piezoelectric materials, leading to greater electromechanical coupling. Materials with 0 indicate that it is in the form of particles, whereas composites with 0–3 contain piezoelectric particles randomly dispersed in polymer phases; in addition, this connectivity has been examined for sensor appliances. It is advantageous to fabricate more complicated shapes using 0–3 connectivity since it has a relatively high piezoelectric voltage coefficient, allowing for increased flexibility [77]. In order to achieve acceptable grain sizes, the ceramic is typically ground into powder during the preparation of 0–3 connectivity. The polymer determines how the materials can be combined; for example, if the polymer is epoxy, mixing can be done by hand at room temperature with an appropriate quantity of resin, hardener, and ceramic [74]. The most complex pattern is the 3–3 pattern, in which two phases combine to produce 3D networks. These connectivity patterns can be found in living systems such as bone, wood, and coral, where organic tissue and inorganic skeleton interpenetrate one another [78]. Additionally, there exist connectivity patterns for more than two phases, including 20 connectivity patterns for a three-phase system and 35 connectivity patterns for a four-phase system [75].

1.6 Fabrication and Characterization of Hybrid Piezoelectric Materials

Hybrid piezoelectric materials can be fabricated by using different techniques, such as ES, melt blending, coating/deposition, 3D printing, and hybridization via surface modification.

1.6.1 Cold-pressing and Curing-molding Method

Reinforcing materials and catalyzed or uncatalyzed resin are positioned between corresponding male and female press-actuated molds in cold press molding for forming and curing. With the addition of pigment to the resin system, the part with two smooth, dimensionally precise surfaces can be colored.

The cold-pressing and curing-molding method allows easy preparation of the composites and formation of flexible shapes. This method has many advantages, such as simple fabrication, formation of flexible shapes, low-temperature treatment, high production rate, and low cost [79]. The first step is usually to mix the piezoelectric powders; the piezoelectric powders usually consist of ceramic material particles or nanoparticles and a polymer material powder, and sometimes to attain specific properties, additives or fillers can be used. After mixing the powders thoroughly to ensure homogeneity, the powder mixture is placed in a mold, and pressure is applied to shape the composite. When the desired shape has been achieved for the curing-molding method, the composite material is subjected to a curing process involving the application of heat and pressure to initiate and complete polymerization or cross-linking. However, the disadvantage of this approach is that the molds are expensive and it also requires high pressure, unlike solution/film casting. Wu et al. [79] fabricated PVDF/PZT composite using cold-pressing and curing-molding method; the PZT nanoparticles were dispersed in ethyl alcohol, and then PVDF powder was added to the solution and the mixture was stirred for two hours at 60 °C. The solvent was then dried in the oven at 70 °C for three hours. The product that was a powder was filled into the mold and pressed into pellets by using cold pressing at 10 MPa and curing molding at 170 °C, as shown in Figure 1.11.

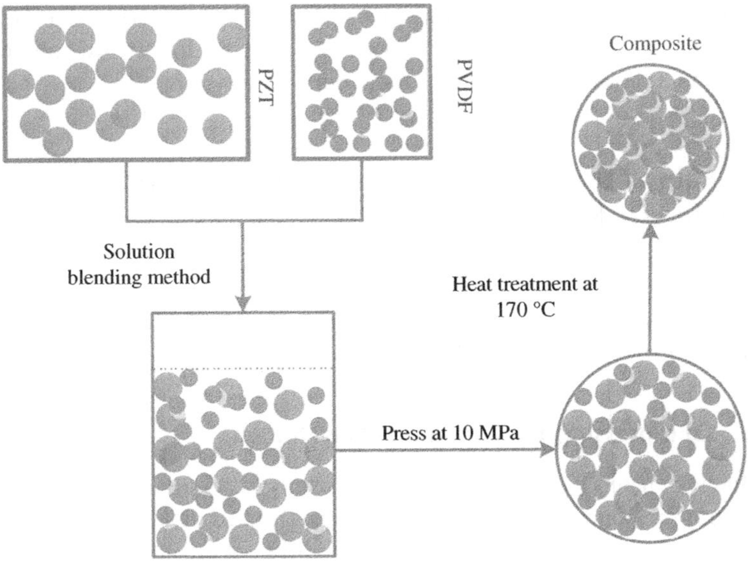

Figure 1.11 Cold-pressing and curing-molding method to fabricate PVDF/PZT composite. Source: Reproduced with permission from Wu et al. [79], Copyright 2016, Wiley Publications.

1.6.2 Cold Sintering Process

Compared to conventional methods that typically require high temperatures for sintering, the cold sintering process (CSP) significantly reduces the sintering temperature. Alternative CSP techniques can produce a high dielectric permittivity, have the ability to integrate temperature-sensitive materials, enable low-temperature densification, have reduced energy consumption, and faster processing times [80]. CSP enables co-sintering of thermoplastic polymers and ceramic materials in one step [81]. The first step involves preparing a powder mixture, usually consisting of ceramic material particles or nanoparticles along with a polymer matrix material. Using various techniques, the powder mixture is shaped into a certain form, such as a pellet. The samples are placed in a chamber or furnace, where they are subjected to pressure as well as heat [82]. Densification often occurs with uniaxial pressure and temperature applied to assist the kinetics of dissolution and evaporation. With evaporation, a supersaturated condition is created, which precipitates the solutes. This permits densification and grain growth, enabling the CSP [82]. Vinoth and Wang [80] fabricated $BaTiO_3$/PVDF nanocomposite using CSP, as shown in Figure 1.12. $BaTiO_3$ powder (0.99 wt%) and PVDF powder (0.01 wt%) were mixed with di-H_2O

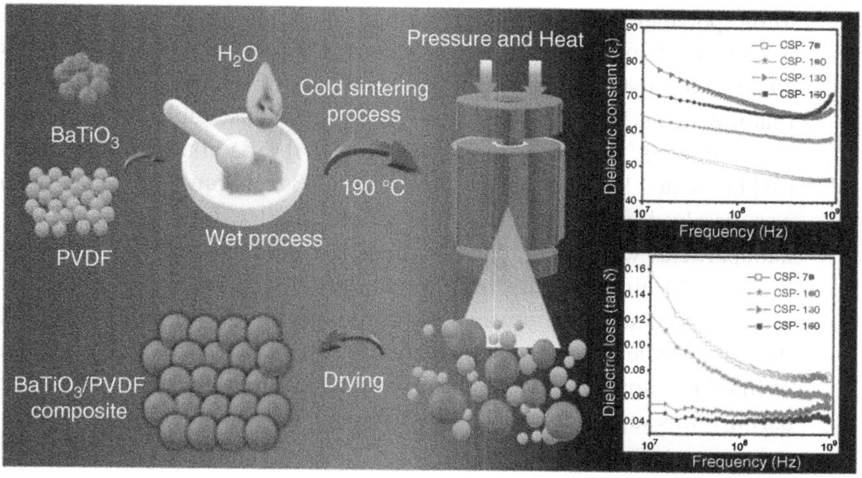

Figure 1.12 Cold sintering method to fabricate $BaTiO_3$/PVDF nanocomposite. Source: Reproduced with permission from Vinoth and Wang [80], Copyright 2023, ACS Publications.

(10 wt%) in a mortar and pestle for 15–20 minutes to get a homogeneous mixture. The mixture was then poured into a 12.7 mm uniaxial die, which was subsequently heated to 190 °C, and a pressure of 350 MPa was then applied. The sintering temperature was gradually lowered to room temperature, and the sintered samples were dried at 120 °C for 12 hours in the oven to obtain the final product ($BaTiO_3$/PVDF composite/hybrid).

1.6.3 Electrospinning (ES)

ES is a versatile and efficient approach for generating continuous nanofibers, as the fiber diameter can be adjusted from micrometers to nanometers [83]. The ceramic/polymer electrospun composite nanofibers exhibit precise control over fiber morphology, composition, and alignment, and they also exhibit high surface-to-volume ratios, making them suitable for various applications [83]. ES techniques can be used to create fibers with intricate structures, such as ribbon-shaped, porous, core-shell, or hollow [83]. ES process operates on the idea that when a solution is subjected to high voltage, the ohmic current evenly distributes the charges among the molecules. The ohmic current becomes predominantly convective as the solution is ejected from the spinneret tip; through fiber deposition, charges are transported from the ES tip to the target [84]. In ES, the first step is to prepare the polymer solution, which is attained by dissolving the polymer in a desired solvent, and then the ceramic material particles are added to the polymer solution. Then a syringe containing the polymer solution is connected to a high-voltage power supply [85]. A metallic needle is attached to the syringe, acting as the spinneret. Ground collectors, such as rotating drums or stationary plates, are placed at a suitable distance from the spinnerets, as shown in Figure 1.13a [86]. Sharafkhani and Kokabi [87] fabricated PVDF/$BaTiO_3$ nanofibers by ES technique. PVDF (15 wt%) was dissolved in a mixture of DMF/acetone at 30 °C for 24 hours. $BaTiO_3$ nanoparticles were then added to the PVDF solution and was ultrasonicated for 30 minutes, and stirred for 24 hours to obtain stable suspension. The prepared suspension was poured into a syringe with a metal needle, then the needle tip was kept at a constant distance of 10 cm, and a voltage of 16 kV was applied between the tip and target, and a constant flow rate of 0.6 ml/h was supplied through a syringe pump, as shown in Figure 1.13b.

1.6.4 Solvent-casting Method

Due to its simplicity, solvent casting has been used for years to prepare thin films or membranes of piezoelectric composites [88]. Using the solvent-casting method, thin and flexible piezoelectric composite films can be prepared with controlled thicknesses and compositions. Solution/solvent casting is a cost-effective way to produce composite films with piezoelectric properties [89]. Its simple application, flexible film thickness control, and compatibility with different substrates make it a viable choice for various applications, such as energy-harvesting systems. The first step involves the preparation of a polymer solution by dissolving a polymer matrix material in an appropriate solvent, and then ceramic particles are added to the polymer solution [90]. The solution is stirred or sonicated until it becomes homogeneous; the particles must be evenly distributed throughout the solution in order to form a uniform composite film. The homogeneous solution is then cast onto a flat substrate, such as a glass plate or a Petri dish, and it is spread evenly on the substrate. After casting, the solvent is allowed to evaporate, then the nanocomposite is obtained as a membrane or a film [90]. Jain et al. [91] prepared PVDF–PZT composite films using the solvent-casting method. PVDF was dissolved in DMF, and the solution was stirred at a high temperature in a microwave oven until everything was dissolved. PVDF–PZT composite film was obtained after casting the solution on the glass plate and heating it at 65 °C until the solvent evaporated. Marycz et al. [92] also used solvent casting to prepare a PU/PLA/ZnO nanocomposite film. The polymer solution was prepared, and polymer granules (PLA and PU 20/80 w/w) were dissolved in DMF by magnetic stirring at 40 °C for three days. Then ZnO nanoparticles were

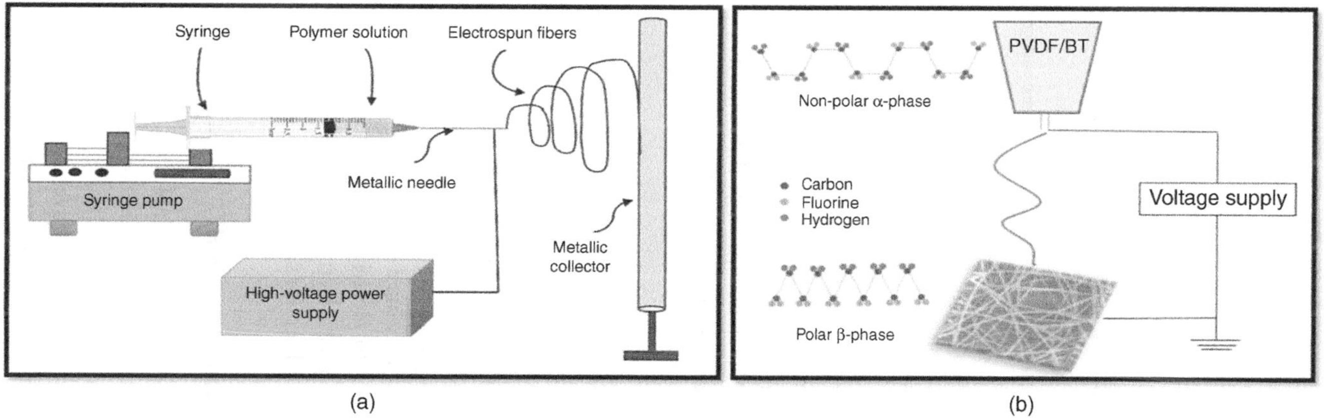

Figure 1.13 (a) The schematic representation of electrospinning technique. Source: Reproduced with permission from Sharma et al. [86], Copyright 2023, IntechOpen. (b) Schematic representation of electrospinning technique to form PVDF/$BaTiO_3$ nanofibers. Source: Reproduced with permission from Sharafkhani and Kokabi [87], © 2018/AIP Publishing.

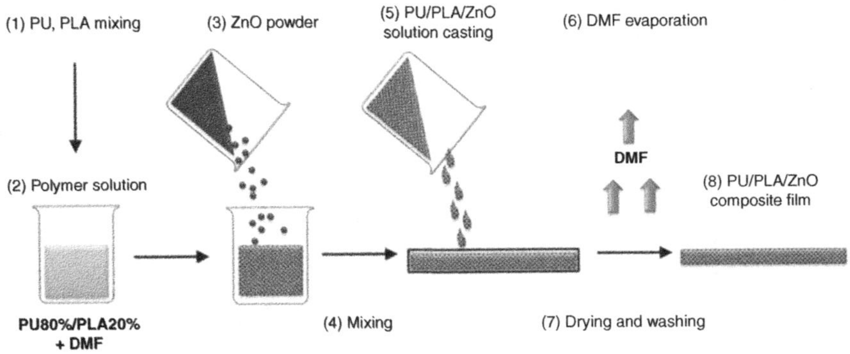

Figure 1.14 Schematic representation of solvent-casting method to prepare PU/PLA/ZnO film. Source: Reproduced with permission from Marycz et al. [92], Copyright 2016, MDPI Publications.

added to the polymer solution, and the mixture was stirred at 40 °C for 24 hours to obtain a uniform suspension. The solution was then stirred with an ultrasonic homogenizer for 15 minutes, and the solution was cast into a glass Petri dish. PU/PLA/ZnO nanocomposite film was then removed from the glass mold after DMF was evaporated, as shown in Figure 1.14.

1.7 Piezoelectric Energy Harvesters (PEHs) System

PEH is made up of an energy conversion system, energy storage, a power control circuit, as well as mechanical vibration equipment. First, the structure in the surrounding environment transforms the vibration energy into mechanical energy. By utilizing a piezoelectric device, this mechanical energy is then converted into electrical energy and stored as electricity. There are several types of piezoelectric energy reapers' configurations as discussed in the following subsections.

1.7.1 Cantilever Configuration of PEH

Many structural configurations are available for PEHs devices, such as cymbal type, circular diaphragm, as well as stack configurations; however, most PEHs use a cantilever-based configuration, as depicted in Figure 1.15. According to the number of piezoelectric layers attached to the metal-based layer, a cantilever shaft structure can be categorized into two forms: a unimorph structure having just single active piezoelectric layer and a bimorph

Figure 1.15 A cantilever configuration of PEH: (a) unimorph type, (b) bimorph type.

structure with two thin layers of active piezoelectric material. The bimorph structure has received noteworthy attention within PEH systems due to its ability to significantly improve electrical energy output without causing any significant volume changes [28]. The cantilever beam-type piezoelectric generator is the most extensively researched as well as constructed structure because of its straightforward structural geometry and capacity to generate relatively significant stresses for a given force input. Because the resonant frequency of the cantilever beam structure is frequently and significantly higher than external excitation, a metallic mass is typically placed at the end of the beam to lower it. An example of this configuration would be combining a thin layer of piezoelectric material with a conductive metal layer and fixing the resultant structure at one end in order to achieve its flexural function. By applying mechanical vibrations to the cantilever's mass system, the piezoelectric layer produces an AC voltage. PZT is a commonly used bimorph/unimorph piezoelectric material for cantilevers in resonant modes. The piezoelectric layer deforms when subjected to external force, creating an electric potential as the cantilever oscillates. A variety of other beam shapes have also been investigated to increase the output performance, including triangular, taper, S-shaped, and arc-shaped.

1.7.2 Circular Diaphragm Configuration of PEH

As shown in Figure 1.16, a circular diaphragm is typically constructed from a thin disk-shaped piezoelectric material fastened to a metal shim fixed to the edges of the clamping ring. A mass known as "proof mass" is fixed to the core of the diaphragm under low-frequency operation to enhance low-frequency performance and power output. An affixed proof mass is fixed to the center of the circular diaphragm of piezoelectric materials. Through the production of the piezo-metallic composite, prestress can also

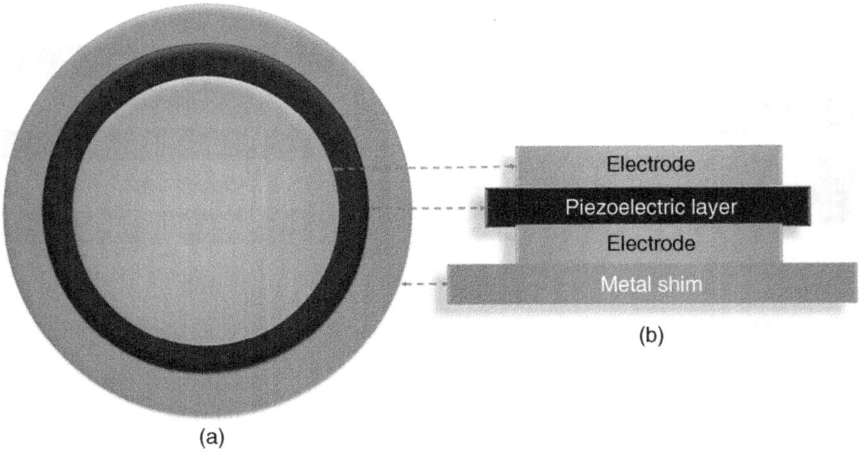

Figure 1.16 Circular diaphragm PEH: (a) top view, (b) side/edge view.

be introduced into the piezoelectric ceramic. A cymbal transducer can produce more energy than a cantilever energy harvester since the cymbal construction can resist more impact, but due to its robustness, it can only be used with powerful vibration sources.

1.7.3 Cymbal Configuration of PEH

A piezoelectric layer is sandwiched stuck between the two metallic end caps on either side of the cymbal transducer (Figure 1.17). Typically, steel is used since it has a higher yield strength than brass or aluminum, resulting in a higher force loading capability. The piezoelectric layer experiences radial tension when axial stress is applied to the metal end caps. A cymbal transducer can produce more energy than a cantilever energy harvester since the cymbal construction can resist more impact, but due to its robustness, it can only be used with powerful vibration sources. A cymbal transducer

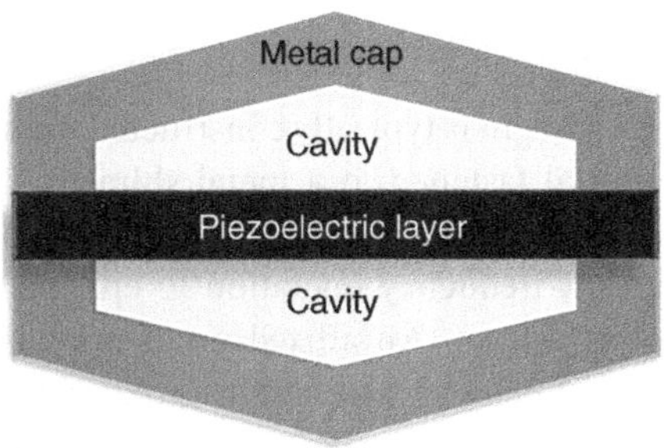

Figure 1.17 Cymbal configuration of PEH.

is designed to be used in applications with high-impact forces. Cymbal and other structures of PEH can be combined to obtain maximum output energy, such as multilayer stacks, cantilevers, or integrating them under high loads.

1.7.4 Stacked Configuration of PEH

Multiple piezoelectric layers are piled on top of one another in the stack configuration of PEHs, so that their polarities line up with the pressure being applied (Figure 1.18). The stacked configuration is created by mixing ceramic powders, dispersants, binders, plasticizers, and solvents to prepare a homogenous slurry. In the next step, tape casting is used to obtain ceramic films, and electrodes are screen-printed on dried tapes. The ceramic layers are then stacked and sintered [93]. In order to improve the output power, PEH stack configuration is always electrically arranged in series or parallel. When piezoelectric stacks are connected in series, they produce a higher open circuit voltage capable of igniting lighter devices, whereas when they are connected in parallel, they generate a higher electric current, as well as their power could be utilized for electricity supply with energy harvesting [93].

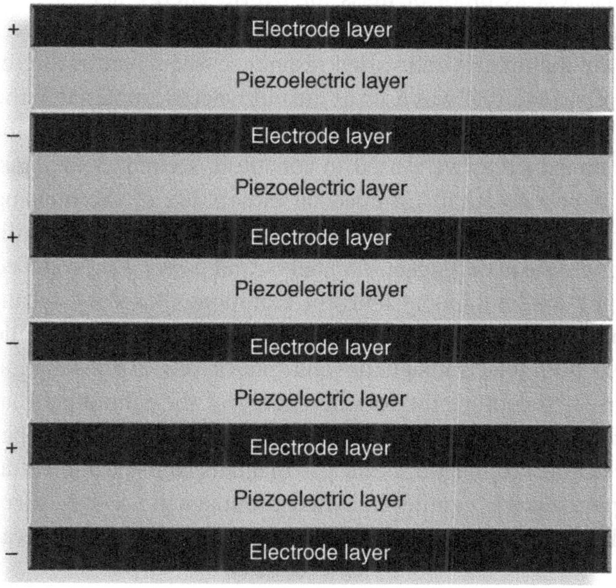

Figure 1.18 Stacked configuration of PEH.

1.8 Application of Hybrid Materials for Hybrid Energy-Harvesting Systems

A hybrid energy harvester is a new class of harvesters that integrate multiple harvester technologies into a single unit. The energy deficiency of a single harvester has led to the suggestion of hybrid energy-harvesting devices. Strong optical absorption, photoconductivity, electroluminescence, solution-based processing ease, electrical and ionic conductivity, high charge mobility, adjustable bandgaps, high photoconversion efficiency, as well as light-harvesting properties for solar devices are just a few benefits of these materials, according to a recent review.

1.8.1 Piezoelectric and Electromagnetic Hybrid Systems

Piezoelectric power generation as well as electromagnetic power production could be combined to harvest more vibration energy. Multi-fidelity nonlinear coupling systems of force, electricity, and magnetism are capable of operating at higher frequencies. In order to design and optimize a piezoelectric–electromagnetic hybrid harvester, theoretical analysis, and Finite Element Method (FEM) simulation are conducted first. It has been successfully applied to use PZT cantilever arrays, induction coils, as well as permanent magnets as proof of mass for hybrid energy harvester. A vibration-based multimodal piezoelectric–electromagnetic hybrid harvester with various vibrational modes and varied resonance frequencies was created by Toyabur et al. [94]. PZT was used as the piezoelectric material in the system, and NdFeB magnets with conductive coils attached at their tips for electromagnetic transduction were used to constitute secondary systems. The system has a primary shaft clamped on either side of the housing harvester with four lateral symmetric piezoelectric beams. Without significantly expanding the size of the gadget, a higher output power was produced from modest input accelerations at ambient vibrations. Based on a cantilever beam structure, Yang created a unique piezoelectric–electromagnetic hybrid harvester. Both electromagnetic and piezoelectric transduction methods were used to capture the kinetic energy of the vibrations that energized the gadget [95]. The system was made up of two-layer copper coils arranged close to the magnets, and the magnets were positioned at the end of the piezoelectric cantilever. When the magnets' poling orientation was parallel to the plane of the coils, a higher output voltage was attained [95].

1.8.2 Piezoelectric–Triboelectric Hybrid Systems

Some potential solutions have been developed by combining piezoelectric and triboelectric systems in order to maximize energy production. Numerous mechanical energy sources, including vibration, compression, and deflection, are well reacted to the two mechanisms. The triboelectrification and piezoelectric phenomena do not overlap during electricity generation because triboelectrification occurs on the surfaces of materials and the piezoelectric phenomenon occurs below the surface whenever an external pressure is applied to a suitable configuration. By developing various components to simultaneously increase the piezoelectric and triboelectric effects, a number of groundbreaking experiments have shown the practicality of merging the triboelectric and piezoelectric effects into one integrated system for mechanical energy harvesting. Han et al. created an r-shaped piezoelectric–triboelectric hybrid nanogenerator system [96]. The upper part was the piezoelectric nanogenerator, consisting of a Polyethylene Terephthalate (PET) film, polarized PVDF film, and surface-nanostructured Al electrodes. The triboelectric nanogenerator was located at the bottom and was made up of a bottom surface-nanostructured Al electrode, PET/Indium Tin Oxide (ITO) electrode, and micro/nano dual-scale (MNDS) polydimethylsiloxane electrode [96]. In the system, pressing the hybrid nanogenerator causes the PVDF to bend, and the gap distance decreases, producing piezoelectric as well as triboelectric output (Figure 1.19a), and then the piezoelectric and triboelectric outputs decrease to zero shortly after the hybrid nanogenerator is pressed (Figure 1.19b). PVDF then recovers its original shape and the gap distance increases, producing reversed piezoelectric as well as triboelectric effects, upon releasing the hybrid nanogenerator (Figure 1.19c). Lastly, a movement cycle is achieved when the hybrid nanogenerator is released, the charges are equilibrated, and no electrical output is produced (Figure 1.19d).

1.8.3 Triboelectric, Piezoelectric, and Electromagnetic Hybrid System

Research is being conducted on triple-hybrid energy-harvesting systems, integrating electromagnetic, piezoelectric, as well as triboelectric mechanisms in one system. This might be an effective way to raise output performance. All three types of energy can be harvested as well as utilized simultaneously because they operate under the same conditions to produce electricity. Several hybrid triboelectric–electromagnetic–piezoelectric

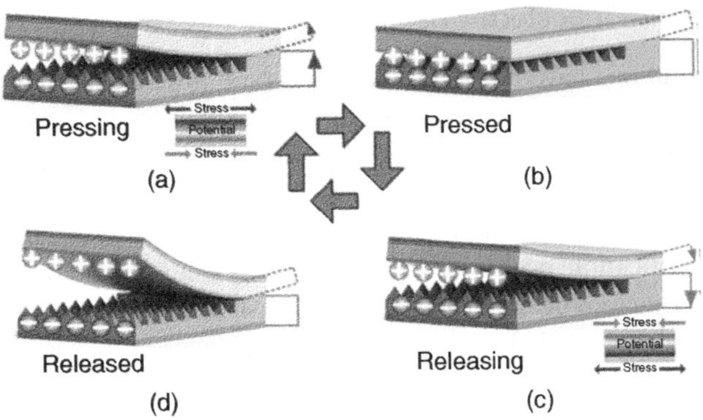

Figure 1.19 The working principle of the piezoelectric–triboelectric hybrid nanogenerator. (a) PVDF is bent in the pressed state, resulting in an increase in piezoelectric and triboelectric output. (b) The charges reach equilibrium when pressed and no electrical output is produced. (c) Upon releasing, PVDF is recovered, resulting in reversed piezoelectric output and triboelectric output. (d) During the released state, the charges reach equilibrium and do not produce any electrical output. Source: Reproduced with permission from Mengdi Han et al. [96], Copyright 2013, ACS publications.

devices have been reported. Ma et al. developed an integrated triboelectric/electromagnetic/piezoelectric hybrid energy harvester centered on rotational motion. The device consisted of three sections: the electromyography (EMG), a piezoelectric nanogenerator (PENG), and triboelectric nanogenerators (TENGs), which produced electricity. An acrylic cylinder and shell were used to integrate three different generator components into a hybrid energy harvester [97]. The generator was mounted on a rotating disk for rotation. A three-component generator was produced by reciprocating Magnet I in an acrylic cylinder to harvest energy. As a result, the energy-harvesting efficiency was improved, and the energy conversion efficiency was higher. To scavenge mechanical vibrational energy, the technology was combined with transformers and rectifiers to create a hybridized triboelectric–electromagnetic–piezoelectric nanogenerator by Du et al. [98]. The nanogenerator could also be transformed into direct current through power management. The maximum output power density was up to 100 W [98]. As depicted in Figure 1.20A, Li et al. [100] proposed a hybrid flutter-based energy harvester (HFEH) system consisting of a piezoelectric and electromagnetic component connected by magnetic forces [99]. The piezoelectric component was made up of a tip magnet A, two rigid airfoil components that were hinged at the end of the shaft, and a piezoelectric cantilever beam. For synchronous deflection, the two airfoil pieces were bonded to the spinning shaft. The piezoelectric component

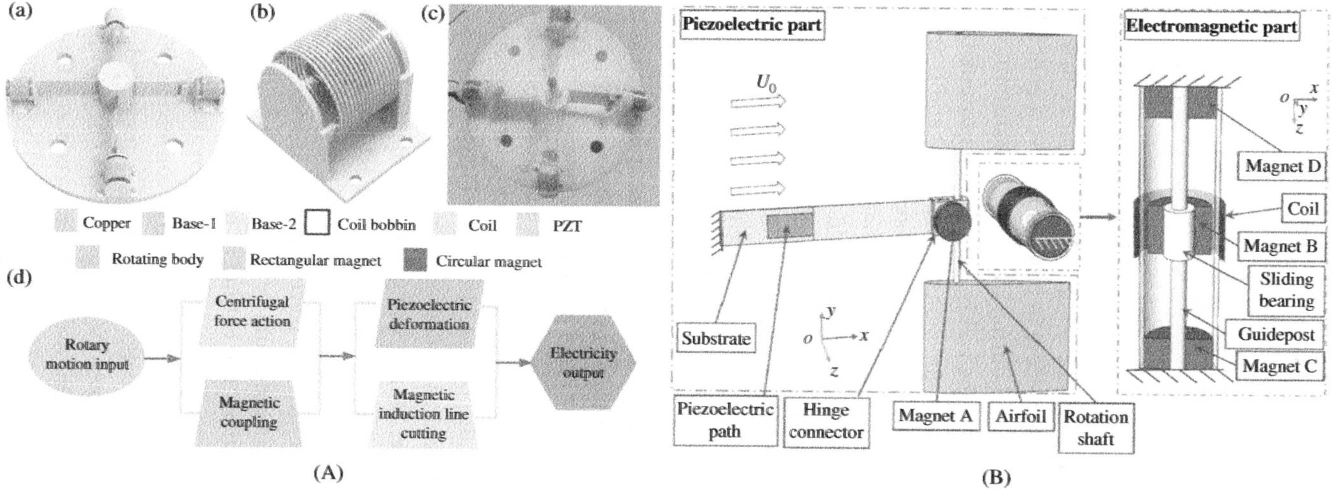

Figure 1.20 (A) Prototype rendering and physical schematic. (a) Prototype rendering diagram, (b) enlarged schematic of the electromagnetic unit, and (c) the physical drawing of the entire prototype. Source: Reproduced with copyright permission from Zhou et al. [99]/Elsevier. (B) Schematic diagram of the HFEH. Source: Reproduced with permission from Li et al. [100], Copyright 2022, Elsevier Science Ltd.

is affected by both the aeroelastic lift and the moment in the flow field. The piezoelectric cantilever beam and the airfoil structure both produce periodic pitching motions when the negative aerodynamic dampening is greater than the structural dampening. In order to transform the vibrational energy into electrical energy, the piezoelectric patch is linked to the external resistance R_1. The magnet B, which is slidably connected to a stationary guidepost by a sliding bearing, the coil, a sleeve fixed at both ends, magnets C and D installed at both ends of the sleeve, and the coil make up the electromagnetic component. Magnet A may drive the movement of magnet B in the sleeve in order to achieve a periodic change in the magnetic flux in the coil during vibration. To transform vibrational energy into electrical energy, the coil is coupled to an external resistance R_2. Because the magnetization directions of magnets C and D are opposite to that of magnet B, the repellent magnetic force functions as a magnetic spring to create a nonlinear restoring force that is comparable to magnet B [100]. A piezoelectric electromagnetic hybrid energy harvester based on magnetic coupling suction–repulsion motion along with centrifugal action was therefore explored by Zhou et al. in a different report, as depicted in Figure 1.20B [99]. The hybrid piezoelectric–electromagnetic energy harvester is mainly composed of piezoelectric cantilever beams, base-1, base-2, coil bobbin, coil, rotating body, rectangular magnets, and circular magnets. The electromagnetic component of base-2 is attached to base-1 with the use of bolts. The input excitation consists of a rotating body with a piezoelectric cantilever beam arranged around it. The piezoelectric cantilever beam also has a rectangular magnet attached to the end of it that fulfills the magnetic coupling by means of the circular magnet inside the coiled bobbin. A physical display, like the one in Figure 1.20A(c), makes it possible to see the represented structure more clearly. The structure was first constructed by these authors using 3D modeling software, and the assembled structure was then materialized using 3D printing technology. In summary, Figure 1.20A(d) depicts the transformation of mechanical energy into electrical energy. The piezoelectric cantilever shaft deforms as a result of magnetic coupling and centrifugal force, with the input of external rotational excitation. The cutting of magnetic induction lines by electromagnetic coil is now complete, enabling the realization of electrical energy output and the accomplishment of energy conversion [99].

Zhou et al. described the working principle as presented. Thus, the motion state diagram of the prototype is provided to help with the introduction and understanding of the overall operating concept. Here, it is easy to separate the entire motion state into three states. The magnets on top of the piezoelectric cantilever shaft have polarities that are the exact opposite of those in the coil frame. Additionally, the magnet's poles facing outward that are mounted at the piezoelectric cantilever beam's end are likewise in opposition. Using the examples of PEH1 and EMH1, under the influence of attraction, the two magnets are closest to one another, and as the spinning body continues to rotate, the attraction between them decreases. Because the magnets at the ends of PEH2 and EMH1 have identical magnetic poles, the repulsive force will steadily grow. This finalizes the slicing of the magnetic induction field lines necessary to produce electricity as well as the reciprocating motion of the circular magnet inside the coil bobbin. The piezoelectric–electromagnetic energy compound is finished by the simultaneous deformation of the piezoelectric cantilever shaft under the influence of centrifugal as well as magnetic force.

1.9 Present Development Challenges and Future Perspectives

Hybrid energy-harvesting via piezoelectric systems is still a relatively new field, and there is currently very limited research on hybrid energy-harvesting systems in comparison to single ones. Since input quantities have a substantial impact on the system's behavior and inputs and outputs do not have a unified specification standard, it is still impossible to determine which structure or mechanism is ideal. To develop successfully hybridized energy harvesters, it is also necessary to understand more fully the influence of the hybrid method on the communal efficiency of the system. In addition to the device's design, it must be compatible with advanced technologies and meet performance requirements to last for a long time. It will take more investigation to ascertain how much hybrid energy-harvesting technology can increase power density and energy conversion efficiency. Researchers should also be aware of the threat of creating devices with additional structural complexity and greater volume in pursuing hybrid effects, which have no effect on improved performance.

Due to their application-dependent nature, piezoelectric energy-harvesting devices have limited commercialization potential. Scientists and engineers have been actively researching this area for many years. According to a report by Zervos [101], piezoelectric harvesters' power efficiency could increase within a short period of time, from the order of microwatts (W) to the order of several milliwatts (mW). The major improvement in PEHs' output, dependability, and related technologies has increased demand for them on the international market, fostering further growth. According to the estimates from IDTechEx, piezoelectric energy-harvesting technology investment will increase to $145 million in 2018 in order to successfully create a $667 million industry by 2022 [28]. According to a recent report by Technavio [102], the market for piezoelectric smart materials was estimated to reach $42 billion in 2020. According to research analysts, the aerospace and military industries will have good growth due to rising demand, the building industry will experience growth, and the research and development industries will increase their efforts [102]. Due to their superior qualities, a different Technavio analysis claimed that piezo-ceramics supplied significant shares to the piezoelectric smart materials market in 2016 [103]. Throughout the forecast period (2017–2021), the United States will be the market's largest revenue contributor from a geographic perspective. Piezoelectric systems having improved functional behaviors have found applications in the aerospace and defense industries for controlling engine vibration, ice formation on wings, turbulence-caused flow separation, high cabin noise levels, cold climates and control surfaces, smart sensors, and smart nano-robotics, among other things [28].

The performance of PEHs and their use with respect to social lives are both the subject of several research initiatives. Many piezoelectric devices, suited for sensor and actuator applications, use thick piezoelectric films made using the aerosol deposition process [28]. The National Institute of Aerospace (NIA), USA, exhibited a hybrid PEH ("33"-longitudinal mode) with an innovative structure that produced four times more power than a conventional "31" harvester [28]. Stewart et al. (2012) discovered a device's maximal power output obtained by filling the cantilever beam's middle two-thirds with active piezoelectric materials. A unique soccer field made of piezoelectric components has been built in Rio de Janeiro, Brazil, by the British energy business PAVEGEN to harvest electrical energy by harnessing the kinetic energy of people walking and running on the pitch. Six LED lights are powered by the captured energy at night to illuminate the slum regions [28]. By applying the piezoelectric energy-harvesting approach, the Korea Institute of Machinery and Materials (KIMM) created the first artificial basilar membrane in history. The cochlea in human ears can be mimicked by this technology in order to improve hearing. The frequency of the signal received from the basilar membrane is converted into an electric signal by the cochlea using a trapezoid-shaped nano-piezoelectric substance. This generated electric signal helps to stimulate the auditory nerves by permitting the operator to hear sound through it [104].

Again, growing technological breakthroughs lead to higher R&D spending, supporting a wide range of firms and industries in thriving and surviving in a cutthroat market. The worldwide chemical producer Arkema has created ultra-high value-added electroactive polymers such as PVDF and

PVDF films that could be utilized in the future for a variety of applications in the electrical, robotic, automotive, and aerospace industries. These kinds of polymers, which are an essential component of textiles, can also make garments fashionable and well-liked. These textiles with various sensors and sensing capabilities are effective enough to transmit a person's biomedical information. The Arkema group company Piezotech offers an exclusive line of piezo-polymer-based electroactive inks that fulfill the requirements of the actuator and sensor industries. Thus, by printing basic electronic components such as transistors and sensors with conductive inks and thin layers of electroactive polymers, it is now possible to fabricate electronic devices on flexible substrates such as paper, cloth, or plastic [105].

Through these years, there has been an increase in the market demand for polymer/ceramic composites/hybrids and piezoelectric polymers. Given their extensive potential, the quantity of scientific research articles in the field of piezoelectric energy harvesting has been rising dramatically over the past 10 years, as demonstrated earlier in Figure 1.1. Electrical energy, material science, applied physics, and mechanical engineering are just a few of the fields that PEH's multidisciplinary techniques are addressing. Due to the wide range of applications for hybrid, composite, or hybrid piezoelectric actuators and motors in the automotive, robotics, IT sector, biomedical, ecology, and energy sectors, the market for energy-harvesting devices made of ceramics, polymers, and ceramic/polymer composites/hybrids is still expanding globally [106, 107].

We would like to point out that there is either intentional or unintentional negligence on the part of researchers within the niche of piezoelectric hybrid materials to study/investigate converse piezoelectric effects. This along with the direct effect need to be reported together for proper validation of research reports.

1.10 Conclusion

Although batteries are considered conventional sources of energy, they tend to suffer from numerous constraints, including limited lifespan, power efficiency, as well as a limited capacity for storing energy, which necessitates frequent recharging. Energy harvesting from ambient sources is becoming a major focus of researchers who aim to eliminate or extend the life of batteries. The ability to harvest energy can extend the working lifespan of electronics, making it a particularly useful technique for inaccessible electronics or expensive maintenance, such as sensor nodes in isolated locations, biomedical systems/components such as implanted health trackers, as well as a large-scale network of sensors. A piezoelectric generator has several advantages over other methods, including its inherent transduction capability, ability to maintain efficiency regardless of scale, higher power density, low cost, capability to operate at high frequencies, and ease of application. There has been significant interest in piezoelectric hybrid such as piezoelectric transducers due to their electrical coupling properties and higher power density in comparison to electromagnetic as well as electrostatic transducers. A piezoelectric generator, however, has some disadvantages, including depolarization, brittleness, and poor coupling in piezo-film. As a result of poor adhesion between PVDF film and electrode materials, the fabrication of devices is difficult. To overcome the limitations of conventional PEHs, hybrid energy-harvesting systems have been proposed as a way to solve the energy-insufficiency problem of a single harvester. Hybrid energy harvesting is a newly emerging field that is also being extensively researched, as discussed in this chapter. In hybrid harvesting, energy is not only scavenged from manifold sources but also converted into electricity through various devices. Integrating manifold conversion systems in one design, such as hybrid energy harvesters, offers a wide range of applications. Hybrid nanogenerators' integrated construction enables them to be used anytime, anywhere, and to generate energy via a variety of energy sources. To boost output currents or voltages, the various components of the device unit can also be coupled in parallel or series. In the next decade, hybrid energy-harvesting systems are expected to experience rapid growth in a variety of applications due to a growing interest in hybrid energy-harvesting technologies for application in advanced electrical/electronics, machines, biomedical materials, etc.

References

1. Zeadally, S., Karim, F., Talpur, A., and Sheng, Q.Z. (2020). Design architectures for energy harvesting in the Internet of Things. *Renewable Sustainable Energy Rev.* 128: 109901.
2. Alvarado, U., Juanicorena, A., Adin, I. et al. (2012). Energy harvesting technologies for low-power electronics. *Trans. Emerging Telecommun. Technol.* 23: 728–741. https://doi.org/10.1002/ett.2529.
3. Wang, H. and Jasim, A. (2020). Chapter 14: Piezoelectric energy harvesting from pavement. In: *Eco-Efficient Pavement Construction Materials* (ed. F. Pacheco-Torgal, S. Amirkhanian, H. Wang, and E. Schlangen), 367–382. Woodhead Publishing https://doi.org/10.1016/B978-0-12-818981-8.00014-X.

4 Abdelkefi, A. (2016). Aeroelastic energy harvesting: a review. *Int. J. Eng. Sci.* 100: 112–135. https://doi.org/10.1016/j.ijengsci.2015.10.006.

5 Trolier-Mckinstry, S., Zhang, S., Bell, A.J., and Tan, X. (2018). High-performance piezoelectric crystals, ceramics, and films. *Annu. Rev. Mater. Res.* 48: 191–217. https://doi.org/10.1146/annurev-matsci-070616-124023.

6 Laboratories, P. and Manor, B. (1979). A comprehensive optical second harmonic generation study of the non-centrosymmetric character of biological structures. *J. Biol. Phys.* 6: 105–117.

7 Roy, M.S., Mirunalini, A.R., Krishna, P.S. et al. (2021). Harnessing energy from everyday life. *J. Phys. Conf. Ser.* 2115 (1): https://doi.org/10.1088/1742-6596/2115/1/012032.

8 Habib, M., Lantgios, I., and Hornbostel, K. (2022). A review of ceramic, polymer and composite piezoelectric materials. *J. Phys. D: Appl. Phys.* 55 (42): https://doi.org/10.1088/1361-6463/ac8687.

9 Aleksandrova, M., Tudzharska, L., Nedelchev, K., and Kralov, I. (2023). Hybrid organic/inorganic piezoelectric device for energy harvesting and sensing applications. *Coatings* 13 (2): 464.

10 Hossain, I.Z., Khan, A., and Hossain, G. (2022). A piezoelectric smart textile for energy harvesting and wearable self-powered sensors. *Energies* 15 (15): 5541.

11 Fath, A., Xia, T., and Li, W. (2022). Recent advances in the application of piezoelectric materials in microrobotic systems. *Micromachines* 13 (9): 1–18. https://doi.org/10.3390/mi13091422.

12 Thakur, V.K., Thakur, M.K., and Gupta, R.K. (2017). *Hybrid polymer composite materials: structure and chemistry*. Woodhead publishing.

13 Vijayakanth, T., Liptrot, D.J., Gazit, E. et al. (2022). Recent advances in organic and organic–inorganic hybrid materials for piezoelectric mechanical energy harvesting. *Adv. Funct. Mater.* 32 (17): 1–49. https://doi.org/10.1002/adfm.202109492.

14 Vicars, G.R. (1878). Acoustical effects of atmospheric pressure. *Nature* 17: 1–2.

15 Curie, J. and Curie, P. (2019). Développement par compression de l ' électricité polaire dans les cristaux hémièdres à faces inclinées. *Bull. de minéralogie* 3: 90–93.

16 Bailato, A. (1996). Piezoelectricity: history and new thrusts. In: *1996 IEEE Ultrasonics Symposium. Proceedings*, 575–583. IEEE.

17 Lippmann, G. (1881). Principe de la conservation de lélectricité, ou second principe de la théorie des phénomènes électriques. *J. Phys. Theor. Appl.* 10: 381–394.

18 Voigt, W. (1910). *Lehrbuch der kristallphysik:(mit ausschluss der kristalloptik)*, vol. 34. BG Teubner.

19 Kelvin (1893). Note by Lord Kelvin on electric molecules for the explanation of the piezo-electric and pyro-electric properties of crystals. *Lond. Edinb. Dublin Philos. Mag. J. Sci.* 36: 384–384. https://doi.org/10.1080/14786449308620495.

20 Zimmerman, D. (2018). 'A more creditable way': the discovery of active sonar, the Langevin–Chilowsky patent dispute and the Royal Commission on Awards to Inventors. *War in History* 25 (1): 48–68. https://doi.org/10.1177/0968344516651308.

21 Zhang, S., Li, F., Jiang, X. et al. (2015). Advantages and challenges of relaxor-$PbTiO_3$ ferroelectric crystals for electroacoustic transducers – a review. *Prog. Mater Sci.* 68: 1–66. https://doi.org/10.1016/j.pmatsci.2014.10.002.

22 Uchino, K. (2015). Glory of piezoelectric perovskites. *Sci. Technol. Adv. Mate.* 16 (4): 046001. https://doi.org/10.1088/1468-6996/16/4/046001.

23 Shirane, G., Kazuo, S., and Akitsu, T. (1952). Phase transitions in solid solutions of $PbZrO_3$ and $PbTiO_3$ (II) X-ray study. *J. Phys. Soc. Japan* 7: 12–18.

24 Tripathy, S.S. (1969). The piezoelectricity of poly(vinylidene fluoride). *Jpn. J. Appl. Phys.* 8 (7): 975.

25 Caliò, R., Rongala, U.B., Camboni, D. et al. (2014). Piezoelectric energy harvesting solutions. *Sensors* 4755–4790. https://doi.org/10.3390/s140304755.

26 Shepley, P.M., Stoica, L.A., Burnell, G., and Bell, A.J. (2019). Effects of poling and crystallinity on the dielectric properties of $PbTiO_3$ at cryogenic temperatures. *Sci. Rep.* 1–8. https://doi.org/10.1038/s41598-019-38995-9.

27 X. Zhu, "Piezoelectric Ceramic Materials: Processing, Properties, Characterization, and Applications," Nova Science Publishers, 2010, pp. 1-34

28 Mishra, S., Unnikrishnan, L., Nayak, S.K., and Mohanty, S. (2019). Advances in piezoelectric polymer composites for energy harvesting applications: a systematic review. *Macromol. Mater. Eng.* 304: 1–25. https://doi.org/10.1002/mame.201800463.

29 Guerin, S., Tofail, S.A.M., and Thompson, D. (2019). Organic piezoelectric materials: milestones and potential. *NPG Asia Mater.* 11 (1): 1–5. https://doi.org/10.1038/s41427-019-0110-5.

30 Zhang, S.J., Li, F., and Yu, F.P. (2016). Piezoelectric materials for cryogenic and high-temperature applications. In: *Structural Health Monitoring (SHM) in Aerospace Structures* (ed. F.-G. Yuan), 59–93.

Woodhead Publishing https://doi.org/10.1016/B978-0-08-100148-6.00003-2.

31 Nunes-Pereira, J., Costa, P., and Lanceros-Mendez, S. (2018). Piezoelectric energy production. *Compr. Energy Syst.* 3–5: 380–415. https://doi.org/10.1016/B978-0-12-809597-3.00324-2.

32 Priya, S., Song, H., Zhou, Y. et al. (2017). A review on piezoelectric energy harvesting: materials, methods, and circuits. *Energy Harvesting Syst.* 4 (1): 3–39. https://doi.org/10.1515/ehs-2016-0028.

33 Chen, Q., Sun, Y., Wang, Y. et al. (2013). ZnO nanowires-polyimide nanocomposite piezoresistive strain sensor. *Sens. Actuators, A* 190: 161–167. https://doi.org/10.1016/j.sna.2012.11.006.

34 Chowdhury, A.R., Jaksik, J., Hussain, I. et al. (2019). Surface-modified nanostructured piezoelectric device as a cost-effective transducer for energy and biomedicine. *Energy Technol.* 7 (5): 1–8. https://doi.org/10.1002/ente.201800767.

35 Wang, Z., He, L., Gu, X. et al. (2021). Rotational energy harvesting systems using piezoelectric materials: a review. *Rev. Sci. Instrum.* 92 (4): https://doi.org/10.1063/5.0039730.

36 Zhou, Q., Lam, K.H., Zheng, H. et al. (2014). Piezoelectric single crystal ultrasonic transducers for biomedical applications. *Prog. Mater. Sci.* 66: 87–111. https://doi.org/10.1016/j.pmatsci.2014.06.001.

37 Chen, J. and Panda, R. (2005). Review: Commercialization of piezoelectric single crystals for medical imaging applications. *Proc. IEEE Ultrason. Symp.* 1 (c): 235–240. https://doi.org/10.1109/ULTSYM.2005.1602839.

38 Xue, G., Geng, W., Fu, W. et al. (2022). Integrated fabrication and ferroelectric domain adjustment of lithium niobate single crystal films based on silicon substrate. *Mater. Des.* 215: 110447. https://doi.org/10.1016/j.matdes.2022.110447.

39 Carville, N.C., Collins, L., Manzo, M. et al. (2015). Biocompatibility of ferroelectric lithium niobate and the influence of polarization charge on osteoblast proliferation and function. *J. Biomed. Mater. Res. Part A* 103 (8): 2540–2548. https://doi.org/10.1002/jbm.a.35390.

40 Schweinler, H.C. (1952). Ferroelectricity in the ilmenite structure. *Phys. Rev.* 87 (1): 5–11. https://doi.org/10.1103/PhysRev.87.5.

41 Uchino, K. (2017). The development of piezoelectric materials and the new perspective. In: *Advanced Piezoelectric Materials* (ed. K. Uchino), 1–92. Woodhead Publishing https://doi.org/10.1016/b978-0-08-102135-4.00001-1.

42 Xu, M., Kang, H., Guan, L. et al. (2017). Facile fabrication of a flexible $LiNbO_3$ piezoelectric sensor through hot pressing for biomechanical monitoring. *ACS Appl. Mater. Interfaces* 9 (40): 34687–34695. https://doi.org/10.1021/acsami.7b10411.

43 Bartasyte, A., Margueron, S., Baron, T. et al. (2017). Toward high-quality epitaxial $LiNbO_3$ and $LiTaO_3$ thin films for acoustic and optical applications. *Adv. Mater. Interfaces* 4 (8): 1–36. https://doi.org/10.1002/admi.201600998.

44 Uršič, H., Santo Zarnik, M., and Kosec, M. (2011). $Pb(Mg_{1/3}Nb_{2/3})O_3$–$PbTiO_3$ (PMN-PT) material for actuator applications. *Smart Mater. Res.* 2011: 1–6. https://doi.org/10.1155/2011/452901.

45 Park, S.E. and Shrout, T.R. (1997). Ultrahigh strain and piezoelectric behavior in relaxor based ferroelectric single crystals. *J. Appl. Phys.* 82 (4): 1804–1811. https://doi.org/10.1063/1.365983.

46 Gamboa, B., Bhalla, A., and Guo, R. (2020). Assessment of PZT (soft/hard) composites for energy harvesting. *Ferroelectrics* 555 (1): 118–123. https://doi.org/10.1080/00150193.2019.1691389.

47 Algueró, M., Cheng, B.L., Guiu, F. et al. (2001). Degradation of the d_{33} piezoelectric coefficient for PZT ceramics under static and cyclic compressive loading. *J. Eur. Ceram. Soc.* 21 (10–11): 1437–1440. https://doi.org/10.1016/S0955-2219(01)00036-X.

48 Lay, R., Deijs, G.S., and Malmström, J. (2021). The intrinsic piezoelectric properties of materials – a review with a focus on biological materials. *RSC Adv.* 11 (49): 30657–30673. https://doi.org/10.1039/d1ra03557f.

49 Banerjee, S., Bairagi, S., and Wazed Ali, S. (2021). A critical review on lead-free hybrid materials for next generation piezoelectric energy harvesting and conversion. *Ceram. Int.* 47 (12): 16402–16421. https://doi.org/10.1016/j.ceramint.2021.03.054.

50 Sood, A., Desseigne, M., Dev, A. et al. (2022). A comprehensive review on barium titanate nanoparticles as a persuasive piezoelectric material for biomedical applications: prospects and challenges. *Small* 19 (12): 1–29. https://doi.org/10.1002/smll.202206401.

51 Karaki, T., Yan, K., Miyamoto, T., and Adachi, M. (2007). Lead-free piezoelectric ceramics with large dielectric and piezoelectric constants manufactured from $BaTiO_3$ nano-powder. *Jpn. J. Appl. Phys., Part 2* 46 (4–7): 44–46. https://doi.org/10.1143/JJAP.46.L97.

52 Takahashi, H., Numamoto, Y., Tani, J., and Tsurekawa, S. (2006). Piezoelectric properties of $BaTiO_3$ ceramics with high performance fabricated by microwave sintering. *Jpn. J. Appl. Phys., Part 1* 45 (9 B): 7405–7408. https://doi.org/10.1143/JJAP.45.7405.

53 Wada, S., Takeda, K., Muraishi, T. et al. (2007). Preparation of [110] grain oriented barium titanate ceramics by templated grain growth method and their piezoelectric properties. *Jpn. J. Appl. Phys., Part 1* 46 (10 B): 7039–7043. https://doi.org/10.1143/JJAP.46.7039.

54 Wu, J., Xiao, D., and Zhu, J. (2015). Potassium-sodium niobate lead-free piezoelectric materials: Past, present, and future of phase boundaries. *Chem. Rev.* 115 (7): 2559–2595. https://doi.org/10.1021/cr5006809.

55 Saito, Y., Takao, H., Tani, T. et al. (2004). Lead-free piezoceramics. *Nature* 432 (7013): 84–87.

56 Guo, Y., Kakimoto, K.I., and Ohsato, H. (2004). Phase transitional behavior and piezoelectric properties of $(Na_{0.5}K_{0.5})NbO_3$-$LiNbO_3$ ceramics. *Appl. Phys. Lett.* 85 (18): 4121–4123. https://doi.org/10.1063/1.1813636.

57 Yang, L., Chi, S., Dong, S. et al. (2020). Preparation and characterization of a novel piezoelectric nanogenerator based on soluble and meltable copolyimide for harvesting mechanical energy. *Nano Energy* 67: 104220. https://doi.org/10.1016/j.nanoen.2019.104220.

58 Kong, E.S.-W. (1983). Developments in crystalline polymers edited by DC Bassett. *J. Appl. Crystallogr.* 16 (3): 361–362. https://doi.org/10.1107/s0021889883010626.

59 Katsouras, I., Asadi, K., Li, M. et al. (2016). The negative piezoelectric effect of the ferroelectric polymer poly(vinylidene fluoride). *Nat. Mater.* 15 (1): 78–84. https://doi.org/10.1038/nmat4423.

60 Sencadas, V., Gregorio, R., and Lanceros-Méndez, S. (2009). α to β phase transformation and microestructural changes of PVDF films induced by uniaxial stretch. *J. Macromol. Sci., Part B: Phys.* 48 (3): 514–525. https://doi.org/10.1080/00222340902837527.

61 Branciforti, M.C., Sencadas, V., Lanceros-Mendez, S., and Gregorio, R. (2007). New technique of processing highly oriented poly(vinylidene fluoride) films exclusively in the β phase. *J. Polym. Sci., Part B: Polym. Phys.* 45 (19): 2793–2801. https://doi.org/10.1002/polb.21239.

62 Uchino, K. (2017). Piezoelectro composites. *Compr. Compos. Mater.* II: 613–624. https://doi.org/10.1016/B978-0-12-803581-8.03939-4.

63 Li, H., Tian, C., and Deng, Z.D. (2014). Energy harvesting from low frequency applications using piezoelectric materials. *Appl. Phys. Rev.* 1 (4): https://doi.org/10.1063/1.4900845.

64 Chang, J., Shen, Y., Chu, X. et al. (2015). Large d_{33} and enhanced ferroelectric/dielectric properties of poly(vinylidene fluoride)-based composites filled with $Pb(Zr_{0.52}Ti_{0.48})O_3$ nanofibers. *RSC Adv.* 5 (63): 51302–51307. https://doi.org/10.1039/c5ra07932b.

65 Chamankar, N., Khajavi, R., Yousefi, A.A. et al. (2020). A flexible piezoelectric pressure sensor based on PVDF nanocomposite fibers doped with PZT particles for energy harvesting applications. *Ceram.*

Int. 46 (12): 19669–19681. https://doi.org/10.1016/j.ceramint.2020.03.210.

66 Kumar, M., Kulkarni, N.D., and Kumari, P. (2022). Fabrication and characterization of PVDF/$BaTiO_3$ nanocomposite for energy harvesting application. *Mater. Today Proc.* 56: 1151–1155. https://doi.org/10.1016/j.matpr.2021.11.133.

67 Dashtizad, S., Alizadeh, P., and Yourdkhani, A. (2021). Improving piezoelectric properties of PVDF fibers by compositing with $BaTiO_3$-Ag particles prepared by sol-gel method and photochemical reaction. *J. Alloys Compd.* 883: 160810. https://doi.org/10.1016/j.jallcom.2021.160810.

68 Kong, H., Jin, Y., Li, G. et al. (2023). Design and fabrication of a hierarchical structured pressure sensor based on $BaTiO_3$/PVDF nanofibers via near-field electrospinning. *Adv. Eng. Mater.* 25: 1–9. https://doi.org/10.1002/adem.202201660.

69 Patra, S., Ray, A., Roy, A. et al. (2018). ZnO polymer composite based visible blind UV photo detector. *Mater. Res. Bull.* 101: 240–245. https://doi.org/10.1016/j.materresbull.2018.01.041.

70 Liu, H., Fu, H., Sun, L. et al. (2021). Hybrid energy harvesting technology: from materials, structural design, system integration to applications. *Renewable Sustainable Energy Rev.* 137: 110473. doi: 10.1016/j.rser.2020.110473.

71 Bhunia, R., Das, S., Dalui, S. et al. (2016). Flexible nano-ZnO/polyvinylidene difluoride piezoelectric composite films as energy harvester. *Appl. Phys. A* 122 (7): 1–13. https://doi.org/10.1007/s00339-016-0161-1.

72 Peihai, J., Ling, W., Lizhu, L., and Xiaorui, Z. (2016). Preparation and characterisation of Al-doped ZnO and PVDF composites. *High Voltage* 1 (4): 166–170. https://doi.org/10.1049/hve.2016.0057.

73 Fu, Y., Cheng, Y., Chen, C. et al. (2022). Study on preparation process and enhanced piezoelectric performance of pine-needle-like ZnO@PVDF composite nanofibers. *Polym. Test.* 108: 107513. https://doi.org/10.1016/j.polymertesting.2022.107513.

74 Newnham, R.E. (1986). Composite electroceramics. *Ferroelectrics* 68 (1): 1–32. https://doi.org/10.1080/00150198608238734.

75 Britz, D. (1978). iR elimination in electrochemical cells. *J. Electroanal. Chem.* 88: 309–352.

76 Dang, Z.M., Yuan, J.K., Zha, J.W. et al. (2012). Fundamentals, processes and applications of high-permittivity polymer-matrix composites. *Prog. Mater. Sci.* 57 (4): 660–723. https://doi.org/10.1016/j.pmatsci.2011.08.001.

77 Lee, H.J., Zhang, S., Bar-Cohen, Y., and Sherrit, S. (2014). High temperature, high power piezoelectric composite transducers. *Sensors* 14 (8): 14526–14552. https://doi.org/10.3390/s140814526.

78 Binner, J., Chang, H., and Higginson, R. (2009). Processing of ceramic-metal interpenetrating composites. *J. Eur. Ceram. Soc.* 29 (5): 837–842. https://doi.org/10.1016/j.jeurceramsoc.2008.07.034.

79 Wu, Z., Yang, Z., Zhang, J., and Qu, X. (2016). Fabrication and characterization of the piezoelectric ceramic–polymer composites. *Int. J. Appl. Ceram. Technol.* 13 (4): 690–696. https://doi.org/10.1111/ijac.12550.

80 Vinoth, S. and Wang, S.F. (2023). Cold sintering process for a $BaTiO_3$/poly(vinylidene difluoride) ceramic-polymer composite: evaluation of the structural and microwave dielectric properties. *Inorg. Chem.* https://doi.org/10.1021/acs.inorgchem.3c00838.

81 Guo, J., Berbano, S.S., Guo, H. et al. (2016). Cold sintering process of composites: bridging the processing temperature gap of ceramic and polymer materials. *Adv. Funct. Mater.* 26 (39): 7115–7121. https://doi.org/10.1002/adfm.201602489.

82 Guo, J., Zhao, X., Herisson De Beauvoir, T. et al. (2018). Recent progress in applications of the cold sintering process for ceramic–polymer composites. *Adv. Funct. Mater.* 28 (39): 1–15. https://doi.org/10.1002/adfm.201801724.

83 Esfahani, H., Jose, R., and Ramakrishna, S. (2017). Electrospun ceramic nanofiber mats today: synthesis, properties, and applications. *Materials* 10 (11): https://doi.org/10.3390/ma10111238.

84 Wu, H., Pan, W., Lin, D., and Li, H. (2012). Electrospinning of ceramic nanofibers: fabrication, assembly and applications. *J. Adv. Ceram.* 1 (1): 2–23. https://doi.org/10.1007/s40145-012-0002-4.

85 Xu, Y., Ndayikengurukiye, J., Akono, A.T., and Guo, P. (2022). Fabrication of fiber-reinforced polymer ceramic composites by wet electrospinning. *Manuf. Lett.* 31: 91–95. https://doi.org/10.1016/j.mfglet.2021.07.017.

86 Sharma, G.K. and James, N.R. (2022). Electrospinning: the technique and applications. In: *Recent Developments in Nanofibers Research.* Intechopen.

87 Sharafkhani, S. and Kokabi, M. (2018). Tailoring favor crystalline structure via electrospun PVDF/$BaTiO_3$ nanofibers. *AIP Conf. Proc.* 1920: https://doi.org/10.1063/1.5018933.

88 Gayathiri, S., Gowdhaman, P., and Annamalai, V. (2017). Piezoelectric ceramic-polymer composites as smart materials: an overview of preparation methods. *J. Environ. Nanotechnol.* 6 (3): 89–89. https://doi.org/10.13074/jent.2017.09.173275.

89 Hintermueller, D. and Prakash, R. (2022). Comprehensive characterization of solution-cast pristine and reduced graphene oxide composite polyvinylidene fluoride films for sensory applications. *Polymers* 14 (13): https://doi.org/10.3390/polym14132546.

90 Salit, M.S., Jawaid, M., Bin Yusoff, N., and Hoque, M.E. (2015). *Manufacturing of Natural Fibre Reinforced Polymer Composites*, 1–383. Springer https://doi.org/10.1007/978-3-319-07944-8.

91 Jain, A., Kumar, S.J., Kumar, M.R. et al. (2014). PVDF-PZT composite films for transducer applications. *Mech. Adv. Mater. Struct.* 21 (3): 181–186. https://doi.org/10.1080/15376494.2013.834094.

92 Marycz, K., Marędziak, M., Grzesiak, J. et al. (2016). Polyurethane/polylactide-blend films doped with zinc ions for the growth and expansion of human olfactory ensheathing cells (OECs) and adipose-derived mesenchymal stromal stem cells (ASCs) for regenerative medicine applications. *Polymers* 8 (5): https://doi.org/10.3390/polym8050175.

93 Li, L., Xu, J., Liu, J., and Gao, F. (2018). Recent progress on piezoelectric energy harvesting: structures and materials. *Adv. Compos. Hybrid Mater.* 1: 478–505.

94 Toyabur, R.M., Salauddin, M., Cho, H., and Park, J.Y. (2018). A multimodal hybrid energy harvester based on piezoelectric- electromagnetic mechanisms for low-frequency ambient vibrations. *Energy Convers. Manage.* 168: 454–466. https://doi.org/10.1016/j.enconman.2018.05.018.

95 Yang, B. (2022). Hybrid energy harvester based on piezoelectric and electromagnetic mechanisms. *J. Micro/Nanolithogr. MEMS MOEMS* 9: 1–10. https://doi.org/10.1117/1.3373516.

96 Han, M., Zhang, X.S., Meng, B. et al. (2013). r-Shaped hybrid nanogenerator with enhanced piezoelectricity. *ACS Nano* 7 (10): 8554–8560.

97 Ma, T., Gao, Q., Li, Y. et al. (2020). An integrated triboelectric–electromagnetic–piezoelectric hybrid energy harvester induced by a multifunction magnet for rotational motion. *Adv. Eng. Mater.* 22 (2): 1–9. https://doi.org/10.1002/adem.201900872.

98 Du, X., Zhao, S., Xing, Y. et al. (2018). Hybridized nanogenerators for harvesting vibrational energy by triboelectric–piezoelectric–electromagnetic effects. *Adv. Mater. Technol.* 3 (6): 3–9. https://doi.org/10.1002/admt.201800019.

99 Zhou, J., He, L., Liu, L. et al. (2022). Design and research of hybrid piezoelectric-electromagnetic energy harvester based on magnetic couple suction-repulsion motion and centrifugal action. *Energy Convers. Manage.* 258: 115504. https://doi.org/10.1016/j.enconman.2022.115504.

100 Li, Z., Zhou, S., and Li, X. (2022). A piezoelectric–electromagnetic hybrid flutter-based wind energy harvester: modeling and nonlinear analysis. *Int. J. Non-Linear Mech.* 144: 104051. https://doi.org/10.1016/j.ijnonlinmec.2022.104051.

101 Zervos, H. Piezoelectric energy harvesting: developments, challenges, future. *Energy Harvest J.* https://www.idtechex.com/en/research-article/piezoelectric-energy-harvesting-developments-challenges-future/5074.

102 Jaganathan, C.B. (2016). R&D to boost global piezoelectric smart materials market, the global piezoelectric smart materials market is expected to exceed $42 billion by 2020. https://www.ceramicindustry.com/articles/95444-rd-to-boost-global-piezoelectric-smart-materials-market/ (accessed 13 September 2022).

103 TechNavio (2017). Global piezoelectric smart materials market 2017–2021. https://www.technavio.com/report/global-piezoelectric-smart-materials-market (accessed 13 September 2022).

104 Stewart, M., Weaver, P.M., and Cain, M. (2012). Charge redistribution in piezoelectric energy harvesters. *Appl. Phys. Lett.* 100 (7): 073901.

105 Yip, M., Jin, R., Nakajima, H.H. et al. (2014). A fully-implantable cochlear implant SoC with piezoelectric middle-ear sensor and arbitrary waveform neural stimulation. *IEEE J. Solid-State Circuits* 50 (1): 214–229.

106 Jesse Maida, T. (2017). Top 5 vendors in the global piezoelectric smart materials market: Technavio.

107 Abraham, T. (2014). Expanding markets for piezoelectrics, piezoelectric ceramics, polymers and ceramic/polymer composite devices are increasingly in demand for multiple high-tech applications. https://www.ceramicindustry.com/articles/93845-expanding-markets-for-piezoelectrics/

2

KNN-Based Hybrid Piezoelectric Materials

S. Wazed Ali[1,3], Swagata Banerjee[1], Chirantan Shee[1], and Mayuri Srivastava[2]

[1] *Department of Textile and Fibre Engineering, Indian Institute of Technology Delhi, Hauz Khas, New Delhi, India*
[2] *School of Interdisciplinary Research, Department of Textile and Fibre Engineering, Indian Institute of Technology Delhi, New Delhi, India*
[3] *School of Interdisciplinary Research (SIRe), Indian Institute of Technology Delhi, New Delhi, India*

2.1 Introduction

Since the discovery of piezoelectricity over a century ago, more than a thousand piezoelectric materials have been created. It is well known that nanoscale piezoelectricity is still in its infancy to comprehend piezoelectric behavior at the nanoscale level [1]. Both lead-containing and lead-free piezoelectric materials can convert mechanical energy into electrical energy. $Pb_xZr_{1-x}TiO_3$, often known as lead zirconate titanate (PZT), with around 60% lead content, holds great promise as a piezoelectric material. A wide variety of polymeric and ceramic compounds can be classified as lead-free piezoelectric materials. Some examples of these materials are polyvinylidene fluoride (PVDF), barium titanate ($BaTiO_3$), and zinc oxide (ZnO) [2, 3]. This category offers several benefits, including a straightforward structure, low manufacturing costs, straightforward synthesis, excellent piezoelectricity, and ease of mass manufacture. The synthesis (thermal treatment) of a combination of various metal oxides, carbonates, and salts produces ferroelectric ceramics, which may be customized to meet the needs of a particular application. Advanced piezoelectric materials are built upon customized solid solutions with various compositions and dopants. PZT, a ceramic material with exceptional piezoelectric capabilities, offers a variety of dopants that can change its structure and properties. They are widely used in sensors, transducers, actuators, and electronic parts [4, 5]. However, their

Hybrid Materials for Piezoelectric Energy Harvesting and Conversion, First Edition.
Edited by S. Wazed Ali, Satyaranjan Bairagi, and Shahid Ul Islam.
© 2024 John Wiley & Sons, Inc. Published 2024 by John Wiley & Sons, Inc.

main disadvantage is that these lead-based piezoelectric devices contain more than 60% lead by weight. Lead remains in the environment long after its disposal. This leads to an increased probability of lead accretion in living beings causing lead toxicity. Thus, there is a need to curb the consumption of lead-based substances and find a suitable alternative to them [6]. The desire for materials that are environmentally friendly and nonhazardous to human health has been the driving force behind the development of novel lead-free piezoelectric substances. Innumerable research endeavors are being carried out in the field of lead-free piezoelectric ceramics to produce new materials that are competent enough to serve the purpose of effective energy-harvesting materials [7]. Devices based on lead-free piezoelectric materials are eco-friendly, suitable for high-temperature applications, and biocompatible. Based on their structural characteristics, lead-free piezoelectric materials can be divided into perovskite, tungsten–bronze, layered bismuth, and pyrochlore categories. Compared to other structures, the perovskite structure has drawn the most attention due to its ease of preparation and basic structure. It has the generic formula ABX_3, where X is an anion that connects the cations, and A and B are the two cations of different sizes, often with A bigger than B. The main advantage of the perovskite structure is that different anions and cations can be switched on the A and B sites without impacting the overall structure [8]. Lead-free materials with perovskite structures were created in the 1940s through the 1960s [9]. Examples include barium titanate (BT), modified bismuth sodium titanate (BNT), potassium sodium niobate (KNN), and its derivatives [9] (Figure 2.1). Compared to their lead-based counterparts, the non-lead-based materials as single systems failed to demonstrate the desired piezoelectric capabilities. They are changed and doped appropriately so that they can acquire exceptional piezoelectric capabilities [10]. The lead-free systems based on $K_xNa_{(1-x)}NbO_3$ (KNN), $Na_{0.5}Bi_{0.5}TiO_3$, and $BaTiO_3$ have received the most attention among the numerous options. A number of studies addressing the state of research into these lead-free piezoelectric compositions have recently been published [6, 11–13]. With a high Curie temperature (T_C) and strong piezoelectric characteristics, KNN is regarded as the most promising lead-free choice among these alternatives [14, 15]. Numerous groups have explored the crystal structure of KNN; however, the results are inconsistent [16–18]. Due to its higher ferroelectric and piezoelectric characteristics, the composition around $x = 0.5$ is more interesting. A polymorphic phase boundary (PPB) occurring between two orthorhombic phases on each side is frequently cited as the cause for the superior qualities of this composition.

A component of this chapter includes a full discussion of ceramics without lead content, especially the KNN-based hybrid material.

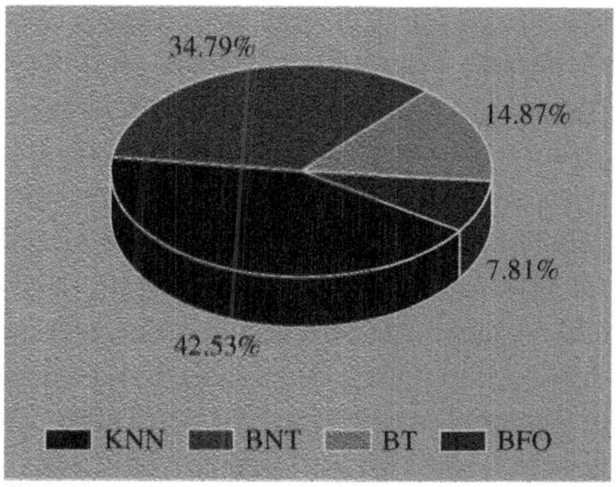

Figure 2.1 Publications on lead-free piezo/ferroelectric KNN, BNT, BT, and BFO-based material from the time period of 2000–2017. Source: Zheng et al. [19]/Elsevier.

2.2 Lead-Free Ceramics

The global market for electronic devices has been dominated by lead-based piezoelectric materials, specifically PZT [20]. However, due to the adverse effects of lead on people and their surroundings, lead-based piezoelectric materials are being replaced with lead-free piezoelectric materials [21]. As a result, many researchers invest time and energy in investigating substitute materials to satisfy environmental ceramic criteria for new uses [22]. The piezoelectric characteristics of these substituted materials were on par with, and occasionally even better than, those of the real PZT. Additionally, because of their biocompatibility and environmental friendliness, they can take the place of lead-based piezoelectric materials in the near future. It is the attempt of many years of research that has resulted in significant advances in the electrical characteristics, physical processes, and associated applications of lead-free materials. Many researchers have suggested a general approach of multiple phases coexisting in lead-free perovskite materials to improve their piezoelectric characteristics [23–25]. It was Jaffe et al. [26] who initially introduced the idea of a "morphotropic phase boundary" (MPB) after observing the highest piezoelectric constant in the best piezoelectric material, PZT, for a composition near to MPB. This acquaints us with the importance of phase boundaries in the electrical advancement of piezoelectric materials, lead-based or lead-free. Research development in lead-free piezoelectric perovskite materials from various angles has been substantially accelerated over the last 15 years by reviews of developments [27, 28]. Numerous lead-free piezoelectric devices have been created; thanks to technological advancements, and they have been found useful in fields, including bioengineering, the military, optoelectronic information, and energy [29]. Prof. Ding-Quan Xiao's group has been studying lead-free ceramics since 1978, China's first attempt to identify substitutes for PZT ceramic materials. The rules for Waste Electrical and Electronic Equipment (WEEE) and Restriction of the Use of Certain Hazardous Substances in Electrical and Electronic Equipment (RoHS) prompted interest in lead-free piezoceramics research [30]. There are two types of lead-free materials: organic and inorganic [31]. PVDF and its copolymers are included in organic materials, while the aurivillius materials (ferroelectrics based on bismuth layered structure), tungsten–bronze, and members of perovskite families of inorganic lead-free materials may be divided into these groups. Excellent piezoelectric capabilities, cost-effective production, straightforward fabrication processes, and the capacity to be rendered into various forms are only a few benefits of ceramic piezoelectric materials. Because of this, ceramic piezoelectric materials are often used in transducers, sensors, and actuators [32]. It was the time of the year 1880 when Curie brothers initially observed piezoelectricity in a quartz crystal. Following this, the first non-lead-based ceramic, that is, $BaTiO_3$ was discovered in 1946. Smolenskii and coworkers found strong ferroelectricity of lead-free BNT materials in 1960.

It was not until after the twenty-first century that the non-lead-based perovskite materials received significant research attention. Lead-free ceramic families such as KNN, $BaTiO_3$, BNT, and $BiFeO_3$ have shown notable progress in the last decade as a result of such scientific vigor [33].

2.2.1 Inorganic Piezoelectric Ceramics

This segment describes some of the inorganic lead-free ceramics. Table 2.1 describes few piezoelectric properties of some of the lead-free ceramics.

Table 2.1 Assets of a few of the lead-free piezoelectric ceramic materials [34].

S. No.	Material	Piezoelectric constant, d_{33} (pC/N)	Electro-mechanical coupling factor	Curie temperature (°C)
1	KNN	80–160 [5]	0.51 [14]	400 [5]
2	KNN-$LiSbO_3$	283 [35]	0.46 [35]	237 [36]
3	$BaTiO_3$	191 [37]	0.49 [38]	130 [37]
4	BNT	75 [39]	Approx. 0.21 [40]	<290 [41]
5	AIN	5 [42]	0.23 [43]	>2000 [44]
6	$LiNbO_3$	6 [45]	0.23 [46]	1150 [45]
7	BCZT	>500 [47]	—	102 [48]
8	ZnO	9.2, 12.4 [49, 50]	—	—

The history began with a ceramic system built on BT with an ABO_3 perovskite structure, which is the first known lead-free piezoelectric system, where A represents basically the structures of barium, calcium, and strontium, and B includes titanium, niobium, and tantalum [51]. It is a well-established ferroelectric that has undergone extensive research in both ceramic and single-crystal forms. In this ceramic system, three ferroelectric phases are found to coexist at the nanoscale: tetragonal (T), orthorhombic (O), and rhombohedral (R). Particularly, the relative mobility of cations (O^{2-}) and anions (Ba^{2+}, Ti^{4+}) leads to spontaneous polarization (Ps) even when no external electric field or mechanical force is applied. At 90 °C (Rhombohedral–Orthorhombic), 0 °C (Orthorhombic–Tetragonal), and 125 °C (Curie temperature), the sequential transition of ferroelectric phases in BT is seen, accordingly [52]. The main ABO_3 perovskite structure, such as $BaTiO_3$ (also labeled BT), was the first ferroelectric substance with a perovskite structure, and it found widespread usage in sonar detection and phonograph needles after its discovery by experts during the World War II in 1941 and 1944 [53]. For field-induced piezoelectric transducers, $BaTiO_3$ is widely recognized as the ultimate ferroelectric material, which is a strong applicant due to its strong dielectric constant, low loss properties, low Curie temperature, large polarizations, large permittivity, and the immense induced strains that can be achieved in these materials [54, 55]. It is observed that there is a certain grain size, which is an effective factor for crystalline BT at the nano level, obtained by the spark plasma sintering (SPS) process, as observed by Wang et al. [56]. Below this grain size, ferroelectricity is lost, and the lattice transforms from tetragonal to cubic. Utilizing hydrothermally created nanoscale particles, SPS was used to create BT ceramics with improved piezoelectric characteristics [30]. BT is the first commercially available piezoelectric material with excellent performance based on polymorphic phase transition (PPT). However, the greatest barrier to its more widespread commercial implementation in actuators and sensors has been the material's poor piezoelectric characteristics (d_{33} = 190 pC/N) and small Curie temperature (T_C = 120 °C). A lot of work has been put into compositional investigations to help BT ceramics become more temperature stable and increase their piezoelectric characteristics. It has been found that the characteristics of BT ceramics can be enhanced by replacing Ti^{4+} with quadrivalency ions such as Zr^{4+} and Hf^{4+}. According to Zheng et al. [57], partially substituting Ti^{4+} with Zr^{4+} and adding a little CuO additive can significantly improve the piezoelectric temperature stability. Exceptional piezoelectric characteristics of d_{33} = 300 pC/N are present in CuO-modified $BaTi_{0.9625}Zr_{0.0375}O_3$ ceramics, and their kp (coupling coefficient) remains above 0.40 over a large temperature range of 43–73 °C and is close to constant between 25 and 55 °C. According to Tian et al. [58], the three-phase transitions in BT are constricted when Ti is replaced with Hf, and the Curie temperatures (T_C) drop as the Hf content of the compounds rises. An excellent d_{33} value of 305 pC/N is achieved between the orthorhombic and rhombohedral phases of $BaHf_{0.05}Ti_{0.95}O_3$.

In 1954, (Ba, Ca) (Ti, Zr)O_3 BCZT was initially studied in order to be used in capacitors. [59]. In 2009, Liu and Ren revealed that BCZT possesses excellent piezoelectric properties, including a piezoelectric coefficient comparable to PZT (the value varies with stoichiometry and grain size) [47]. Unfortunately, BCZT's low Curie temperature (102 °C) is also its limiting factor, making it ineligible for applications requiring a more comprehensive temperature range. The application of doping has also raised the Curie temperature. When BCZT was doped with Bi^{3+}, T_C was shown to rise to 159 °C, as reported by Q. Li et al. [34]. In a different study, Zhang et al. [60] discovered that adding 0.1% $LiTaO_3$ to a material raised the Curie temperature to 129 °C. The best piezoelectric characteristics are demonstrated by BCZT particles at ambient temperature. This is because the BCZT system at this temperature coexists in rhombohedral, orthorhombic, and tetragonal (R–O–T) multiphase. The piezoelectric characteristics of the composite decrease as the temperature rises because the phase changes to tetragonal behavior. Due to its superior dielectric, ferroelectric, and piezoelectric characteristics, BCZT is a strong contender for several applications, including multilayer ceramic capacitors, piezoelectric actuators, and positive temperature coefficient resistors. Additionally, its characteristics may be altered by adjusting the Ba/Ca and Zr/Ti ratio [1]. To improve the piezoelectric and dielectric characteristics with various A-site and B-site dopants (where, generally, A = Ca, Sr, La; B = Nb, Ta, Zr), a variety of $BaTiO_3$-based solid solutions have been created. Numerous papers have focused on specially formulated BCZT $Ba_{1-x}Ca_xTi_{1-y}Zr_yO_3$ after a high piezoelectric coefficient ($d_{33} = 620$ pC/N) for $Ba_{0.85}Ca_{0.15}Ti_{0.90}Zr_{0.10}O_3$was observed [47].

The other perovskite material, BNT, was initially discovered in 1960 by Smolensky et al. BNT has a perovskite structure and rhombohedral (R) symmetry at ambient temperature [61]; it possesses exceptional piezoelectric properties. For instance, it has a d_{33} value of 75 pC/N. BNT can depolarize at temperatures as high as 290 °C [41]. Its strong strain response has made it valuable in actuator applications. There is evidence that the addition of dopants improves BNT's piezoelectric characteristics. For instance, $[Bi_{0.5}(Na_{1-x}Li_x)_{0.5}]0.84Sr_{0.16}TiO_3$ produced by concurrently doping BNT with Li^+ and Sr^{2+}, has a greater piezoelectric constant (173 pC/N) and a higher working temperature (>300 °C) [62]. BNT with a rhombohedral perovskite structure has a relatively high Curie temperature ($T_C = 320$ °C) and a d_{33} of 73 pC/N [63]. It has long been investigated as a possible replacement for lead-based equivalents. BNT has certain limitations, including a strong coercive field and a high conductivity, which provide challenges which provide challenges in the poling process and have very poor piezoelectric characteristics for pure BNT [64]. To address these disadvantages, modifications to the BNT systems' solid solution composition and the addition of various dopants have been made. Due to their exceptional piezoelectric performance close to the rhombohedral and tetragonal MPB, the binary systems of BNT–BKT and BNT–BT have received the most attention among the BNT-based lead-free piezoelectric systems. Also reported [65, 66] are BNT-based compositions that have been altered using $(Bi_{0.5}Li_{0.5})TiO_3$, $Ba(Cu_{0.5}W_{0.5})O_3$, $NaNbO_3$, and Er_2O_3.

Later, BKT was established in the twentieth century. At room temperature, it features tetragonal symmetry, a comparatively high T_C of 380 °C, and a stronger piezoelectric response than BNT. However, because it is challenging to manufacture high-dense ceramics, BKT has received far less research attention than BNT material. Due to their low density and tendency to generate secondary phases when heated, these materials are incredibly challenging to handle. These materials have used sintering aids to increase their density, such as treating BKT with extra bismuth oxide, which enhanced the material's piezoelectric and ferroelectric characteristics (e.g. $d_{33} = 101$ pC/N) [67].

Interest in piezoelectric nanomaterials and structures has grown during the past few years [68]. Hence, ZnO, a common example of a nanomaterial, was developed. It is a naturally occurring metal oxide that is plentiful, inexpensive, chemically stable in the air, and harmless for biological use. The most prevalent phase is wurtzite with a hexagonal shape. ZnO is not a ferroelectric material [68]. The low-temperature approach may make the nanostructure reasonably quickly on various flexible and stiff substrates without poling [69]. ZnO crystallization produces wurtzite, a wide band gap semiconductor with a tetrahedrally coordinated structure. Strong piezoelectric property is produced by the nonexisting center of symmetry and a significant electro-mechanical connection. Its piezoelectric tensor is the greatest among the tetrahedrally linked semiconductors, and it is a typical II–VI combination with a large band gap [70]. Because of this capacity, ZnO is mostly used in actuators and sensor applications that need a strong electro-mechanical connection. Because of this, experimental and theoretical investigations [71, 72] have thoroughly examined the piezoelectricity of ZnO. Many functional ZnO nanostructures have been produced. These nanostructures might be employed in biological studies, sensors, transducers, optoelectronics, and more [73]. The latest research suggests nanoscale piezoelectric ZnO nanowire arrays may produce electrical energy from mechanical energy [74]. The term "piezoelectric nanogenerators" was first established by Wang et al. [75], and they subsequently identified the piezoelectricity in ZnO nanowires (d_{33} = 14–26 pm/V) in 2006. The benefit of employing nanowires is their ability to be excited by very small physical motions with an excitation frequency that ranges from a few Hz to several MHz, making them perfect for harvesting environmental random energy from vibrations, body motion, and moderate airflow [76]. As a result, the scientific community has paid close attention to piezoelectric nanogenerators. Electric power outputs have risen from a few millivolts over time [70]. It has notable applications not just in electronics and photonics but also as a favored filler in piezoelectric nanocomposite applications [21]. ZnO, alike aluminum nitride (AlN), can also be used in MEMS (micro electromechanical system) devices as a thin film or as nanostructures in NEMS (nano electromechanical system) devices. The electromechanical coupling of ZnO is 0.19 [77], which is lower than that of PZT, but it has the advantage of being biocompatible.

Apart from these materials, AlN, sintered by processes such as sputtering, MOCVD (metal organic chemical vapor deposition), PLD (pulse laser deposition), and others, is the most notable non-lead material [78]. AlN now offers a different MEMS device scaling option for the industrial market. AlN may be used for high-temperature applications and does not require poling. This is because it exhibits spontaneous polarization due to the crystal structure of its c-axis. AlN has a relatively low dielectric permittivity ($\varepsilon^{T}_{r,33}$ = 9.5) and a piezo constant for the transverse mode on the order of d_{31} = 2 Pc/N, which results in an electro-mechanical coupling of k_{31} = 0.12 [79].

2.3 Potassium Sodium Niobate (KNN): A Piezoelectric Material

The material most renowned for its piezoelectric property is PZT. It has shown great potential as a piezoelectric material since its discovery. However, recent research is more inclined toward searching for alternative lead-free materials owing to the issues of lead toxicity. Hence, lead-free materials are slowly gaining the limelight. The class of alkali niobates has shown promising potential to replace lead-based materials with established potential in piezoelectricity. KNN has emerged to be one of the forerunners among the alkali niobates, as a potential lead-free alternative.

KNN is a solid solution chemistry of two chemical compounds, that is, potassium niobate and sodium niobate, which are ferroelectric and antiferroelectric, respectively. These two compounds form solutions over a wide composition range, yielding KNN with different stoichiometric ratios of potassium and sodium. However, the piezoelectric properties of the ceramic peak at a particular K/Na ratio of 52/48. This is referred to as the MPB [80]. Potassium niobate also exhibits polymorphism. It is found to be rhombohedral at very low temperatures, orthorhombic at room temperature, tetragonal at higher temperatures, and cubic above its Curie temperature (~435 °C). The phases of sodium niobate are, however, complex and difficult to identify. Apart from MPB, there is also a phenomenon that describes the temperature dependence of polymorphic phases of KNN (Figure 2.2). This is known as polymorphic phase transition (PPT). KNN is found to be orthorhombic at room temperature. At temperatures above 400 °C, KNN converts to cubic form, losing its piezoelectric properties and

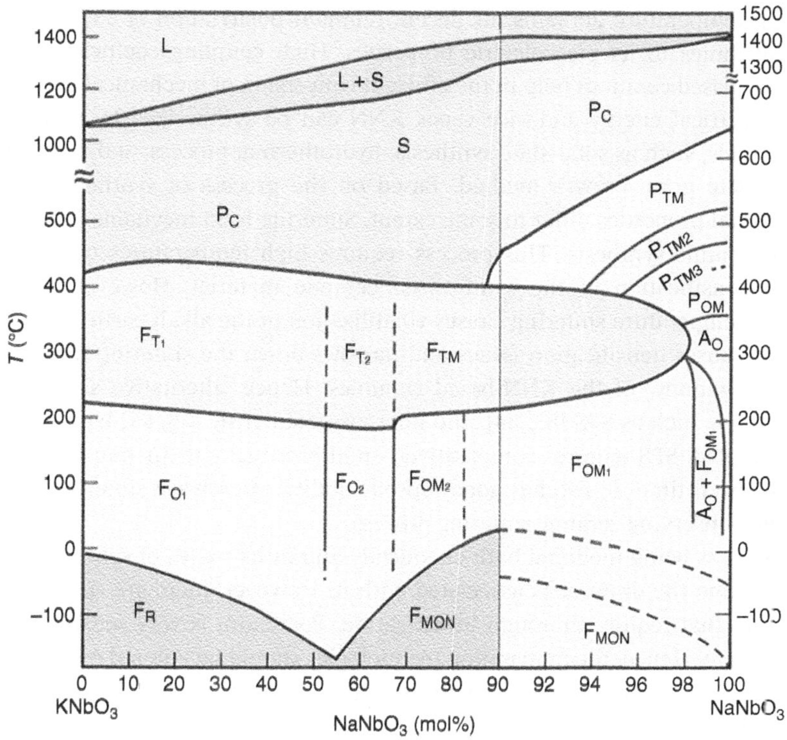

Figure 2.2 Phase transition of Kthethe symbols in the figure stand as follows. L – Liquid, S – Solid, P – Paraelectric, C – Cubic, T – tetragonal, O – Orthorhombic, MON – Monoclinic, TM – Pseudotetragonal, and R – Rhombohedral. Source: Koruza et al. [81]/Public domain/CCBY 4.0/Materials Research Society 2020.

achieving a paraelectric state. Orthorhombic to tetragonal phase transitions of KNN take place above 210 °C. At very low temperatures of −123 °C, the rhombohedral phase of KNN is obtained. Interestingly, crystallographic studies indicate the presence of a monoclinic perovskite-type primary cell of KNN at room temperature, while its unit cell possesses orthorhombic symmetry [82]. Nanostructures of KNN have also been reported to have an extra triclinic symmetry owing to the cationic ordering of the alkali metal ions in the small number of cells in the particles. The piezoelectric properties are maximum at phase boundaries owing to the ease of dipole orientation. Further, the two-phase existence occurring at the phase boundaries also contributes to enhancing the piezoelectric properties of KNN.

KNN is being considered as a potential alternative to its lead-based counterparts, owing to its appreciable piezoelectric properties. It has a high Curie temperature of around 420 °C, which allows extending its applications to high-temperature domains. Its decent remnant polarization of 33 $\mu C/cm^2$ contributes to its piezoelectric properties. High coupling coefficients of KNN-based ceramics help in the efficient conversion of mechanical energy to electrical energy and vice versa. KNN can be synthesized by various methods, such as solid-state synthesis, hydrothermal process, and reactive template grain growth method. Based on the process of synthesis, the material properties differ to some extent. Sintering is an inevitable process for ceramic synthesis. This process requires high temperatures to attain the densification of the synthesized ceramic material. However, such high-temperature sintering causes volatilization of the alkali earth metals. This causes densification issues and narrows down the sintering temperature window of the KNN-based ceramics. Hence, alternative sintering methods such as SPS [83, 84], and microwave sintering [85, 86] have been attempted. SPS requires comparatively mild processing temperatures with less dwell time. It also has good reproducibility. Microwave sintering is a time-conserving ceramic sintering process.

KNN is being modified both chemically and in its routes of synthesis to overcome the drawbacks associated with it. However, there are still challenges that require thorough investigation. Potassium is very sensitive to moisture. Hence, the preparation technologies should be tailored considering the reactivity of potassium with moisture. Achieving comparable piezoelectric coefficients as those of lead-based ceramics is not enough. Electrical properties also should not be compromised at the same time. The poor thermal stability of the KNN-based ceramics and their low-quality factor, all sum up to the unbalanced development of electrical properties. Moreover, the realization of the piezoelectric properties to their potential in the form

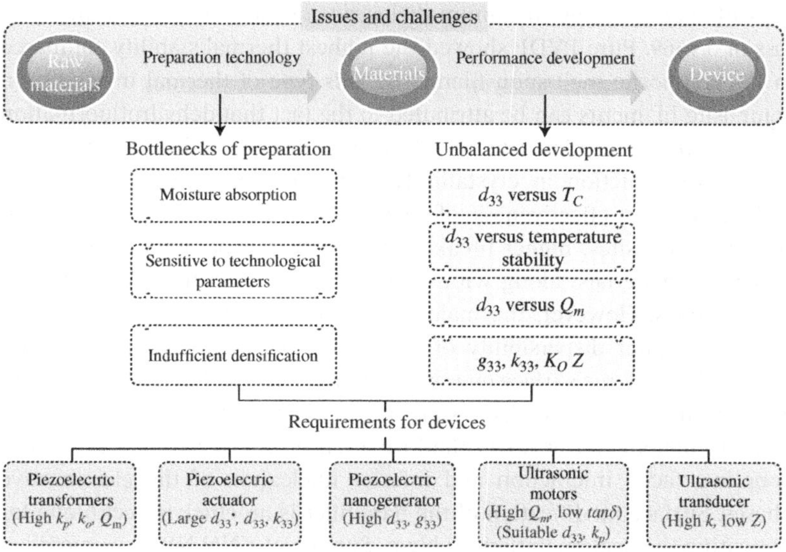

Figure 2.3 Challenges and bottlenecks of KNN-based ceramic piezoelectric material. Source: Wu [87]/American Institute of Physics.

of devices is also a matter of concern. Properties excluding piezoelectricity also come into play. The robustness and low fragility of the prepared devices are essential for their mechanical stability. Figure 2.3 illustrates the challenges and bottlenecks encountered by the KNN-based ceramic piezoelectric materials [87].

2.4 Potassium Sodium Niobate (KNN)-Based Hybrid Piezoelectric Materials

Due to nontoxic nature and excellent piezoelectricity, several research studies [88, 89] have been published, finding new strategies to further improve properties of KNN-based materials. Many researches deal with modification of KNN by other metal ions, whereas numerous publications have been revealed depicting properties of composite materials containing KNN as a piezoelectric filler. Bairagi and Ali [88] investigated the effect of KNN nanorod (NR) content on the piezoelectric property of KNN/PVDF melt-spun composite filament. They prepared composite filaments with 0, 2, 4, and 6 wt% of KNN. 4% KNN-loaded melt-spun filament showed the highest piezoelectric voltage of 3.7 V and 0.326 μA current, due to obtaining maximum β-fraction (26%) at this particular filler loading.

This filament had the maximum dielectric constant of 175 and dielectric loss of 0.9669. Pure PVDF showed the highest thermal stability compared to the composite melt-spun filaments. This type of thermal instability in composite filaments can be attributed to the fact that dehydrofluorination of PVDF occurs at higher temperatures in alkaline conditions, ultimately leading to reduction in crystallinity of the polymer [90]. The authors highlighted the effectiveness of using nanocomposite-based materials. In nanocomposites, higher breakdown voltage can be obtained from the polymeric materials, along with gaining higher dielectric constant from the ceramics. However, the major problem associated with this type of material is poor dispersibility of the nanostructure within the polymer matrix. Due to large surface energy of nanostructures, they tend to agglomerate, resulting in poor dispersion of the filler within the matrix. This type of phenomenon leads to the formation of interfacial defects such as poor interfacial interaction and inferior nucleation of the electroactive phases. Surface modification of the nanofiller is an efficient technique for mitigating the agglomeration problem. Surface of KNN NRs was modified using three different chemicals, that is, 3-aminopropyltrimethoxysilane (APS), polyvinylpyrrolidone (PVP), and polyaniline (PANI). Maximum beta fraction and gamma fraction were obtained for APS and PVP-modified KNN NRs incorporated into the PVDF film, which is due to higher molecular weight of these modifiers, leading to better interaction between the filler and the matrix materials. Polarization for PVP-modified KNN incorporated PVDF and APS-modified KNN incorporated PVDF were found to be 0.15 and 0.14 $\mu C/cm^2$, respectively, which were almost 3.5 times than the polarization value of pure PVDF. Remnant polarization was found to be maximum for silane-modified KNN NR incorporated PVDF, which indicates best performance in terms of output voltage generation. The dielectric constant increases almost two times for the surface-modified KNN-rich PVDF nanocomposites compared to the pure PVDF film. This is due to the more homogeneous distribution of modified KNN NR within the matrix, thereby improving the interfacial polarization in particular. Dielectric loss values were also lowered when KNN NR was surface-modified, which is desirable for getting higher piezoelectric output voltage. In further work, the authors incorporated APS-modified KNN NRs into PVDF matrix using electrospinning technique. They found that a 3% surface-modified KNN-loaded PVDF electrospun membrane produces maximum output voltage of 21 V, maximum output current Of 22 μA, and the highest current and power density values compared to the other specimens under this particular study. In spite of having the lowest β-fraction value, 3% filler-decorated electrospun membrane showed the best piezoelectric property due to maximum incorporation and superior distribution of surface-modified KNN NRs within the PVDF matrix [91].

Teka et al. [89] synthesized KNN nanostructures and, for the first time, incorporated the same into PVDF matrix during electrospinning. They stated that filling volume, temperature, surfactant, and coupling agent have significant roles in the properties of synthesized nanostructures. FTIR (Fourier infrared spectroscopy) spectra and XRD (X-ray diffraction) patterns proved that the addition of KNN helps in the generation of β phase of PVDF, ultimately improving the piezoelectric effect of the nanocomposite structure. Energy-harvesting capability of the composite material was evaluated by digital oscilloscope using thumb pressure as stimuli. A 1.9 V output voltage was obtained for KNN-loaded PVDF nanocomposite structure, whereas the output voltage ranged from 50 to 100 mV for the pristine PVDF electrospun nanofibrous membrane. The higher output voltage in nanocomposite can be attributed to the formation of higher electroactive phase in filler-nucleated KNN/PVDF membrane and in-situ poling of incorporated KNN along the fiber axis. Electrospinning process offers in situ poling, thereby providing the same piezoelectric response, such as film or melt-extruded material at lower filler loading. On the contrary, higher filler loading is required to obtain comparable piezoelectric response in normal melt-extruded material or solution-casted composite film, which simultaneously makes the composite structure mechanically unstable [89].

Bairagi and Ali further incorporated KNN NR of 8.5 aspect ratio within PVDF matrix in order to form electrospun nanocomposites. They have prepared pure PVDF electrospun membrane along with nanocomposites of PVDF using 1%, 3%, and 5% KNN NR. Surprisingly, for 3% KNN NR-loaded electrospun membrane, they obtained the least β-fraction along with the maximum piezoelectric outputs in terms of voltage and current. At this particular concentration, the dipoles within the materials get self-oriented, which subsequently obsolete the need for further poling operation. The nanocomposite with 5% KNN shows a decrease in piezoelectric properties compared to that with 3% loading. This can be attributed to the agglomeration of KNN NRs at higher concentrations, thereby clogging the needle during electrospinning process and finally resulting in lesser presence of KNN NRs within the nanocomposite material [92]. The authors studied the effect of KNN NR concentration on the electrical properties of KNN/PVDF nanocomposite film. They found optimum piezoelectric property for 10% KNN-loaded PVDF composite film, as the maximum β-fraction was achieved at this particular concentration. This particular sample also offered maximum dielectric constant, energy density, and remnant polarization values [93]. In further research, they discussed the effect of CNT (carbon nanotubes) addition within PVDF/3% KNN nanocomposite. The incorporation of 0.1% CNT leads to increase in β-fraction of PVDF, whereas further enhancement in CNT loading reduces the β-fraction. The addition of 0.1% CNT improves the conductivity of the solution, thus increasing the mechanical stretching during electrospinning, providing better orientation of crystal phase. On the contrary, at higher concentrations (0.3% and 0.5%) of CNT, viscosity of the solution enhances, thereby causing problems in ejection of the solution from the needle tip during electrospinning. This phenomenon results in poor crystal orientation, thus β-fraction decreases. Piezoelectricity was also found to be maximum for the 0.1% CNT-loaded electrospun membrane due to maximum β-fraction of PVDF within this particular sample. The presence of CNT also uniformly distributes KNN NRs. Energy harvesting is improved due to the conductive path provided by the CNT within the nanocomposite structure [94]. The same research group also revealed that PVDF/KNN/ZnO electrospun nanocomposite membrane offers higher output voltage compared to PVDF/KNN and PVDF/ZnO electrospun membranes due to synergistic effect [95].

Shoorangiz et al. studied the effect of KNN and CNT contents on the dielectric, piezoelectric, and ferroelectric properties of KNN and CNT-incorporated PVDF composite films. Dielectric constant value of a composite material can be theoretically calculated using Eq. (2.1):

$$\log \varepsilon = \sum V_{\mathrm{i}} \log \varepsilon_{\mathrm{i}} \tag{2.1}$$

where V_{i} is the volume fraction of each constituent and ε_{i} is the respective dielectric constant. With an increase in both the contents of KNN and CNT, the dielectric constant of the composite enhances at the expense of an increase in dielectric loss of the material. Increase in dielectric loss is more prominent for enhancing CNT content due to the formation of more conductive paths within the composite. With rising KNN content, the piezoelectric property of the composite film improves as KNN is a better piezoelectric material than PVDF. At higher KNN loading, formation of 3D network of KNN across the PVDF matrix improves the polarization process during poling, ultimately resulting in superior piezoelectric properties. Addition of CNT, while keeping the KNN percentage at constant level, enhances the piezoelectric property of the composite up to a certain concentration of CNT (percolation threshold) followed by deterioration in piezoelectricity. Up to percolation threshold, conductive CNT increases the local electric field during poling, causing superior piezoelectricity. On the other side, above the percolation threshold, electric field applied during poling partially canceled due to more electronic movement, causing deterioration in poling efficacy, and thus piezoelectric voltage constant reduces. Change in ferroelectric property of composite material with increasing CNT loading is quite interesting (Figure 2.4(a)). With an increase in CNT percentage to 0.75 wt%, the remnant polarization of the composite material enhances, along with a decrease in the magnitude of required electric field for polarization. With the addition of CNT, a conductive path is created within the composite, thereby causing effective DC poling, and the alignment process of the dipoles also becomes easier. Up to 0.75 wt% of CNT loading, the nature of the ferroelectric curve tends toward typical circular nature. Beyond this critical concentration of CNT, coercive field of the composite increases, and achievable polarization reduces. These characteristics had been described by two means. Beyond critical concentration, agglomeration of CNT can occur, resulting in lower electrical conductivity enhancement. Otherwise, if agglomeration can be prevented, the CNTs can create power-consuming path through the composite material, offering electrical breakdown of the material at lower electrical field [96].

Apart from this type of composite structure, many research studies have dealt with KNN-based hybrid filler materials. Vivekananthan et al. prepared $(1-x)$ KNN-x $BaTiO_3$ nanoparticle (NP), where $x = 0.02, 0.04, 0.06, 0.08$. Thereafter, this NP has been incorporated within PDMS (polydimethyl siloxane) matrix with the help of stirring. Next, this solution has been spin-coated onto an ITO (indiun tin oxide)-coated PET substrate. Then another ITO-coated PET (polyester) substrate was used as the top electrode Copper wires were attached for the connection, and the device was finally encapsulated within the Kapton tape. Lastly, the device was poled under 5 kV voltage for a lapse of 12 hours in order to get maximum orientation of the dipoles within the composite structure. Among KNN, KNN-0.02 $BaTiO_3$, and $BaTiO_3$-loaded composites, KNN-0.02 $BaTiO_3$-loaded material offered maximum voltage output of 58 V and current output of 450 nA. The effect of applied force on the electrical output was also studied, and 10 N force was found to give maximum electrical outputs. A 12 N force further reduces the voltage output, which is due to the depolarization of the dipoles at higher pressure [97].

Ichangi et al. reported that Li and Ta modification of KNN gives a very high piezoelectric output voltage due to superior piezoelectricity and polarization originating from desirable lattice distortion derived from difference in ionic radii of Li^+ and Ta^{5+}. They prepared modified KNN nanofibers using electrospinning technique and finally incorporated these within the PVDF matrix in order to characterize its piezoelectric behavior [98].

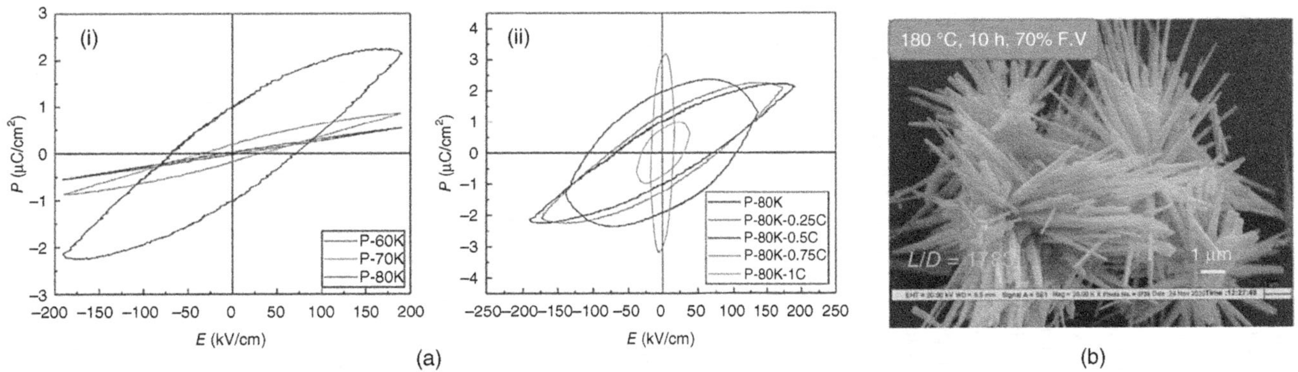

Figure 2.4 (a) P–E hysteresis loops of (i) PVDF/KNN and (ii) PVDF/KNN/CNT composite. Source: Shoorangiz et al. [96]/Elsevier, and (b) chemically modified KNN nanorods. Source: Banerjee et al. [99]/with permission from Elsevier.

Banerjee et al. synthesized Zn- and Sn-doped hybrid KNN NR-like structure (Figure 2.4(b)). They found very high piezoelectric coefficient (d_{33}) of almost 95 pC/N for the new hybrid material (KNN–ZS) compared to 13 pC/N for pure KNN. The authors prepared electrospun nanofibrous membrane using optimized KNN–ZS synthesized material by incorporating it within a PVDF matrix. A 3% filler-loaded membrane exhibited highest beta fraction and finer diameter due to higher stretching experienced during electrospinning. This device had been tested for piezoelectricity, and also triboelectric characterization had been conducted with PTFE (polytetrafluoroethylene) as another surface. KNN–ZS filler-loaded nanofibrous membrane offered higher output voltage compared to pure KNN-loaded membrane and pristine PVDF membrane. Piezoelectric-cum-triboelectric output voltage and output current were found as 25 V and 2.11 μA, respectively [99].

Xue et al. prepared $0.9525(K_{0.5}Na_{0.5}NbO_3)$–$0.0475LiTaO_3$ (KNN–LTS) nanopowder-based nanogenerator, which offered d_{33} of 330 pC/N along with providing output voltage and output current of 53 V and 15 μA, respectively. The authors stated that the output voltage grows linearly with the applied force. The voltage also enhances with increase in thickness of nanogenerator (Figure 2.4) [100].

2.5 Applications

Most of the reported applications of piezoelectric materials focus on converting wasted mechanical energy from environment into electrical energy in order to mitigate the purpose of "green technology." KNN-based materials are no exception. There are many research studies investigating the possibilities of enhancing the output electrical signals with the help of piezoelectric materials. Many case studies have already been described in the abovementioned section.

Sound pollution is a great threat to today's technology-driven lifestyle. To define noise, sound pressure level (SPL) of more than 80 dB are detrimental to our physical and environmental conditions. Fibrous materials are well known for their sound-insulating behavior. However, low-frequency sound wave insulation is still a challenge using normal fibrous structures. Piezoelectric material can absorb lower-frequency sound waves using piezoelectric shunt damping phenomenon. Mohamed et al. prepared composite foam with the help of KNN nanofibers and PVDF matrix. KNN nanofiber having positive longitudinal piezoelectric coefficient (d_{33}), produces charges with opposite polarity compared to the same produced by PVDF, which has a negative d_{33} value. Due to simultaneous production of opposite charges, application of sound pressure causes electrical discharge, ultimately helping in sound absorption [101].

Zhang et al. pointed out the potential application of KNN in degrading organic pollutants by piezocatalytic activity. The authors dispersed appropriate amount of KNN powder in the rhodamine B (RhB) solution. After stirring for 60 minutes, ultrasonication was done to investigate the dye degradation performance of the piezoelectric material. Effects of different annealing temperatures and poling voltages were revealed. Under the experimental conditions, 900 °C annealing temperature and maximum value of poling voltage offered the best piezocatalytic activity. KNN powders annealed at 1000 and 1100 °C generated agglomeration. Agglomerated form can cause recombination of positive and negative charges generated by piezoelectricity, which, in turn, reduces the piezocatalytic activity. At mechanical equilibrium, due to ferroelectricity of KNN, obviously opposite charges are adsorbed onto the KNN powders to neutralize the spontaneous polarization of the ferroelectric KNN. While any force imbalance occurs, the polarity of the material reduces due to decrease in dipole orientation. As the amount of opposite surface charge is constant, some of the heterogeneous charges react with the hydroxide ions to dissolve oxygen within the solution. Later, free radicals with oxidizing ability were formed, which subsequently perform the degradation of dye [102].

Vivekananthan et al. reported that $(1-x)$ KNN-x $BaTiO_3$ NP reinforced PDMS composite material can be used in sleep monitoring and neurological disorder recognition applications [97]. Moreover, KNN-based material has been used in structural health monitoring applications [103]. Xue et al. reported $0.9525(K_{0.5}Na_{0.5}NbO_3)$–$0.0475LiTaO_3$ (KNN–LTS) nanopowder-based nanogenerator can be used as a motion detection sensor [100].

2.6 Conclusion

KNN is a competent lead-free piezoelectric ceramic that has the potential to replace its lead-based counterparts. The high Curie temperature of this ceramic extends its applications to high-temperature domains. The balance of d_{33} with a high Curie temperature is still under consideration by various research groups. Chemical doping of the pristine ceramic to design new phase transition boundaries has been carried out extensively, while some novel compositions are still underway. These new phase boundaries aim to reduce the phase transition temperature toward room temperature in order to facilitate the domain reversals easily. The multiple-phase existence has also been sought after to improve the piezoelectric properties of the ceramic. Due to the various lucrative properties of this material, it finds extensive application as an energy storage material. It has also found extensive use in the transportation and civil industries. Treatment of effluents through the piezocatalytic pathway is another emerging application of piezoelectric materials. However, there remains a lot to be explored in order to establish KNN as a suitable counterpart of lead-based PZT.

References

1 Pinheiro, E. and Deivarajan, T. (2019). A concise review encircling lead free porous piezoelectric ceramics. *Acta Phys. Pol. A* 136: 555–565.

2 Hiruma, Y., Aoyagi, R., Nagata, H., and Takenaka, T. (2004). Piezoelectric properties of $BaTiO_3$-$(Bi_{1/2}K_{1/2})TiO_3$ ferroelectric ceramics. *Jpn. J. Appl. Phys., Part 1* 43: 7556–7559.

3 Le, A.T., Ahmadipour, M., and Pung, S.-Y. (2020). A review on ZnO-based piezoelectric nanogenerators: synthesis, characterization techniques, performance enhancement and applications. *J. Alloys Compd.* 844: 156172.

4 Polla, D.L. and Francis, L.F. (1998). Processing and characterization of piezoelectric materials and integration into microelectromechanical systems. *Annu. Rev. Mater. Sci.* 28: 563–597.

5 Rödel, J., Jo, W., Seifert, K.T. et al. (2009). Perspective on the development of lead-free piezoceramics. *J. Am. Ceram. Soc.* 92: 1153–1177.

6 Panda, P.K. (2009). Review: Environmental friendly lead-free piezoelectric materials. *J. Mater. Sci.* 44: 5049–5062.

7 Rödel, J., Kounga, A.B., Weissenberger-Eibl, M. et al. (2009). Development of a roadmap for advanced ceramics: 2010–2025. *J. Eur. Ceram. Soc.* 29: 1549–1560.

8 Kumari, P., Rai, R., Shandilya, M. et al. (2015). State-of-the-art of lead free ferroelectrics: a critical review. *Adv. Mater. Lett.* 6: 453–484.

9 Rödel, J., Webber, K.G., Dittmer, R. et al. (2015). Transferring lead-free piezoelectric ceramics into application. *J. Eur. Ceram. Soc.* 35: 1659–1681.

10 Mishra, S., Unnikrishnan, L., Nayak, S.K., and Mohanty, S. (2019). Advances in piezoelectric polymer composites for energy harvesting applications: a systematic review. *Macromol. Mater. Eng.* 304: 1800463.

11 Takenaka, T., Nagata, H., Hiruma, Y. et al. (2007). Lead-free piezoelectric ceramics based on perovskite structures. *J. Electroceram.* 19: 259–265.

12 Malič, B., Otoničar, M., Radan, K., and Koruza, J. (2021). Lead-free piezoelectric ceramics. *Encycl. Mater.: Tech. Ceram. Glasses* 3–3: 358–368.

13 Ringgaard, E., Wurlitzer, T., and Wolny, W.W. (2005). Properties of lead-free piezoceramics based on alkali niobates. *Ferroelectrics* 319: 97–107.

14 Egerton, L. and Dillon, D.M. (1959). Piezoelectric and dielectric properties of ceramics in the system potassium—sodium niobate. *J. Am. Ceram. Soc.* 42: 438–442.

15 Tellier, J., Malic, B., Dkhil, B. et al. (2009). Crystal structure and phase transitions of sodium potassium niobate perovskites. *Solid State Sci.* 11: 320–324.

16 Kodre, A., Tellier, J., Arčon, I. et al. (2009). Extended x-ray absorption fine structure study of phase transitions in the piezoelectric perovskite $K_{0.5}Na_{0.5}NbO_3$. *J. Appl. Phys.* 105: 113528.

17 Shirane, G., Newnham, R., and Pepinsky, R. (1954). Dielectric properties and phase transitions of $NaNbO_3$ and $(Na,K)NbO_3$. *Phys. Rev.* 96: 581–588.

18 Wada, S., Seike, A., and Tsurumi, T. (2001). Poling treatment and piezoelectric properties of potassium niobate ferroelectric single crystals. *Jpn. J. Appl. Phys., Part 1* 40: 5690–5697.

19 Zheng, T., Wu, J., Xiao, D., and Zhu, J. (2018). Recent development in lead-free perovskite piezoelectric bulk materials. *Prog. Mater. Sci.* 98: 552–624.

20 Jaffe, H. (1958). Piezoelectric ceramics. *J. Am. Ceram. Soc.* 41: 494–498.

21 Banerjee, S., Bairagi, S., and Wazed Ali, S. (2021). A critical review on lead-free hybrid materials for next generation piezoelectric energy harvesting and conversion. *Ceram. Int.* 47: 16402–16421.

22 Li, T. and Lee, P.S. (2022). Piezoelectric energy harvesting technology: from materials, structures, to applications. *Small Struct.* 3: 2100128.

23 Liu, X., Tan, X., Liu, X., and Tan, X. (2016). Giant strains in non-textured $(Bi_{1/2}Na_{1/2})TiO_3$-based lead-free ceramics. *Adv. Mater.* 28: 574–578.

24 Lee, M.H., Kim, D.J., Park, J.S. et al. (2015). High-performance lead-free piezoceramics with high Curie temperatures. *Adv. Mater.* 27: 6976–6982.

25 Zheng, T., Wu, H., Yuan, Y. et al. (2017). The structural origin of enhanced piezoelectric performance and stability in lead free ceramics. *Energy Environ. Sci.* 10: 528–537.

26 Jaffe, B., Roth, R.S., and Marzullo, S. (2004). Piezoelectric properties of lead zirconate-lead titanate solid-solution ceramics. *J. Appl. Phys.* 25: 809.

27 Jo, W., Dittmer, R., Acosta, M. et al. (2012). Giant electric-field-induced strains in lead-free ceramics for actuator applications – status and perspective. *J. Electroceram.* 29: 71–93.

28 Cross, E. (2004). Lead-free at last. *Nature* 432: 24–25.

29 Maeder, M.D., Damjanovic, D., and Setter, N. (2004). Lead free piezoelectric materials. *J. Electroceram.* 13: 385–392.

30 Lu, Y.-Q. and Li, Y.-X. (2011). A review on lead-free piezoelectric ceramics studies in China. *J. Adv. Dielectr.* 01: 269–288.

31 Bai, S., Xu, Q., Gu, L. et al. (2012). Single crystalline lead zirconate titanate (PZT) nano/micro-wire based self-powered UV sensor. *Nano Energy* 1: 789–795.

32 Heywang, W., Lubitz, K., and Wersing, W. (2008). *Piezoelectricity: Evolution and Future of a Technology*. Springer Science & Business Media.

33 Xiao, D. (2011). Progresses and further considerations on theresearch of perovskite lead-free piezoelectric ceramics. *J. Adv. Dielectr.* 1: 33–40.

34 Li, Q., Zhang, Q., Cai, W. et al. (2020). Enhanced ferroelectric and piezoelectric responses of $(Ba_{0.85}Ca_{0.15})(Zr_{0.1}Ti_{0.9})O_3$ ceramics by Tm^{3+} amphoteric substitution. *Mater. Chem. Phys.* 252: 123242.

35 Vaish, R. (2013). Piezoelectric and pyroelectric materials selection. *Int. J. Appl. Ceram. Technol.* 10: 682–689.

36 Li, P., Zhai, J., Shen, B. et al. (2018). Ultrahigh piezoelectric properties in textured $(K,Na)NbO_3$-based lead-free ceramics. *Adv. Mater.* 30: 1705171.

37 Gao, J., Xue, D., Liu, W. et al. (2017). Recent progress on $BaTiO_3$-based piezoelectric ceramics for actuator applications. *Actuators* 6: 24.

38 Xue, D., Zhou, Y., Bao, H. et al. (2011). Elastic, piezoelectric, and dielectric properties of $Ba(Zr_{0.2}Ti_{0.8})O_3$-$50(Ba_{0.7}Ca_{0.3})TiO_3$ Pb-free ceramic at the morphotropic phase boundary. *J. Appl. Phys.* 109.

39 Raj, N.P.M.J., Abisegapriyan, K.S., and Khandelwal, G. (2020). A lead-free ferroelectric $Bi_{0.5}Na_{0.5}TiO_3$ based flexible, lightweight nanogenerator for motion monitoring applications. *Sustainable Energy Fuels* 4: 5636–5644.

40 Taghaddos, E., Charalambous, H., Tsakalakos, T., and Safari, A. (2019). Electromechanical properties of flash sintered BNT-based piezoelectric ceramic. *J. Eur. Ceram. Soc.* 39: 2882–2888.

41 Reichmann, K., Feteira, A., and Li, M. (2015). Bismuth sodium titanate based materials for piezoelectric actuators. *Materials* 8: 8467–8495.

42 Guy, I.L., Muensit, S., and Goldys, E.M. (1999). Extensional piezoelectric coefficients of gallium nitride and aluminum nitride. *Appl. Phys. Lett.* 75: 4133–4135.

43 Tadigadapa, S. and Mateti, K. (2009). Piezoelectric MEMS sensors: state-of-the-art and perspectives. *Meas. Sci. Technol.* 20: 092001.

44 Hou, R., Hutson, D., and Kirk, K.J. (2013). Development of sputtered AlN thin-film ultrasonic transducers for durable high-temperature applications. *Insight Non-Destr. Test. Condition Monit.* 55: 302–307.

45 Boes, A., Chang, L., Langrock, C. et al. (2023). Lithium niobate photonics: Unlocking the electromagnetic spectrum. *Science* 379 (6627): eabj4396.

46 Tressler, J.F., Alkoy, S., and Newnham, R.E. (1998). Piezoelectric sensors and sensor materials. *J. Electroceram.* 2: 257–272.

47 Liu, W. and Ren, X. (2009). Large piezoelectric effect in Pb-free ceramics. *Phys. Rev. Lett.* 103: 257602.

48 Coondoo, I., Panwar, N., Alikin, D. et al. (2018). A comparative study of structural and electrical properties in lead-free BCZT ceramics: influence of the synthesis method. *Acta Mater.* 155: 331–342.

49 Broitman, E., Soomro, M.Y., Lu, J. et al. (2013). Nanoscale piezoelectric response of ZnO nanowires measured using a nanoindentation technique. *Phys. Chem. Chem. Phys.* 15: 11113.

50 Cardoso, J., Oliveira, F., Proenca, M., and Ventura, J. (2018). The influence of shape on the output potential of ZnO nanostructures: sensitivity to parallel versus perpendicular forces. *Nanomaterials* 8: 354.

51 Yasseri, M., Abdizadeh, H., Shakeri, A., and Golobostanfard, M.R. (2014). Low temperature synthesis of barium titanate powder by a modified sol-gel method. *Adv. Mater. Res.* 829: 727–731.

52 Zhao, C., Huang, Y., and Wu, J. (2020). Multifunctional barium titanate ceramics via chemical modification tuning phase structure. *InfoMat* 2: 1163–1190.

53 Uchino, K. (2010). The development of piezoelectric materials and the new perspective. In: *Advanced Piezoelectric Materials* (ed. K. Uchino), 1–85. Elsevier https://doi.org/10.1533/9781845699758.1.

54 Simon-Seveyrat, L., Hajjaji, A., Emziane, Y. et al. (2007). Re-investigation of synthesis of $BaTiO_3$ by conventional solid-state reaction and oxalate coprecipitation route for piezoelectric applications. *Ceram. Int.* 33: 35–40.

55 Sonia, R.K., Patel, P., Prakash, C. et al. (2012). Low temperature synthesis and dielectric, ferroelectric and piezoelectric study of microwave sintered $BaTiO_3$ ceramics. *Ceram. Int.* 38: 1585–1589.

56 Wang, X., Deng, X., Wen, H., and Li, L. (2006). Phase transition and high dielectric constant of bulk dense nanograin barium titanate ceramics. *Appl. Phys. Lett.* 89: 162902.

57 Zheng, P., Zhang, J.L., Shao, S.F. et al. (2009). Piezoelectric properties and stabilities of CuO-modified $Ba(Ti,Zr)O_3$ ceramics. *Appl. Phys. Lett.* 94: 032902.

58 Tian, H.Y., Wang, Y., Miao, J. et al. (2007). Preparation and characterization of hafnium doped barium titanate ceramics. *J. Alloys Compd.* 431: 197–202.

59 McQuarrie, M. and Behnke, F.W. (1954). Structural and dielectric studies in the system (Ba, Ca) (Ti, Zr)O_3. *J. Am. Ceram. Soc.* 37: 539–543.

60 Zhang, L., Ma, J., Zhang, F. et al. (2022). High-temperature stability of $Ba_{0.99}Ca_{0.01}Zr_{0.02}Ti_{0.98}O_3$-doped $LiTaO_3$ lead-free ceramics. *J. Mater. Sci.: Mater. Electron.* 33: 5878–5885.

61 Zhou, C.R. and Liu, X.Y. (2007). Dielectric properties and relaxation of $Bi_{0.5}Na_{0.5}TiO_3$-$BaNb_2O_6$ lead-free ceramics. *Bull. Mater. Sci.* 30: 575–578.

62 He, H., Lu, W., Oh, J. et al. (2020). Probing the coexistence of ferroelectric and relaxor states in $Bi_{0.5}Na_{0.5}TiO_3$-based ceramics for enhanced piezoelectric performance. *ACS Appl. Mater. Interfaces* 12: 30556.

63 Hiruma, Y., Nagata, H., and Takenaka, T. (2009). Thermal depoling process and piezoelectric properties of bismuth sodium titanate ceramics. *J. Appl. Phys.* 105: 084112.

64 Chen, W., Li, Y., Xu, Q., and Zhou, J. (2005). Electromechanical properties and morphotropic phase boundary of $Na_{0.5}Bi_{0.5}TiO_3$-$K_{0.5}Bi_{0.5}TiO_3$-$BaTiO_3$ lead-free piezoelectric ceramics. *J. Electroceram.* 15: 229–235.

65 Wang, X.X., Chan, H.L.W., and Choy, C.L. (2005). $(Bi_{0.5}Na_{0.5})_{0.94}Ba_{0.06}TiO_3$ lead-free ceramics with simultaneous addition of CeO_2 and La_2O_3. *Appl. Phys. A* 80: 333–336.

66 Sasaki, A., Chiba, T., Mamiya, Y., and Otsuki, E. (1999). Dielectric and piezoelectric properties of $(Bi_{0.5}Na_{0.5})TiO_3$-$(Bi_{0.5}K_{0.5})TiO_3$ systems. *Jpn. J. Appl. Phys., Part 1* 38: 5564–5567.

67 Hlruma, Y., Nagata, H., and Takenaka, T. (2007). Grain-size effect on electrical properties of $(Bi_{1/2}K_{1/2})TiO_3$ ceramics. *Jpn. J. Appl. Phys., Part 1* 46: 1081–1084.

68 Mahapatra, S.D., Mohapatra, P.C., Aria, A.I. et al. (2021). Piezoelectric materials for energy harvesting and sensing applications: roadmap for future smart materials. *Adv. Sci.* 8: 2100864.

69 Briscoe, J. and Dunn, S. (2015). Piezoelectric nanogenerators – a review of nanostructured piezoelectric energy harvesters. *Nano Energy* 14: 15–29.

70 Zheng, Q., Shi, B., Li, Z., and Wang, Z.L. (2017). Recent progress on piezoelectric and triboelectric energy harvesters in biomedical systems. *Adv. Sci.* 4: 1700029.

71 Li, C., Guo, W., Kong, Y., and Gao, H. (2007). First-principles study of the dependence of ground-state structural properties on the dimensionality and size of ZnO nanostructures. *Phys. Rev. B: Condens. Matter Mater. Phys.* 76: 035322.

72 Preissig, F.J.v., Zeng, H., and Kim, E.S. (1998). Measurement of piezoelectric strength of ZnO thin films for MEMS applications. *Smart Mater. Struct.* 7: 396–403.

73 Kuoni, A., Holzherr, R., Boillat, M., and de Rooij, N.F. (2003). Polyimide membrane with ZnO piezoelectric thin film pressure transducers as a differential pressure liquid flow sensor. *J. Micromech. Microeng.* 13 (4): S103.

74 Wang, Z.L. and Song, J. (2006). Piezoelectric nanogenerators based on zinc oxide nanowire arrays. *Science* 1979 (312): 242–246.

75 Wang, X., Song, J., Liu, J., and Zhong, L.W. (2007). Direct-current nanogenerator driven by ultrasonic waves. *Science* 1979 (316): 102–105.

76 Wang, Z.L., Zhu, G., Yang, Y. et al. (2012). Progress in nanogenerators for portable electronics. *Mater. Today* 15: 532–543.

77 Jaffe, H. and Berlincourt, D.A. (1965). Piezoelectric transducer materials. *Proc. IEEE* 53: 1372–1386.

78 Fei, C., Liu, X., Zhu, B. et al. (2018). AlN piezoelectric thin films for energy harvesting and acoustic devices. *Nano Energy* 51: 146–161.

79 Tsubouch, K. and Mikoshiba, N. (1985). Zero-temperature-coefficient SAW devices on AlN epitaxial films. *IEEE Trans. Sonics Ultrason.* 32: 634–644.

80 Dai, Y.J., Zhang, X.W., and Chen, K.P. (2009). Morphotropic phase boundary and electrical properties of $K_{1-x}Na_xNbO_3$ lead-free ceramics. *Appl. Phys. Lett.* 94: 1–4.

81 Koruza, J., Liu, H., Höfling, M. et al. (2020). $(K,Na)NbO_3$-based piezoelectric single crystals: growth methods, properties, and applications. *J. Mater. Res.* 35: 990–1016.

82 Wang, K. and Li, J.F. (2007). Analysis of crystallographic evolution in $(Na,K)NbO_3$-based lead-free piezoceramics by x-ray diffraction. *Appl. Phys. Lett.* 91: 2005–2008.

83 Pinho, R., Tkach, A., Zlotnik, S. et al. (2020). Spark plasma texturing: a strategy to enhance the electro-mechanical properties of lead-free potassium sodium niobate ceramics. *Appl. Mater. Today* 19: 100566.

84 Dubernet, M., Pitcher, M.J., Zaghrioui, M. et al. (2022). Synthesis routes for enhanced piezoelectric properties in spark plasma sintered Ta-doped KNN ceramics. *J. Eur. Ceram. Soc.* 42: 2188–2194.

85 Li, H., Hao, Y., Lin, Z. et al. (2022). (K,Na)NbO_3 lead-free piezoceramics prepared by microwave sintering and solvothermal powder synthesis. *Solid State Commun.* 353: 114871.

86 Feizpour, M., Barzegar Bafrooei, H., Hayati, R., and Ebadzadeh, T. (2014). Microwave-assisted synthesis and sintering of potassium sodium niobate lead-free piezoelectric ceramics. *Ceram. Int.* 40: 871–877.

87 Wu, J. (2020). Perovskite lead-free piezoelectric ceramics. *J. Appl. Phys.* 127: 190901.

88 Bairagi, S. and Ali, S.W. (2020). Flexible lead-free PVDF/SM-KNN electrospun nanocomposite based piezoelectric materials: significant enhancement of energy harvesting efficiency of the nanogenerator. *Energy* 198: 117385.

89 Teka, A., Bairagi, S., Shahadat, M. et al. (2018). Poly(vinylidene fluoride) (PVDF)/potassium sodium niobate (KNN)–based nanofibrous web: a unique nanogenerator for renewable energy harvesting and investigating the role of KNN nanostructures. *Polym. Adv. Technol.* 29: 2537–2544.

90 Bairagi, S. and Ali, S.W. (2019). A unique piezoelectric nanogenerator composed of melt-spun PVDF/KNN nanorod-based nanocomposite fibre. *Eur. Polym. J.* 116: 554–561.

91 Bairagi, S. and Ali, S.W. (2019). Effects of surface modification on electrical properties of KNN nanorod-incorporated PVDF composites. *J. Mater. Sci.* 54: 11462–11484.

92 Bairagi, S. and Ali, S.W. (2019). Influence of high aspect ratio lead-free piezoelectric fillers in designing flexible fibrous nanogenerators: demonstration of significant high output voltage. *Energy Technol.* 7: 1–10.

93 Bairagi, S. and Ali, S.W. (2020). Poly(vinylidine fluoride) (PVDF)/potassium sodium niobate (KNN) nanorods based flexible nanocomposite film: influence of KNN concentration in the performance of nanogenerator. *Org. Electron.* 78: 1–12.

94 Bairagi, S. and Ali, S.W. (2020). Investigating the role of carbon nanotubes (CNTs) in the piezoelectric performance of a PVDF/KNN-based electrospun nanogenerator. *Soft Matter* 16: 4876–4886.

95 Bairagi, S. and Ali, S.W. (2020). A hybrid piezoelectric nanogenerator comprising of KNN/ZnO nanorods incorporated PVDF electrospun nanocomposite webs. *Int. J. Energy Res.* 44: 5545–5563.

96 Shoorangiz, M., Sherafat, Z., and Bagherzadeh, E. (2022). CNT loaded PVDF-KNN nanocomposite films with enhanced piezoelectric properties. *Ceram. Int.* 48: 15180–15188.

97 Vivekananthan, V., Chandrasekhar, A., Alluri, N.R. et al. (2019). A flexible piezoelectric composite nanogenerator based on doping enhanced lead-free nanoparticles. *Mater. Lett.* 249: 73–76.

98 Ichangi, A., Shvartsman, V.V., Lupascu, D.C. et al. (2021). Li and Ta-modified KNN piezoceramic fibers for vibrational energy harvesters. *J. Eur. Ceram. Soc.* 41: 7662–7669.

99 Banerjee, S., Bairagi, S., and Ali, S.W. (2022). A lead-free flexible piezoelectric-triboelectric hybrid nanogenerator composed of uniquely designed PVDF/KNN-ZS nanofibrous web. *Energy* 244: 123102.

100 Xue, Q., Wang, Z., Tian, H. et al. (2015). A record flexible piezoelectric KNN ultrafine-grained nanopowder-based nanogenerator. *AIP Adv.* 5: 017102.

101 Mohamed, A.M., Yao, K., Yousry, Y.M. et al. (2020). Enhanced airborne sound absorption effect in poly(vinylidene fluoride)/$(K_{0.5}Na_{0.5})NbO_3$-nanofiber composite foams. *J. Appl. Polym. Sci.* 137: 1–10.

102 Zhang, A., Liu, Z., Geng, X. et al. (2019). Ultrasonic vibration driven piezocatalytic activity of lead-free $K_{0.5}Na_{0.5}NbO_3$ materials. *Ceram. Int.* 45: 22486–22492.

103 Narita, F., Nagaoka, H., and Wang, Z. (2019). Fabrication and impact output voltage characteristics of carbon fiber reinforced polymer composites with lead-free piezoelectric nano-particles. *Mater. Lett.* 236: 487–490.

3

MoS_2-Based Hybrid Piezoelectric Materials

Sahal Ebrahim[1] and Deepalekshmi Ponnamma[2]

[1] *Department of Nanoscience and Technology, University of Calicut, Kerala, India*
[2] *Materials Science and Technology Program, Department of Mathematics, Statistics and Physics, College of Arts and Sciences, Qatar University, Doha, Qatar*

3.1 Introduction

The burning of fossil-derived fuels exacerbates environmental conditions, and their limited availability turns the world's focus toward renewable energy resources. Nowadays, renewable energy sources such as solar energy, wind energy, and ocean energy have been progressively established [1]. They are abundant in nature and have little effect on the atmosphere during utilization; however, their merits have faded due to large space requirements, economic infeasibility, and nonprogressive technological improvements, which led to a novel idea called energy harvesters that employ piezoelectric materials for their functioning. Piezoelectricity is the property of materials that converts the deformation caused by a mechanical stimulus into electrical energy, and vice versa. In other words, when a material undergoes stress, a change in electric potential is generated (the positive piezoelectric effect), and conversely, strain is produced by the action of applied electrical energy (the inverse piezoelectricity) [2]. Piezoelectric materials easily consume waste mechanical energy using a piezoelectric nanogenerator (PENG), removing the need for frequent charging of battery cells [3], and aiding them in certain attractive conversion applications such as in pacemakers where heat beat converts into electricity for its functioning [4], wind energy conversion in micro air vehicles [5], ocean wave current generation [6], and energy from footsteps[7], and breathing [3]. Various applications of PENG are illustrated in Figure 3.1 [9].

Hybrid Materials for Piezoelectric Energy Harvesting and Conversion, First Edition.
Edited by S. Wazed Ali, Satyaranjan Bairagi, and Shahid Ul Islam.
© 2024 John Wiley & Sons, Inc. Published 2024 by John Wiley & Sons, Inc.

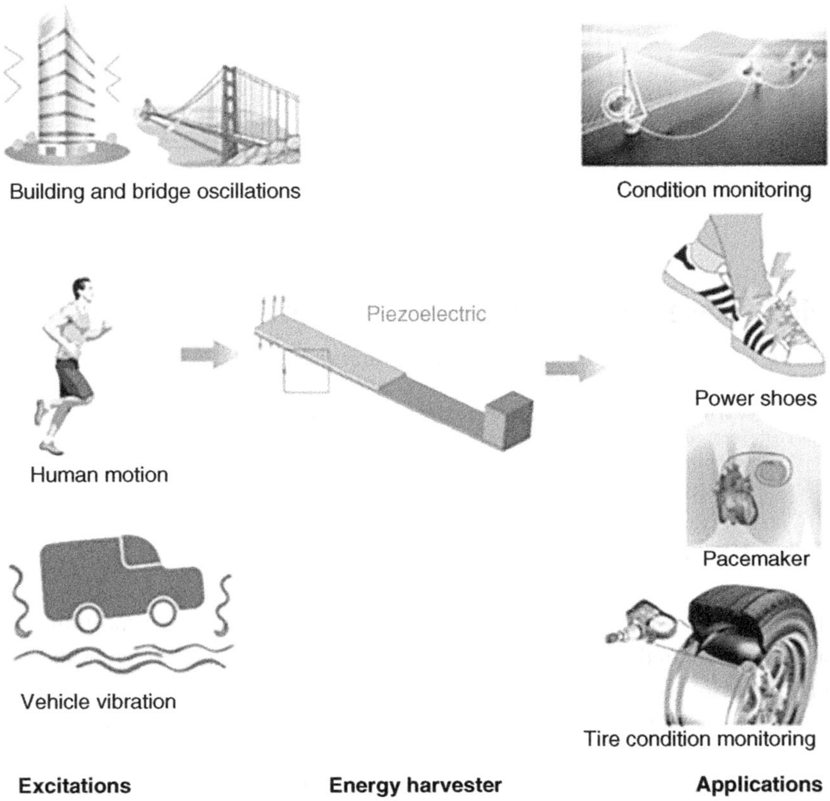

Figure 3.1 Illustration of various applications of the PENG. Source: Adapted from Wang et al. [8]. Reproduced with permission from Yang et al. [9], © 2018/ ELSEVIER.

Most piezoelectric materials are made of synthetic polymers that are nonrenewable and bio-incompatible, and are harmful to nature [10]. Therefore, it is crucial to bring biocompatible, renewable, and biodegradable materials for expanding their applications. Natural polymers, such as cellulose, which are most abundant in nature [11], have flexible, low-cost, renewable, biodegradable, biocompatible, higher strength, and lower thermal expansion characteristics. Furthermore, cellulose is known for robust, straight, and unbranched d-glucose polysaccharide groups linked with β-1,4-glycosidic bonds, and are organized into both crystalline and amorphous structures to form the molecule [12]. The flexibility of cellulose makes it suitable for electrical and electronic gadgets. However, while compared with conventional polymers, cellulose-based polymers have lower piezoelectric properties with limited composite structures, and their internal and external strong hydrogen bonds cause an insoluble behavior in regular solvent or melt systems [11, 13]. Two-dimensional (2D) materials have received a lot of attention because of the presence of van der Waals (vdW) forces of attraction between the monolayers, excellent crystallinity, capacity to bear ultimate strain, extremely fine structure [14], high electromechanical sensitivity [15], and other distinctive physical features. Among them, 2D transition-metal dichalcogenides (TMDs) such as molybdenum disulfide (MOS_2), having similar structure as that of graphene has high elastic behavior and excellent piezoelectric properties. The layered structure of MoS_2 is formed by locating the molybdenum layer between two sulfide surfaces, with both molybdenum and sulfide covalently bonded, and the layers are connected by vdW forces of attraction [16]. The symmetry of the MOS_2 structure is crucial in determining the piezoelectric properties. Intriguingly, bulk MOS_2 material has inversion symmetry with respect to an inversion center due to the opposing geometrical arrangement of its closest atomic layers. Meanwhile, when MoS_2 is exfoliated from the bulk, the atomic scale of MoS_2 layers is formed, causing inversion symmetry to be broken and producing the piezoelectric effect [17]. While considering the number of MoS_2 atomic layers, the presence of an inversion center in an even number of MoS_2 layers does not produce piezoelectricity, as a result of the net zero effect caused by the cancellation of two equal and opposite charges. Even so, the lack of inversion symmetry in MoS_2 with an odd number of layers, due to uneven charge distribution, would result in piezoelectricity when subjected to external stress [18]. Figure 3.2a [19] depicts the variation of piezoelectric behavior of the MoS_2 structure with respect to different views as structural symmetry is changed. If the top view of Figure 3.2a, which has no inversion center, is applied with a stress p (along inner plane d_{11}), polarization occurs ensuing in the uneven distribution of electric dipoles owing to the movement of positive Mo and negative S atoms. Viewed from the side, however, the centrosymmetric structure is restored, and the force acting along the z-plane (outer plane d_{33}) does not produce a piezoelectric effect. Notably, piezoelectric characteristics were bound to a certain orientation [2]. The flexibility of zigzag and armchair structure does not show any disparity during bending (illustrated in Figure 3.2b). After extreme flexure, inextricable bonding of Mo–Mo provides super elasticity to MoS_2 structure [20]. The remarkable flexibility of MoS_2 sheets has been noticed. Polarization causes current generation in the monolayer MoS_2 sheet without tensile stress, there is no polarization in MoS_2, and a strain along the x-axis (ε_{11}) distorts the dipole moment. As a result, domains were uniformly arranged in the

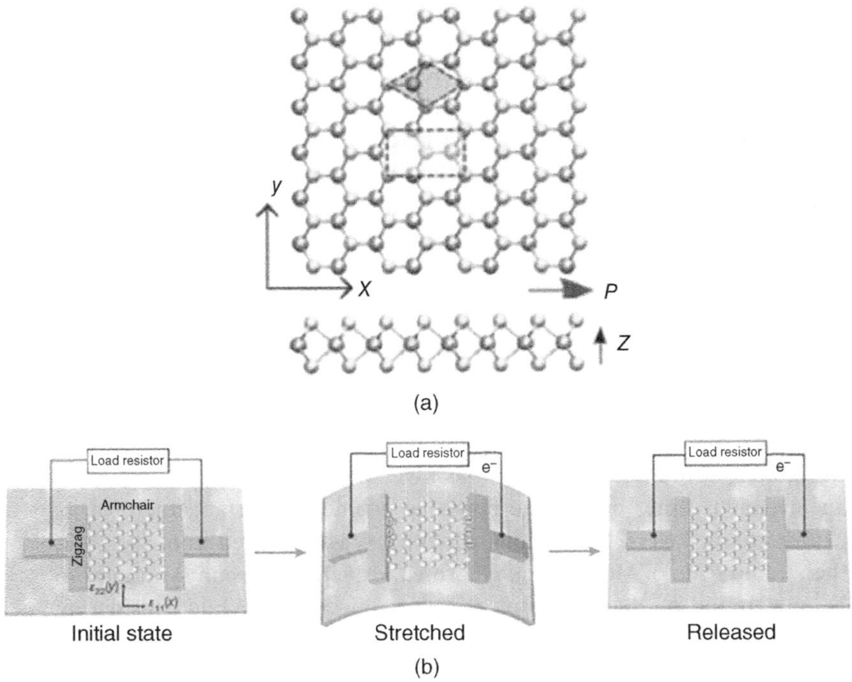

Figure 3.2 (a) Plan view and front view of MoS_2, where no piezoelectric effect in front plane because of centrosymmetric structure, while top view loses its inversion center and has piezoelectric characteristic, (b) charge accumulation at the edges of metal–MoS_2 interface with respect to stretching [19, 20]. Source: Copyright 2012 and 2015. Reprinted with permission from American chemical society.

opposite direction, resulting in a polarization of charge at the two zigzag edges of electrodes. The strain along the *y*-axis (ε_{12}) is a pure shear strain and can be neglected. The sign can be reversed by applying strain parallel to the *y*-axis, at which point the charge forms at the armchair edges. After the withdrawal of force, the material returned to its equilibrium position.

Importantly, the monolayer MoS_2 formed by chemical vapor deposition (CVD) exhibits direction piezoelectricity, and it was discovered that the output power from the armchair direction of MoS_2 was almost twice as great as that from the zigzag direction within the same strain [21].

3.2 Different Methods of MoS_2 Synthesis

3.2.1 Exfoliation Method

In this method, the layers are cut out of the bulk materials. There are two types of exfoliations: (i) mechanical and (ii) chemical exfoliation. Production can be considerably increased by using chemical exfoliation rather than mechanical exfoliation. However, adding sonication to this procedure can lead to flake size reduction to a few thousand nanometers, causing imperfections in the 2D lattice structure [22]. Different exfoliation techniques with their synthesis procedure are included in Table 3.1.

3.2.2 Hydrothermal Method

Thin film MoS_2 is prepared using the hydrothermal method. In general, 15 ml of monohydrated hydrazine and 85 ml of deionized water are mixed with 0.05 mol of $(NH_4)_6Mo_7O_{24}{\cdot}4H_2O$ and 0.1 mol sulfur powder and agitated for 30 minutes. The resulting solution was taken into a Teflon-lined autoclave, heated to 180 °C for 36 hours, then cooled naturally to ambient temperature. The final solution is cleaned and filtered using ethanol and deionized water, and then dried under a vacuum [3].

3.2.3 CVD Method

Generally, there are two types of CVD methods. The first one is thermal vapor sulfurization (TVS) (Figure 3.3a), which involves a couple of steps. Initially, MoS_2 is coated on a SiO_2/Si substrate from a source (commonly used sources are Mo and MoO_3) to synthesize a thin layer structure of MoS_2. Then the as-prepared thin layer undergoes thermal annealing in a sulfur vapor atmosphere at a high temperature to improve the quality of the resulting MoS_2 layer. Importantly, the arrangement of the source and substrate controls the morphology, quality, and yield of the MoS_2 sheet layer. The second one is thermal vapor deposition (TVD) (Figure 3.3b), which utilizes powdered MoO_3 and sulfur as source materials. Figure 3.3c indicates the different steps in the formation of the MoS_2 atomic layer. At first, the powder source is converted into carrier gas through an evaporation or sublimation process. The carrier gas, called the reaction species, transports and diffuses above the substrate. After re-evaporation or desorption, it carries

Table 3.1 Different exfoliation methods utilized to synthesize MoS_2 layers.

Different exfoliation methods	Procedure	References
Mechanical exfoliation		
Scotch tape method	MoS_2 flakes exfoliate by pasting Scotch tape over it. This specific Scotch tape was folded three times after exfoliation, and then put on a new piece of Scotch tape. This resulted in the development of two pieces of Scotch tape with exfoliated MoS_2 flakes.	[23]
Anodic bonding	A fine layer is chopped from each substrate and placed on a glass substrate. Both the substrate and glass are connected to a bonding machine through two electrodes. After applying heat and voltage in the range of 130–200 °C and 200–1500 V, bonding occurs, and the bonded material is detached using mechanical peeling.	[24]
Chemical exfoliation		
Ion intercalation and solvent-based exfoliation	The layers are separated from MoS_2 by introducing ions into the gaps of MoS_2 layers. Commonly, *n*-butyl lithium in hexane is used due to its small size. After Li-ion intercalation, water is used to form a vigorous reaction with the Li ions, which produces hydrogen gas and pushes the MoS_2 layers apart.	[25]
Electrochemical lithiation discharge process	The layered MoS_2 acts as the cathode in a test cell, and the anode is the lithium foil. Moreover, lithium foil acts as a lithium-ion source, and the lithium intercalation occurs in a galvanostatic discharge with a current density of 0.005 mA. The intercalation process creates $LixMoS_2$, which gets washed with acetone and ultrasonically treated with water or ethanol to form fine MoS_2 layers.	[26]
Solvent-based exfoliation also known as Coleman method	MoS_2 is sonicated with solvent having surface tension. The final mixture was centrifuged, and the supernatant was decanted.	[27]

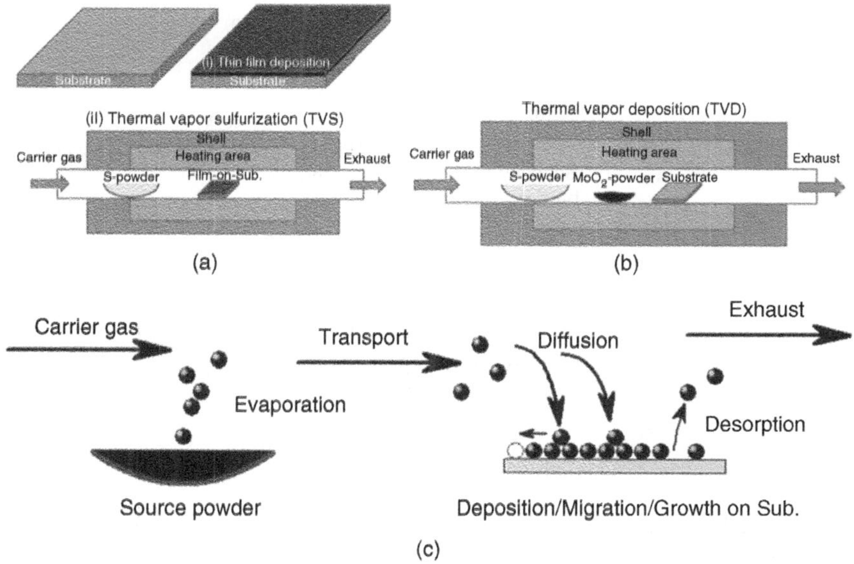

Figure 3.3 (a) CVD–TVS process, (b) CVD–TDS process, and (c) mechanism involved in the TDS film generation process [28]. Source: Copyright 2019. Reprinted with permission from Elsevier.

the byproduct out to the exhaust. In spite of having the crystal quality of atomic layer in TVD, nonuniformity in layers fades its merits. In conclusion, TVD is superior to TVS in generating quality, greater area, and single-crystal single-layer MoS_2 nanosheets [28].

3.3 MoS_2 Working Mechanism

Metal-induced gap (MIG), which is depicted by bars at the band gaps in MoS_2 semiconductor (Figure 3.4), is a gap state that forms at the interface and moves further into the MoS_2 semiconductor as soon as the MoS_2 is linked to the metal electrodes. The Fermi levels of MoS_2 and metal are represented as Φ_{CNL}, and E_F, correspondingly. Electrons from the MoS_2 and metal cross each other to match their Fermi levels Φ_{CNL}, and E_F (Figure 3.4). MIG has a specific energy value and is situated in the MoS_2 bandgap that causes electron transfer between MoS_2 and metal electrodes. Therefore, dipole at the interface is produced by the passage of electrons between the metal electrode and MoS_2. During deformation, the domains align and the infinite dielectric property of metal causes the positive charges

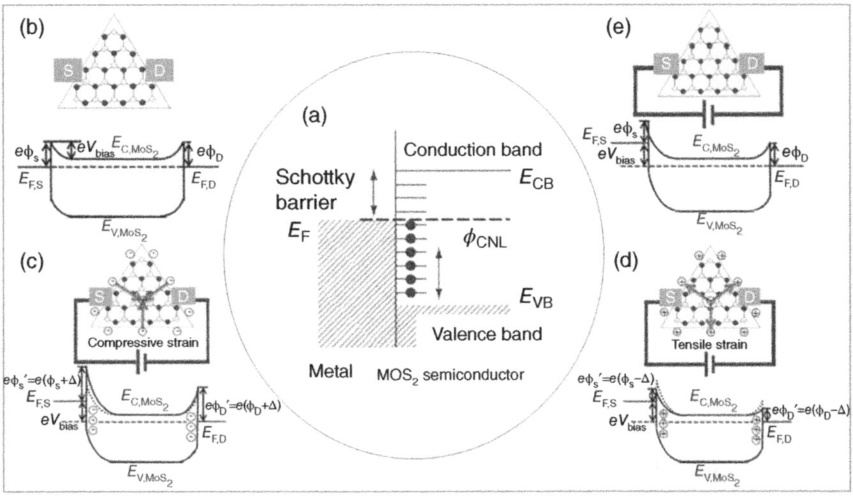

Figure 3.4 Represents (a) metal Fermi level E_F matches with MOS_2 semiconductor, (b) Schottky barrier height differences at different conditions (b) at equilibrium (c) under compression force, (d) tensile force, (e) inverse piezoelectric effect by applying an external supply [29, 30]. Source: Copyright 2019. Reprinted with permission from the American Chemical Society. Copyright. Reprinted with permission from Elsevier 2011 and Nature 2015.

to accumulate at the interface, which attracts the electrons and produces piezoelectric current. In addition, the Schottky barrier is the energy difference between MoS_2's lowest conduction band (CB) state, E_{CB}, and its highest valence-band state, E_{VB}. It serves as a potential barrier for either electrons or holes [29].

The band diagram of triangular piezoelectric MoS_2 material synthesized by CVD is attached to the metal on both edges, which undergoes tensile and compression stress, as shown in Figure 3.4. At equilibrium, there is no force acting on the MoS_2 material; therefore, dipoles are randomly oriented, and the Schottky barrier heights (SBH) of both the source and drain are equal. With the application of a compressive force to the MoS_2 structure, the strain vector is toward the center of the material, so it produces a negative potential at all the zigzag edges and hinders the free electrons from overcoming the Schottky barriers with no piezoelectric effect, that is, SBH increases at the two interfaces (Figure 3.4b). When tensile forces are applied, stress vectors move away from the center of the triangular MoS_2, thus producing positive polarization along the edges, which attracts the electrons toward it and results in a piezoelectric effect. The SBH at the two contact edges are reduced when tensile stress is applied (Figure 3.4c). On the other hand, inverse piezoelectricity is analyzed in Figure 3.4d by connecting it with an external electrical supply, raising the Fermi level of the source contact E_{FS}, and keeping SBH stable, as in an equilibrium condition [30].

3.4 Investigating the Transition of MoS_2 Structure from Bulk to Nanostructured Materials

For bulk MoS_2, MO is surrounded by six S atoms having thickness up to three atoms. The layers are connected by weak vdW forces of attraction; therefore, giving them the ability to glide between the layers. The three distinct polytypes of MoS_2 are 1T, 2H, and 3R with point groups $D6d$, $D6h$, and $C3v$, respectively. The alphabetic letters T, H, and R represent Bravais lattices, such as trigonal, hexagonal, and rhombohedral crystal structures, and numerical value represents the number of atomic layers to define the unit cell. Figure 3.5 shows the stacking of MO and S atoms in the

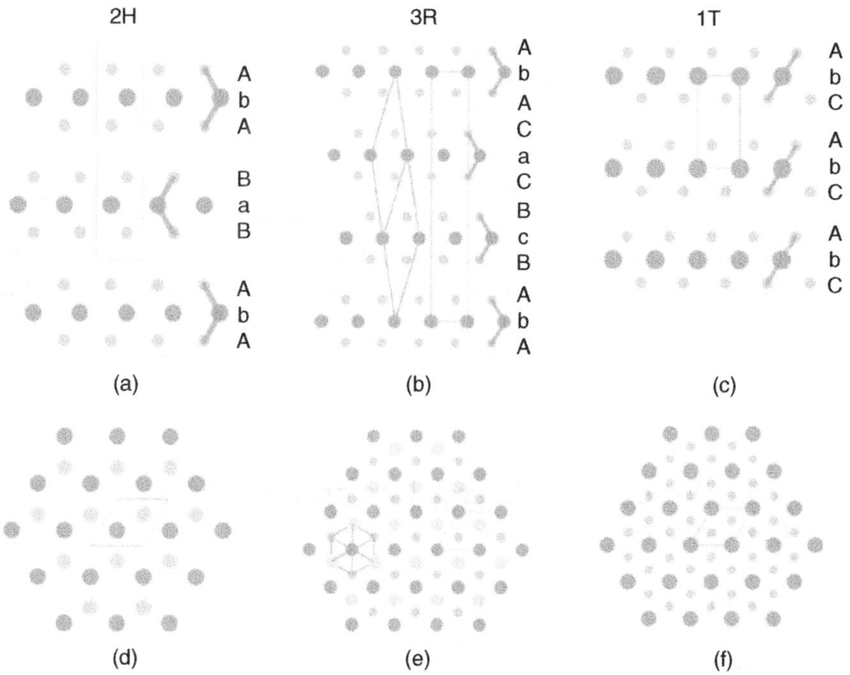

Figure 3.5 The crystal orientation of various polytypes of MoS_2 when viewed along (a–c) the *a*-axis, and (d–f) *b*-axis. Source: Reproduced with permission from Strachan et al. [31]/American chemical Society.

Table 3.2 The different polytypes of MoS_2 with its properties and application.

Polytypes	Properties	Applications
1T Tetragonal	Metastable polymorphs, i.e. 1T′, 1T″, and 1T‴ with stability 1T″ > 1T‴ and 1T′ Octahedral coordination, metallic and paramagnetic characteristics. Lattice parameters $a = 5.60$ Å, $c = 5.99$ Å	Act as source and drain electrodes in FET (field effect transistor) It could be employed for intercalation chemistry; hence, able to function them from N type to P type
2H···MoS_2	Hexagonal coordination Most stable form nonmagnetic semiconductor Space group: *p*63 *mmc* Lattice parameters $a = 3.15$ Å, $c = 12.30$ Å	Dry lubricant, fuel filtration
3R···MoS_2	Stable form Trigonal prismatic coordination Space group *R*3*m* Lattice parameters, $a = 3.17$ Å, $c = 18.38$ Å.	Electronic industry Valley polarization

2H, 3R, and 1T crystal structures. Figure 3.5a–c demonstrates the crystal arrangement when observed along axis, and likewise, Figure 3.5b,c shows the crystal stacking along the *b*-axis. The boxes in the figure represent the unit cell, and the uppercase case and lowercase letters define the position of sulfur and molybdenum atom, respectively [31]. Furthermore, the different polytypes of MoS_2 along with its properties are shown in Table 3.2 [32].

3.5 Piezoelectric Energy Harvesting by MoS_2 Composites

Piezoelectricity has a wide range of uses in actuation, sensing, and energy harvesting since its very discovery in 1880. Nanoscale piezoelectric materials are required due to the constantly expanding need for high-performance and miniaturized devices in micro–electro–mechanical systems (MEMSs) and electronics. Dui et al. experimented with the difference in piezoelectric effects on single crystalline–MoS_2 (SC–MoS_2) and crystals with different orientations, i.e., grain boundaries (GB), in the single-layer MoS_2 (GB–MoS_2). For bulk production of 2D single-layer MoS_2, a CVD technique is implemented that forms GB at different orientations. As a result, some noticeable changes occur in the piezoelectric properties. The formation of defects at the GB–MoS_2 surface altered its symmetry, resulting in a significant increase in its power output by exactly half when compared with SC–MoS_2 (Figure 3.6a). While comparing current density across different orientations, the monolayer MoS_2 has current density outputs in "armchair" directions, which are greater than those across "zigzag" directions. Interestingly, current outputs along parallel and perpendicular directions with respect to GB are larger than those of SC–MoS_2 monolayer flakes, demonstrating improved piezoelectricity of the GB. The inline polarization of GB–MoS_2 is shown in Figure 3.6b, and the magnitude of polarization formed on SC–MoS_2 and GB–MoS_2 under the action of tensile stress is evaluated in Figure 3.6c. [33]. Piezoelectric-induced voltage and current by nanocomposites such as polyvinylidene fluoride (PVDF), MoS_2-reduced graphene oxide (rGO), and PVDF–MoS_2 are compared by Faraz and coworkers (Figure 3.6d,e). PVDF, MoS_2–Rgo, and PVDF–MoS_2 are synthesized using hydrothermal, modified Hummer method, and polymer solution-casting method, respectively. They found that PVDF–rGO–MoS_2-based PENG has an open-circuit voltage of 2.4 V and a maximum short-circuit current of 0.68 A, which is five times better than the other two [3]. Single-crystalline (SC) monolayer triangular MoS_2 shows different piezoelectric energy output when force is applied along the two distinct edges known as armchair and zigzag. The orientation dependency of SC-triangular MoS_2 is caused by the displacement of lattice under the application of external stress which is depicted in Figure 3.6f. The experimental analysis of Kim et al. shows that the armchair and zigzag of MoS_2 produce 3.78 pm/V and 1.38 pm/V due to the anisotropic behavior of monolayer SC triangular MoS_2. Moreover, at 0.48% applied strain, the armchair direction generates power double that of the zigzag direction (Figure 3.6g) [21]. Undoubtedly, MoS_2 structure heavily depends on the orientation, with the armchair orientation producing larger piezoelectric outputs than the zigzag orientation [34].

The 2D vdW heterostructure indium selenide (In_2Se_3)/MoS_2 system is fabricated using the mechanical exfoliation method, and its piezoelectric response was analyzed. When In_2Se_3 and MoS_2 are combined to form a heterostructure, band alignments are altered because of electron transfer between the interfaces of the two materials to match their Fermi levels. This heterostructure has a piezoelectric coefficient (d_{33}) of about 17.5 pm/V; moreover, the reason behind the escalated piezoelectric response is due to

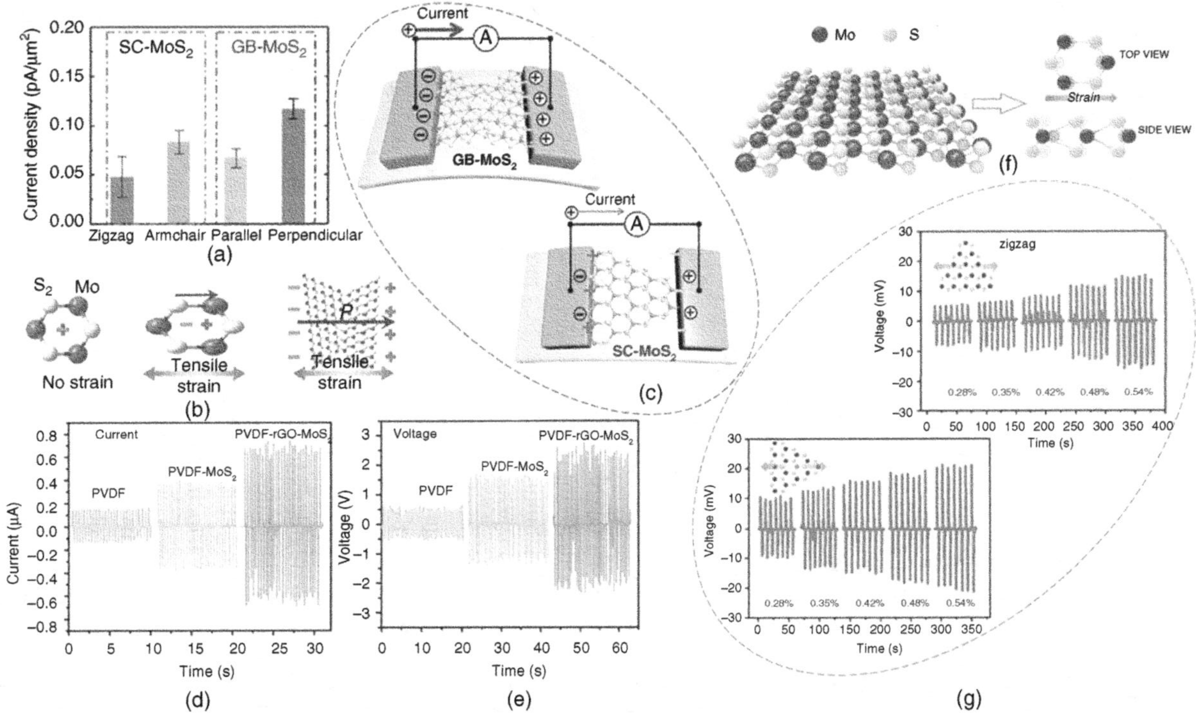

Figure 3.6 (a) Comparison of SC and GB MoS_2 current density under different directions of strain, (b) strain-induced polarization in MoS_2, (c) magnitude of polarization formed on SC–MoS_2 and GD–MoS_2 under the action of tensile stress, (d and e) comparative analysis of voltage and current for a nanocomposite piezoelectric material such as PVDF, PVDF–MoS_2, and PVDF–rGO–MoS_2; hence, PVDF–rGO–MoS_2 has more efficiency for piezoelectric application, (f) lattice distortion of MOS_2 under external stress, and (g) anisotropic piezoelectric characteristics of MoS_2 [21, 33].

the interfacial dipole moment formed by type II staggered band alignment. The large band offset, which is 0.8 eV of the valence band maximum (VBM) and conduction band minimum (CBM), results in enhanced out-of-plane polarization in In_2Se_3/MoS_2 (Figure 3.7a). The inverse piezoelectric output is investigated using piezoresponse force microscopy (PFM). By applying an electric field, the In_2Se_3/MoS_2 structure deforms and is represented as amplitude in Figure 3.7b. The change in amplitude for the change in voltage forms a linear relationship, which is demonstrated in Figure 3.7b. In addition, variations of the piezoelectric coefficient (d_{33}) with the number of layers are shown in Figure 3.7c. Thus, Yuan et al. provided an opportunity for the development of advanced nanoelectronics and multifunctional atomic-scale devices through the development of hetero In_2Se_3/MoS_2 piezoelectric material [35].

The excellent properties of hybrid MoS_2 widened its application in the electronic industry. Huang et al. synthesized a hybrid p-type black phosphorus (BP) and n-type MoS_2 heterostructure (BP/MoS_2, Figure 3.8a) and investigated its physical, transport, electronic, and optical characteristics under compressive strain. When strain is applied, the band gap reduces, and more than 0.85 Å strain initiates a semiconductor-to-metal transition, ensuring an improved piezoelectric response. Figure 3.8a shows the band gap reduction that occurs during the application of compressive strain from 0 to 0.4 Å. Let E_F be the quasi-Fermi level of the BP/MoS_2 structure (Figure 3.8b). The strain creates a movement of electrons from the BP bandgap to the MoS_2 bandgap, and holes in MoS_2 are transferred to the BP bandgap. Due to the change in charge density and interlinkage between BP and MoS_2, the quasi-fermi level of MoS_2 (FM) gets shifted upward; following this, a downward shift of BP Fermi level FBP occurs and confirms the bandgap movement (Figure 3.8c) [36].

The photocatalytic operation is stimulated by investigating piezo-photocatalytic ternary $ZnO/ZnS/MoS_2$ nanocomposites synthesized by Fu and the team. The $ZnO/ZnS/MoS_2$ hybrid structure has a heterojunction with a type II staggered band alignment and is synthesized by a hydrothermal process. The photocatalytic degradation efficiency of methylene blue has skyrocketed up to a proportion of 87.14% for the as-prepared piezo-photocatalytic nanocomposite, which is more than double the efficiency of the non-piezo photocatalytic nanocomposite. The piezoelectric polarization in the composite is generated by stirring. When $ZnO/ZnS/MoS_2$ composite is exposed to sunlight: first, without stirring, no polarization is generated due to the absence of piezoelectricity. The electrons from the VB of ZnO, ZnS, and MoS_2 went into the CB as free electrons, thus creating holes in the VB. At the interfaces, the band alignment is distorted

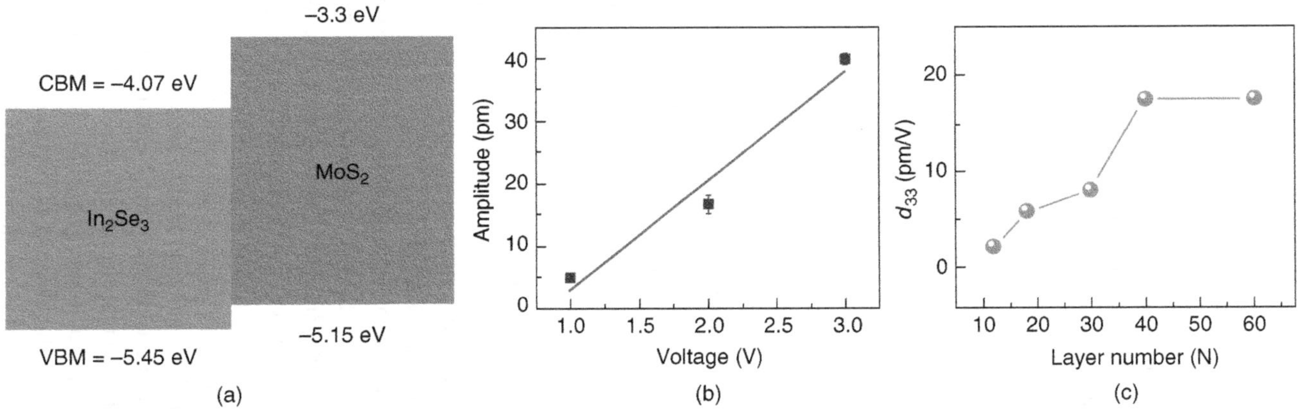

Figure 3.7 (a) Band diagram of In_2Se_3/MoS_2 heterostructure, (b) amplitude versus voltage relationship, (c) variation of piezoelectric coefficient (d_{33}) of In_2Se_3/MoS_2 with number of layers. Source: Reproduced with permission from Yuan et al. [35]/American Chemical Society.

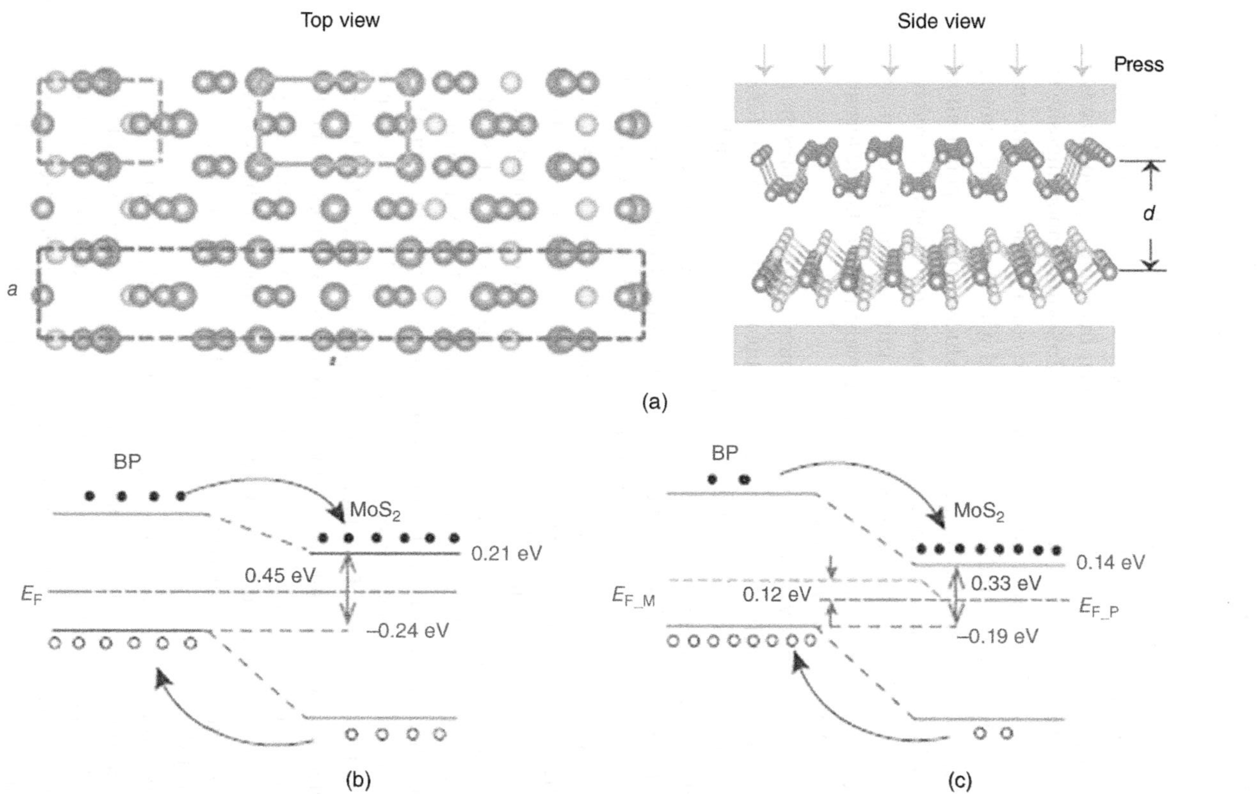

Figure 3.8 Demonstrates (a) the side view and top view of BP/MoS_2 heterostructure, the green and blue rectangular boxes show unit cell of BP and MoS_2, while the red box indicates BP/MoS_2 bilayer. The bandgap shifting of BP/MoS_2 bilayer when applying compressive strain of (b) 0 Å and (c) 0.4 Å. Source: Reproduced with permission from Huang et al. [36]/Springer Nature.

because of the heterojunctions. The electrons from the CB of ZnS and MoS_2 fall into the CB of ZnO (blue arrow), and the holes from the VB of ZnO jump into the VB of MoS_2 and ZnS (red arrow). Hence, the larger potential barrier restricts the effective photocatalytic behavior by confining photogenerated charge carriers, initiated by the hydroxyl-free radical (purple arrow) that forms at the VB of MoS_2 and ZnS, and the oxygen-free radical (yellow arrow) that forms at the CB of ZnO. Second, the ZnO/ZnS/MoS_2 nanocomposite is stirred to create deformation for a piezoelectric output. The electrons and holes move from the VB and CB, as represented with blue and red arrows, respectively. The piezoelectric generation induces charge polarization that detaches electrons and holes in two opposite directions. Therefore, the barrier height decreases, which enables unrestricted movement of photogenerated charge carriers that stimulate piezo-photocatalytic ternary ZnO/ZnS/MoS_2 nanocomposites [37].

Piezoelectric output varies on the basis of number of layers present in MoS_2 and the intensity of applied strain, as depicted in Figure 3.9. First, as mentioned earlier, the piezoelectric effect loses on the even number of layers due to the restoration of the centrosymmetric structure. Despite the voltage induced for odd layers, the intensity of induced voltage decreased when number of odd layers increased. The number of MoS_2 layers $n = 2$, 4, 6, i.e., even to the bulk have no piezoelectric feedback, and $n = 1$, 3, 5; these measure piezoelectric response, which is inversely proportional to n (Figure 3.9a). Therefore, monolayer MoS_2 flakes produce maximum piezoelectric response. Second, the voltage and current generated with respect to the intensity of applied strain is represented in Figure 3.9b. Maximum values of open-circuit voltage and short-circuit current in a single-layer MoS_2 are 18 mV and 27 pA, respectively. The up surging trend in voltage output confirms the conversion of mechanical energy into electricity. In addition, combined piezoelectric output of four monolayer MoS_2 flakes, connected in parallel and series, gained an increased current and voltage accordingly (Figure 3.9c). The total outcomes were the difference between the two individual outputs when connecting devices 1 and 2, whether in series or in parallel [15].

The intensity of emission with respect to piezoelectric response, known as the piezo-phototronic effect, of a monolayer MoS_2 ribbon is studied by Li and the team (Figure 3.10a). A quantum mechanical analysis is made in which a force is varied from 0 to 0.45 N, and the tensile stress causes the charge carriers to pile up on both ends of the ribbon. The charge carries a surge with increasing force, for a fixed force, the emission is maximum when the direct band gap equals to the photon energy and achieves a saturation threshold at high carrier densities. In Figure 3.10b,

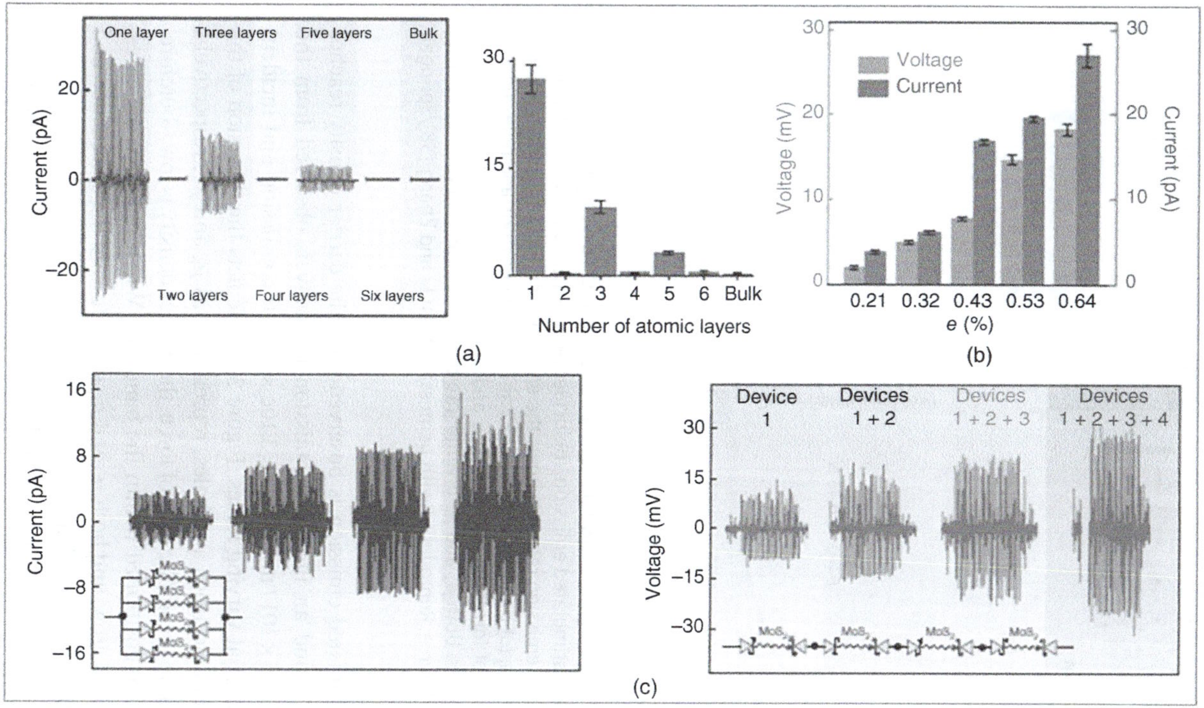

Figure 3.9 Depicts (a) piezoelectric effect in accordance with the number of layers, zero for an even number of layers, maximum at the monolayer, and a decrease when number of odd layers increases, (b) the voltage and current generated with respect to the intensity of applied strain, (c) combined piezoelectric response of a multiple monolayer MoS_2 system connected in series and parallel. Source: Reproduce with permission from Wu et al [15]/Springer Nature.

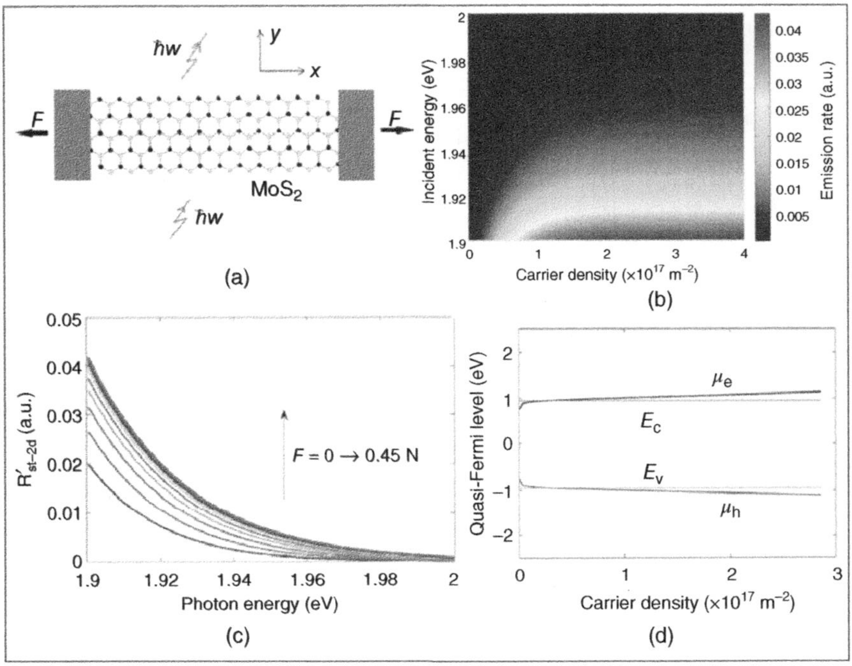

Figure 3.10 (a) Diagram showing how the force F stretched a single layer of MoS_2 and altered the stimulated emission, (b) incidence energy and carrier density in relation to total emission rate, (c) shows the emission and incident photon energy at varying forces, (d) the diversion of the Fermi level of electrons and holes under the application of force, represented as μ_e and μ_h, respectively, when carrier density increases. Both μ_e and μ_h diverge into the conduction and valence bands, respectively. Source: Reproduced with permission from Li and Zhang [38]/Springer Nature.

a stable stimulated emission is observed after the carrier density reaches $2 \times 1017\,m^{-2}$, and an increased emission intensity is observed from the region 0 to $1.5 \times 1017\,m^{-2}$. Figure 3.10c shows the emission and incident photon energy at varying forces. Figure 3.10d depicts the diversion of the Fermi level of electrons and holes, represented as μ_e and μ_h, respectively. Initially, before the application of force, the carrier density is low; therefore, μ_e and μ_h are positioned within the band gap. When the force increases, charge carriers grow, and both μ_e and μ_h divert toward the conduction and valence bands, respectively [38].

A theoretical approach is done by Zhou and team to calculate the piezoelectric response in MoS_2 nanogenerator (NG) that uses square waves to generate strain. They make use of COMSOL software to create mathematical simulations. The experimental setup involves three layers of MoS_2

stacked along z-axis. The two zigzag edges, $(10\bar{1}0)$ and $(\bar{1}010)$, as shown in Figure 3.11a, are connected to the left and right-side electrodes, respectively. MoS_2 acts as an RC circuit with internal resistance R_{IN} (internal resistance), capacitance C_{NG} (nanogenerator's capacitance), and a voltage source to carry the electron flow created by the piezo effect. An external load with resistance R_{EX} (external load resistor) is connected to both ends of the electrode. The results of static characterization, by increasing strain, changing the length, and number of layers and atomic thickness of MoS_2, are illustrated in Figure 3.11b–e, respectively. With the increase in strain, both positive and negative piezo charges (Q) could be produced at the right and left electrode interfaces; therefore, Q and voltage V get uplifted (Figure 3.11b). Despite the width not affecting the generated output V and Q under an applied strain of 0.5%, an increase in length increases the voltage, and is unaffected by the piezocharge Q. The voltage surge meets the saturation level after reaching 200 nm in length (Figure 3.11c). The dependency of voltage output on the number of layers and atomic width (Figure 3.11d) give the same conclusion as mentioned earlier in [15]. The piezocharge has no effect on the number of layers; however, Q is inversely correlated with the number of layers due to the expansion of the edge area. Figure 3.11e shows the relationship between capacitance (CNG) and length, and at 200 nm, CNG reaches its saturation, similar to voltage in Figure 3.11c because of the relation $dV = dQ/C$. Notably, the capacitance of a monolayer 50 nm × 50 nm MoS_2 flake is less in comparison with a ZnO nanowire having a diameter of 2 μm and a length of 200 μm. For this reason, MoS_2's atomic-scale thickness results in a smaller CNG, which is advantageous for collecting high-frequency mechanical impulses. Similarly, dynamic analysis done by applying a high-frequency 5 GHz square wave as strain in MoS_2 gives similar output as in static characterization. The findings depicted in Figure 3.11f illustrate that the conversion efficiency of the monolayer MoS_2 NG remained satisfactory until the signal frequency surpassed 1 GHz, a value significantly exceeding the frequency limitation of the ZnO nanowire NG, which functions within the MHz region. After 1 GHz, MoS_2 conversion efficiency exponentially drops because of the low capacitor discharge time of the NG. The authors have envisaged the potential of the MoS_2 NG for utilization in future device applications and experiments, specifically with regard to high-frequency mechanical stimuli [18].

Zhang and group designed an Au–$MoSe_2$ sensor to detect the presence of ammonia (NH_3). The sensor works by utilizing power from the piezoelectric output of single-layer MoS_2 PENG, because of its excellent energy-harvesting capability. Figure 3.12a illustrates the schematic diagram of Au–$MoSe_2$ sensor operated with a monolayer MoS_2 PENG. The McS_2

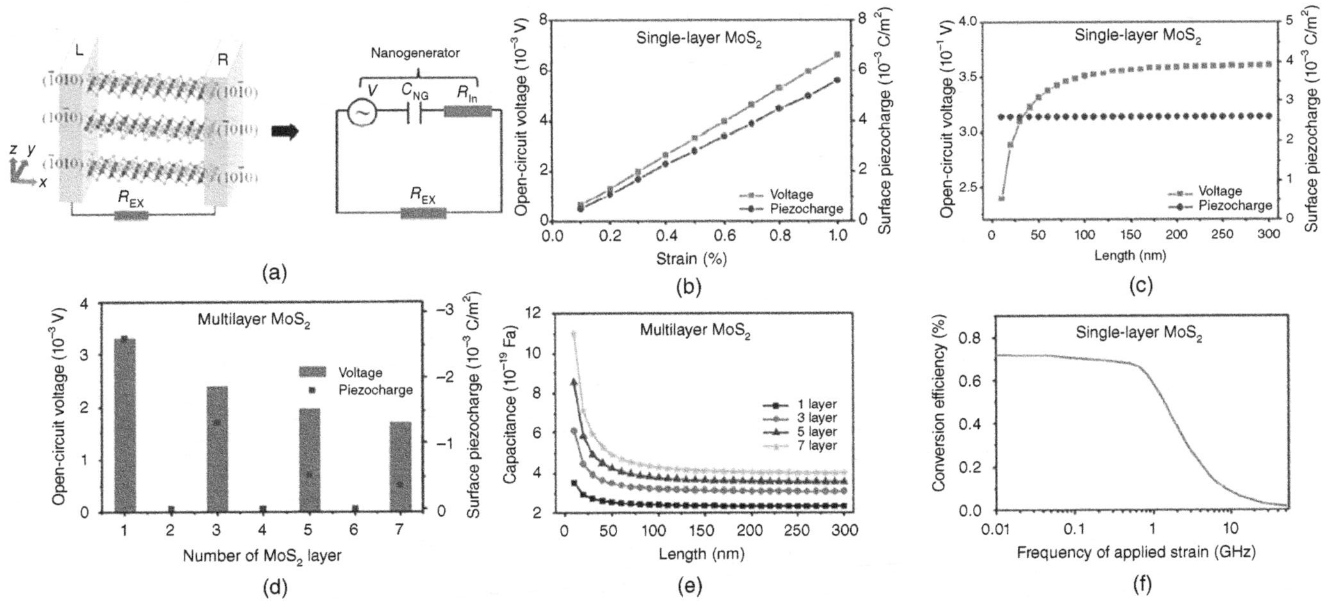

Figure 3.11 (a) Experimental setup involves three layers of MoS_2 stacked along *z*-axis, connected to an external electrode. The relationship between open-circuit voltage (*V*) and surface piezocharge (*Q*) with (b) strain, (c) length, and (d) number of MoS_2 layer, (e) variation of capacitance (C_{NG}) of different MoS_2 layers with respect to length, (f) the conversion efficiency of the monolayer MoS_2 NG for different frequencies. Source: Reproduced with permission from Zhou et al. [18]/Springer Nature.

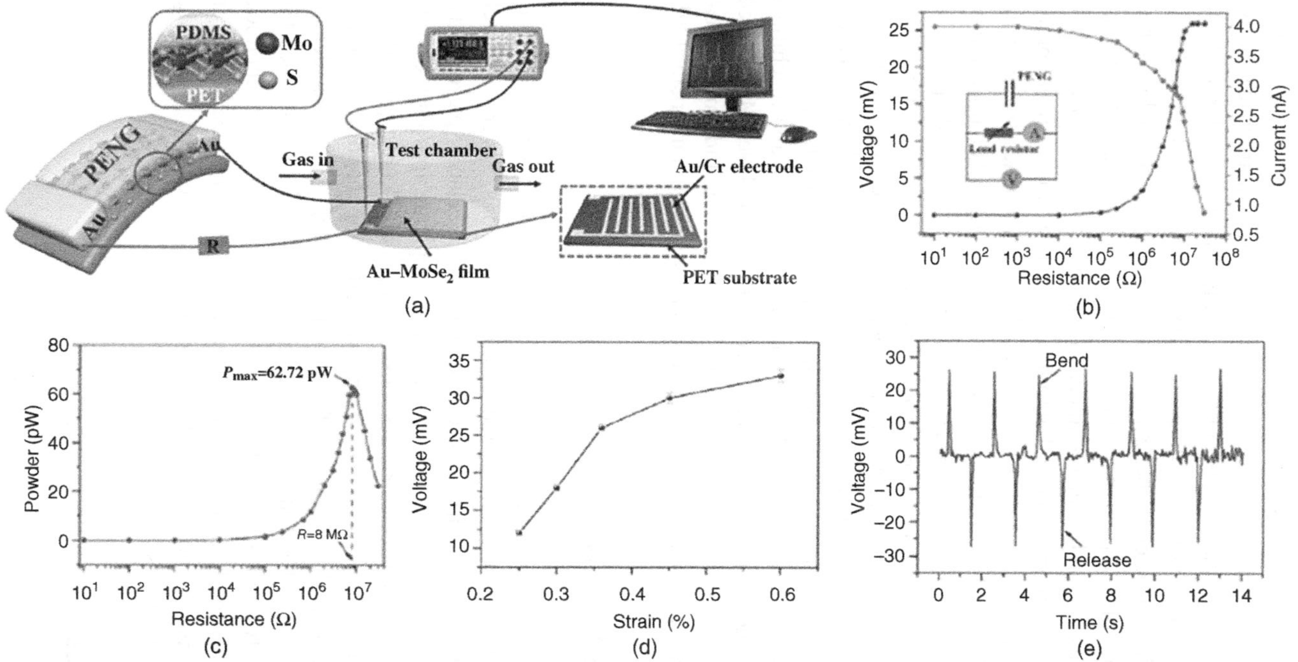

Figure 3.12 (a) The diagram illustrates an Au–$MoSe_2$ sensor to detect NH_3, which is powered by a piezoelectric nanogenerator based on MoS_2 flakes, (b) voltage and current variation with respect to resistance, (c) power output versus resistance, (d) change in strain versus voltage, (e) voltage reaches a maximum of 26 mV when reaching a stable 36% strain then comes down to negative while releasing strain. Source: Reproduced with permission from Zhang [39], © 2019/Elsevier.

flakes in PENG are made by employing atmospheric pressure chemical vapor deposition (APCVD) technique, where both the ends of MoS_2 PENG are connected to Au electrodes (Figure 3.12a), and are placed above polyethylene terephthalate (PET) film substrate. The top portion of MoS_2 is coated with a polydimethylsiloxane (PDMS) film (inset: Figure 3.12a). Therefore, the piezoelectric behavior of MOS_2 PENG functioned under an external strain of 0.36% and a frequency of 0.5 Hz is investigated after connecting it to a rheostat having resistance varied from 10 Ω to 30 MΩ. Despite the voltage does an upward trend with increasing resistance, current plummets after 0.1 MΩ (Figure 3.12b). Moreover, the power reaches its crest value of 62.72 pW toward a resistance of 8 MΩ (Figure 3.12c). The open-circuit voltage reaches a maximum of 26 mV at 36% strain and a frequency of 0.5 Hz, and the voltage drops to negative when releasing MoS_2 layer (Figure 3.12d). Furthermore, the voltage output increases with increase in strain. The polarized charge created during the application of strain on MoS_2 forces the electron to flow through the circuit, hence voltage rises. However, removal of strain leads to the falling of polarized charges, and reaching a negative voltage peak (Figure 3.12e). Hence, the single-layer MoS_2-based PENG exhibits noticeable piezoelectric performance and mechanical energy-harvesting capabilities when placed on distinct human organs [39].

3.6 Conclusions

Since its discovery, piezoelectricity has been used in a wide variety of actuation, sensing, and energy-harvesting applications. Several research studies showed that the armchair orientation produces larger piezoelectric outputs than the zigzag orientation. Therefore, applying appropriate strain can enhance the piezoelectric response. Also, piezoelectric property of MoS_2 could be enhanced such as doping with different dopants and surface functionalization. The impurity formed during doping creates defects that introduce localized states in the bandgap, which can enhance the piezoelectricity, and surface functionalization with different chemical groups would modify its electronic and mechanical properties. Moreover, it has immense application in the fields of catalysis, energy harvesting, actuators, and sensors.

References

1 Ssekulima, E.B., Anwar, M.B., Al Hinai, A., and El Moursi, M.S. (2016). Wind speed and solar irradiance forecasting techniques for enhanced renewable energy integration with the grid: a review. *IET Renew. Power Gener.* 10 (7): 885–989.

2 Cui, C., Xue, F., Hu, W.J., and Li, L.J. (2018). Two-dimensional materials with piezoelectric and ferroelectric functionalities. *npj 2D Mater. Appl.* 2 (1): 18.

3 Faraz, M., Singh, H.H., and Khare, N. (2022). A progressive strategy for harvesting mechanical energy using flexible PVDF-rGO-MoS_2 nanocomposites film-based piezoelectric nanogenerator. *J. Alloys Compd.* 890: 161840.

4 Azimi, S., Golabchi, A., Nekookar, A. et al. (2021). Self-powered cardiac pacemaker by piezoelectric polymer nanogenerator implant. *Nano Energy* 83: 105781.

5 Abdelkefi, A. and Ghommem, M. (2013). Piezoelectric energy harvesting from morphing wing motions for micro air vehicles. *Theor. Appl. Mech. Lett.* 3 (5): 052004.

6 Cui, Y., Wang, H., Li, M., and Sun, K. (2017). Simulation of wave energy harvesting by piezoelectric seaweed. *IOP Conf. Ser.: Mater. Sci. Eng.* 250 (1): 012022. IOP Publishing.

7 Zhu, M., Yi, Z., Yang, B., and Lee, C. (2021). Making use of nanoenergy from human – nanogenerator and self-powered sensor enabled sustainable wireless IoT sensory systems. *Nano Today* 36: 101016.

8 Wang, Z.L. (2008). Towards self-powered nanosystems: from nanogenerators to nanopiezotronics. *Adv. Funct. Mater.* 18 (22): 3553–3567.

9 Yang, Z., Zhou, S., Zu, J., and Inman, D. (2018). High-performance piezoelectric energy harvesters and their applications. *Joule* 2 (4): 642–697.

10 Chen, S., Li, J., Song, Y. et al. (2021). Flexible and environment-friendly regenerated cellulose/MoS_2 nanosheet nanogenerators with high piezoelectricity and output performance. *Cellulose* 28: 6513–6522.

11 Song, Y., Shi, Z., Hu, G.H. et al. (2021). Recent advances in cellulose-based piezoelectric and triboelectric nanogenerators for energy harvesting: a review. *J. Mater. Chem. A* 9 (4): 1910–1937.

12 Ilham, Z. (2022). Biomass classification and characterization for conversion to biofuels. In: *Value-Chain of Biofuels*, 69–87. Elsevier.

13 Zhang, H., Wang, Z.G., Zhang, Z.N. et al. (2007). Regenerated-cellulose/multiwalled-carbon-nanotube composite fibers with enhanced mechanical properties prepared with the ionic liquid 1-allyl-3-methylimidazolium chloride. *Adv. Mater.* 19 (5): 698–704.

14 Wang, Q.H., Kalantar-Zadeh, K., Kis, A. et al. (2012). Electronics and optoelectronics of two-dimensional transition metal dichalcogenides. *Nat. Nanotechnol.* 7 (11): 699–712.

15 Wu, W., Wang, L., Li, Y. et al. (2014). Piezoelectricity of single-atomic-layer MoS_2 for energy conversion and piezotronics. *Nature* 514 (7523): 470–474.

16 Acharya, U., Bober, P., Trchová, M. et al. (2018). Synergistic conductivity increase in polypyrrole/molybdenum disulfide composite. *Polymer* 150: 130–137.

17 Choi, W., Kim, J., Lee, E. et al. (2021). Asymmetric 2D MoS_2 for scalable and high-performance piezoelectric sensors. *ACS Appl. Mater. Interfaces* 13 (11): 13596–13603.

18 Zhou, Y., Liu, W., Huang, X. et al. (2016). Theoretical study on two-dimensional MoS_2 piezoelectric nanogenerators. *Nano Res.* 9 (3): 800–807.

19 Duerloo, K.A., Ong, M.T., and Reed, E.J. (2012). Intrinsic piezoelectricity in two-dimensional materials. *J. Phys. Chem. Lett.* 3 (19): 2871–2876.

20 Casillas, G., Santiago, U., Barrón, H. et al. (2015). Elasticity of MoS_2 sheets by mechanical deformation observed by in situ electron microscopy. *J. Phys. Chem. C* 119 (1): 710–715.

21 Kim, S.K., Bhatia, R., Kim, T.H. et al. (2016). Directional dependent piezoelectric effect in CVD grown monolayer MoS_2 for flexible piezoelectric nanogenerators. *Nano Energy* 22: 483–489.

22 Li, X. and Zhu, H. (2015). Two-dimensional MoS_2: properties, preparation, and applications. *J. Materiomics* 1 (1): 33–44.

23 Budania, P., Baine, P.T., Montgomery, J.H. et al. (2017). Comparison between Scotch tape and gel-assisted mechanical exfoliation techniques for preparation of 2D transition metal dichalcogenide flakes. *Micro Nano Lett.* 12 (12): 970–973.

24 Gacem, K., Boukhicha, M., Chen, Z., and Shukla, A. (2012). High quality 2D crystals made by anodic bonding: a general technique for layered materials. *Nanotechnology* 23 (50): 505709.

25 Ganatra, R. and Zhang, Q. (2014). Few-layer MoS_2: a promising layered semiconductor. *ACS Nano* 8 (5): 4074–4099.

26 Zhiyuan, Z., Zong, Y., Xiao, H. et al. (2011). Single-layer semiconducting nanosheets: high-yield preparation and device fabrication. *Angew. Chem.* 50 (47): 11093–11097.

27 Coleman, J.N., Lotya, M., O'Neill, A. et al. (2011). Two-dimensional nanosheets produced by liquid exfoliation of layered materials. *Science* 331 (6017): 568–571.

28 Liu, H.F., Wong, S.L., and Chi, D.Z. (2015). CVD growth of MoS_2-based two-dimensional materials. *Chem. Vap. Deposition* 21: 241–259.

29 Nakayama, T., Kangawa, Y., and Shiraishi, K. (2011). Atomic structures and electronic properties of semiconductor interfaces, Chap. 1.04. In: *Comprehensive Semiconductor Science and Technology* (ed. P. Bhattacharya, R. Fornari, and H. Kamimura), 113–174. Elsevier.

30 Qi, J., Lan, Y.W., Stieg, A.Z. et al. (2015). Piezoelectric effect in chemical vapour deposition-grown atomic-monolayer triangular molybdenum disulfide piezotronics. *Nat. Commun.* 6 (1): 7430.

31 Strachan, J., Masters, A.F., and Maschmeyer, T. (2021). 3R-MoS_2 2 in review: history, status, and outlook. *ACS Appl. Energy Mater.* 4 (8): 7405–7418.

32 Krishnan, U., Kaur, M., Singh, K. et al. (2019). A synoptic review of MoS_2: synthesis to applications. *Superlattices Microstruct.* 128: 274–297.

33 Dai, M., Zheng, W., Zhang, X. et al. (2019). Enhanced piezoelectric effect derived from grain boundary in MoS_2 monolayers. *Nano Lett.* 20 (1): 201–207.

34 AlTowireb, S.M. and Goumri-Said, S. (2023). Core-Shell structures for the enhancement of energy harvesting in piezoelectric nanogenerators: a review. *Sustainable Energy Technol. Assess.* 55: 102982.

35 Yuan, S., Io, W.F., Mao, J. et al. (2020). Enhanced piezoelectric response of layered In_2Se_3/MoS_2 nanosheet-based van der Waals heterostructures. *ACS Appl. Nano Mater.* 3 (12): 11979–11986.

36 Huang, L., Li, Y., Wei, Z., and Li, J. (2015). Strain induced piezoelectric effect in black phosphorus and MoS_2 van der Waals heterostructure. *Sci. Rep.* 5: 16448.

37 Yongming, F.U., Yujing, W., Hongxing, Z. et al. (2021). Synthesis of ternary $ZnO/ZnS/MoS_2$ piezoelectric nanoarrays for enhanced photocatalytic performance by conversion of dual heterojunctions. *Appl. Surf. Sci.* 556.

38 Li, L. and Zhang, Y. (2017). Controlling the luminescence of monolayer MoS_2 based on the piezoelectric effect. *Nano Res.* 10 (7): 2527–2534.

39 Zhang D, Yang Z, Pang P. Li, M, and Xue Q. Flexible self-powered high-performance ammonia sensor based on Au-decorated $MoSe_2$ nanoflowers driven by single layer MoS_2-flake piezoelectric nanogenerator. *Nano Energy*, 2019; vol. 65,103974.

4

$BaTiO_3$-Based Hybrid Piezoelectric Materials

Chirantan Shee[1], Swagata Banerjee[1], S. Wazed Ali[1,2], and Ramasamy Alagirusamy[1]

[1] *Department of Textile and Fibre Engineering, Indian Institute of Technology Delhi, Hauz Khas, New Delhi, India*
[2] *School of Interdisciplinary Research (SIRe), Indian Institute of Technology Delhi, New Delhi, India*

4.1 Introduction

In the year of 1880, Pierre and Jacques Curie discovered a unique phenomenon, known as piezoelectricity. The Greek word "piezo" means "pressure." As the name implies, electricity originating from pressure is known as piezoelectricity. Piezoelectric materials produce electricity as a result of external mechanical pressure, and vice versa. The generation of electrical output from input mechanical force is known as direct piezoelectric effect. Production of mechanical strain into the piezoelectric material as a result of applied electrical voltage at two ends of the material is known as converse piezoelectric effect. In order to exhibit piezoelectricity, a material must have non-centrosymmetry in its structure. This indicates that centers of positive and negative charges should not coincide when the material is strained. The study of crystalline symmetry enlightens on the actual mechanism of piezoelectricity. Crystalline materials can be distinguished by seven different crystal systems in their basic form. These are cubic, tetragonal, orthorhombic, monoclinic, hexagonal, rhombohedral, and triclinic (Figure 4.1).

Depending upon symmetry, these crystal systems can be classified into 32 point groups. Among these 32 groups, 21 point groups lack a center of symmetry. Twenty point groups, out of these 21, show piezoelectric behavior. Although, the remaining one is also without center of symmetry, other symmetry elements eliminate the piezoelectric property. Half of these 20 non-centrosymmetric point groups show spontaneous polarization. This means, even without any external stimulus, these materials offer some

Hybrid Materials for Piezoelectric Energy Harvesting and Conversion, First Edition.
Edited by S. Wazed Ali, Satyaranjan Bairagi, and Shahid Ul Islam.
© 2024 John Wiley & Sons, Inc. Published 2024 by John Wiley & Sons, Inc.

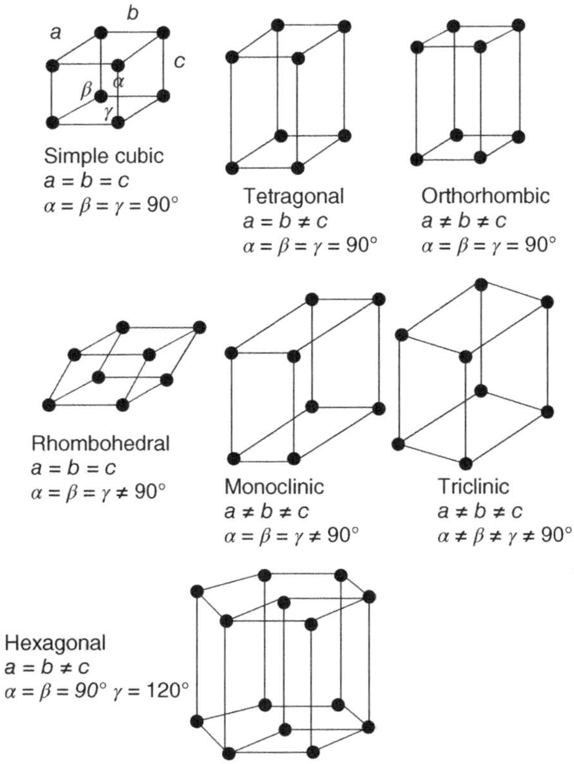

Figure 4.1 Crystallographic lattices in unit cells of crystals. Source: Seto and Romero [1]/Elsevier.

in-built polarization. Materials belonging to these 10 non-centrosymmetric point groups are known as polar or pyroelectric materials. The direction of some of the pyroelectric materials can be changed by applying an electric field without breakdown of the crystalline system. This class of material is termed as ferroelectric material. Ferroelectric material loses its non-centrosymmetry and becomes nonpolar or paraelectric beyond a particular temperature, known as Curie temperature. When the material is cooled through the Curie temperature, it gains non-centrosymmetry [2]. Barium titanate ($BaTiO_3$) is the first-discovered ferroelectric material. After its discovery, it has gained immense importance due to its high piezoelectric coefficient, high dielectric constant (≥1500 at room temperature), and low dielectric loss. It has a Curie temperature of 120 °C [3]. So, it can be used at and above room temperature as a ferroelectric material [4]. Ferroelectric materials exhibit typical polarization versus electric field curves, as shown in Figure 4.2.

Saturation polarization (P_s) is the maximum magnitude of polarization, which can be induced within a ferroelectric material at high electric field. Remnant polarization (P_r) is the magnitude of polarization, which remains

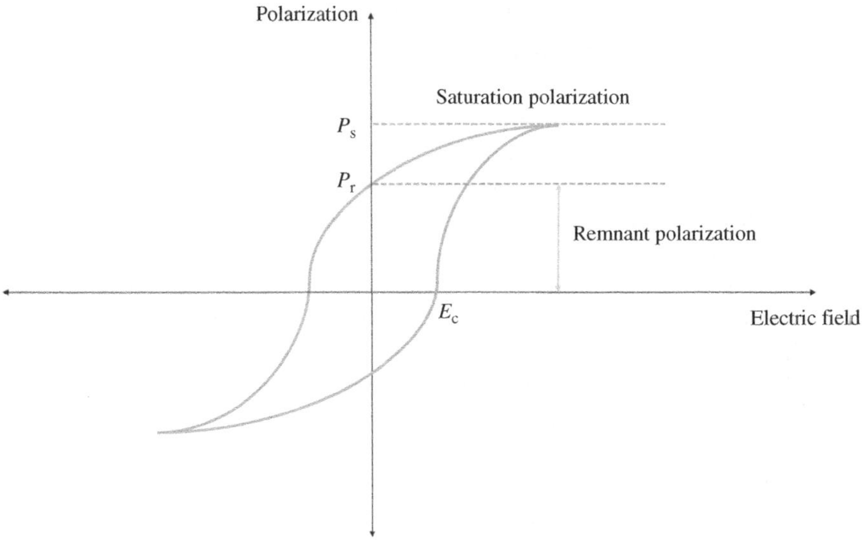

Figure 4.2 Typical polarization versus electric field hysteresis loop of ferroelectric materials.

within a ferroelectric material after the removal of electric field. The coercive field (E_c) is the applied electric field at which the induced macroscopic polarization disappears within a ferroelectric material. Previous research reveals that perovskite $BaTiO_3$ powder was mechanochemically synthesized with crystallite size ranging from 20–50 nm. The polarization versus electric field curve of $BaTiO_3$ followed typical a ferroelectric hysteresis loop with remnant polarization of 2 μC/cm^2 and coercive field of 1060 kV/cm [5]. With the advent of lead zirconate titanate (PZT) in the field of piezoelectricity, $BaTiO_3$ had somewhat lost its popularity. Later, $BaTiO_3$ revived its demand due to awareness regarding potential hazards associated with lead-based materials such as PZT. In order to overcome the limitations associated with pristine $BaTiO_3$, several techniques have been explored to make new kinds of materials. Primarily, different ions were doped in the crystal structure of $BaTiO_3$. Second, composite materials were developed using $BaTiO_3$ as one of the constituents. Finally, physics effects such triboelectricity were utilized to assist the piezoelectricity of $BaTiO_3$ to amplify the electrical output of hybrid nanogenerator.

In this chapter, first the structure and piezoelectric properties of $BaTiO_3$ are explained briefly. Thereafter, several literatures have been reviewed describing piezoelectric properties of hybridized materials based on $BaTiO_3$. Finally, different applications of $BaTiO_3$-based piezoelectric materials have been covered.

4.2 Structure and Piezoelectric Properties of $BaTiO_3$ Perovskite

Perovskite, the name derived from the mineral perovskite $CaTiO_3$, is mainly available in oxide form with the general formula of ABO_3. In a primitive crystal of perovskite, A is the bigger ion positioned in the corners, B is the smaller ion, which is situated within the interstices of the crystal, and oxygen is located at the centers of the faces of the crystal. Both A and B are cations. A is generally monovalent, divalent, and trivalent, whereas B is generally trivalent, tetravalent, and pentavalent. The coordination number of A is 12, and that of B is 6. $BaTiO_3$ is a member of the perovskite family. Perovskite includes not only structures with a strictly cubic perovskite lattice but also derivatives of this ideal structure prepared by lattice distortion or elimination of some atoms. $BaTiO_3$ shows crystal structure transition depending upon various temperature ranges. Above its Curie temperature (120 °C), the crystal structure follows a cubic lattice, and this structure remains stable up to 1460 °C. Further increase in temperature results in a hexagonal crystal structure. If the temperature is lowered through its Curie temperature, a tetragonal structure is obtained, and this structure remains stable from 120 to 5 °C. Lowering the temperature below 5 °C results in transition to orthorhombic structure, which remains stable up to −90 °C. Below −90 °C, rhombohedral structure is obtained. In a $BaTiO_3$ crystal, below its Curie temperature, the Ti cations start to move to a more energetically favorable condition, which can be depicted by distortion of lattice structure, which is lengthening or shortening of the bonds [6]. Perovskite materials can accommodate various donor and acceptor dopants within their structure. Acceptor dopants with valency of 1,2, or 3 can replace Ba^{2+} ions, to make a p-type semiconducting ceramic, whereas replacing Ti^{4+} ions with donor dopants of valency 3,4, or 5 can make an n-type semiconductor in a $BaTiO_3$ perovskite. Using donor dopant at a relatively low concentration makes room temperature semiconducting ceramic, while higher concentration of donor dopant results in the formation of an insulating material [3].

Piezoelectric efficiency can be quantified using piezoelectric coupling coefficient (k), which indicates the fraction of external mechanical energy converted into electrical output. This value is always less than unity, and a higher value implies better piezoelectric performance. $BaTiO_3$ has a very high k value along with having superior thermal and mechanical stability. k_{33}, k_{31}, and k_p are different k factors. k_p is known as planar coupling coefficient. Numerical subscripts denote direction of applied electric field and direction of applied external force. Piezoelectric coefficient (d) indicates the produced electrical output for unit external force ($\times 10^{-12}$ C/N) for direct piezoelectric effect or the resultant deformation for unit electrical input voltage ($\times 10^{-12}$ m/V) for converse piezoelectric effect. A higher *d* value is desirable for higher piezoelectric property. Another parameter, known as open-circuit coefficient is denoted as *g*. *g* value denotes the ability to produce a large amount of electrical voltage while unit external stress is applied to the material. This is generally expressed in *V* (m/N). Ferroelectric ceramics with higher g values indicate ferroelectrically hard ceramics, which do not alter polarization promptly and usually have lower dielectric constant values. The g value is related to the *d* value following the equation below:

$$g = d/K\varepsilon_0$$

where *K* is the relative dielectric constant of the piezoelectric material and ε_0 is the dielectric constant of free space. For $BaTiO_3$, typical values of k_{33}, k_p, d_{33}, d_{31}, and g_{33} are 0.5, 0.36, 190, −78, and 11.4, respectively [3, 7]. Piezoelectric, ferroelectric, and pyroelectric properties of $BaTiO_3$ have been discussed in several kinds of literature [4].

$BaTiO_3$ can be synthesized into various shapes through different synthesis routes. Powdered ceramic obtained through chemical synthesis route had finer grains, and hence better piezoelectric properties. Further, poling of the ceramics possessing a variety of shapes can also help to optimize their piezoelectric performance. A study conducted on the hysteresis loop of this material revealed interesting facts about its piezoelectric behavior. At lower temperatures, the remnant polarization was lower and coercive electric fields were higher compared to those at higher temperatures. This indicates that higher energy is required at lower temperatures to orient the dipoles, increasing the coercive electric field. However, at higher temperatures, the thermal energy provided to the dipoles helps to aid in their orientation, reducing the coercive electric field [3]. Domain engineering can further serve as a method to improve the piezoelectric properties of the $BaTiO_3$ single crystals. It was found that both the piezoelectric and coupling coefficients improved with the decrease in domain size. Finer grains had better piezoelectric coefficients compared to the coarser ones. Thus, in order to achieve lead-free ceramics as a potential alternative to the dominating lead-based materials, domain engineering can be a possible alternative route [8].

Increasing sintering temperature first improves d_{33} and electromechanical coefficient (k_p) of $BaTiO_3$, followed by a decreasing trend is observed. Optimized d_{33} and k_p were found as 355 pC/N and 40%, respectively for the sample sintered at 1190 °C. The value of d_{33} depends on removal of cavities during sintering as well as grain size. Along with having a large relative density and moderate grain size, sample sintered at 1190 °C is close to the transition between tetragonal and orthorhombic lattice structure. This results in easier domain movement, giving higher piezoelectricity. Further enhancement in sintering temperature results in deterioration in the piezoelectric property. The increase in sintering temperature raises the Curie temperature. Dielectric constant also initially enhances with increasing sintering temperature. The highest dielectric constant at the Curie temperature was achieved at 1190 °C sintering temperature. This is due to the fact that at this particular temperature, the densification is maximum, leading to the least energy dissipation [9].

4.3 Synthesis of Barium Titanate

The production processes of $BaTiO_3$ not only influence the cost of the final product but also determine the amount of the formed $BaTiO_3$ powder. In this section, manufacturing routes of $BaTiO_3$ are briefly explained.

(a) Solid-state reaction: This process involves ball milling of $BaCO_3$ or BaO with TiO_2 and calcination of the mixture at higher temperature (800–1300 °C). The resultant powder produced by this technique is prone to agglomeration, and large particle size (2–5 μm) is obtained as an output. The output product also contains impurities due to large processing temperatures and heterogeneous reactions in solid phase. These ultimately lead to inferior electrical properties of the product [6].

(b) Chemical routes of $BaTiO_3$ synthesis: In order to overcome the shortcomings associated with solid-state reactions, several chemical synthesis routes have been evolved. Due to quasi-atomic dispersion into a liquid precursor, submicron-crystallized particles with high purity can be obtained at lower temperatures [6]. Methods of chemical synthesis are sol–gel method, hydrothermal method, coprecipitation method, and polymer precursor method (Figure 4.3).

The hydrothermal method is often used for the synthesis of fine-grained $BaTiO_3$ powders with an appreciable content of the desired phase. A modification of the hydrothermal method, called the microwave-assisted hydrothermal synthesis, is also used to form $BaTiO_3$. This is a faster process with good control and enhancement of the crystallization kinetics of the ceramic is achieved compared to the hydrothermal process. The volumetric heating aids in homogeneous nucleation, which further helps in rapid crystallization. The phases of $BaTiO_3$ form during the first few minutes of the reaction. However, few hours of reaction are carried out to aid in the completion of crystallization. The higher the reaction temperature, the shorter the time required.

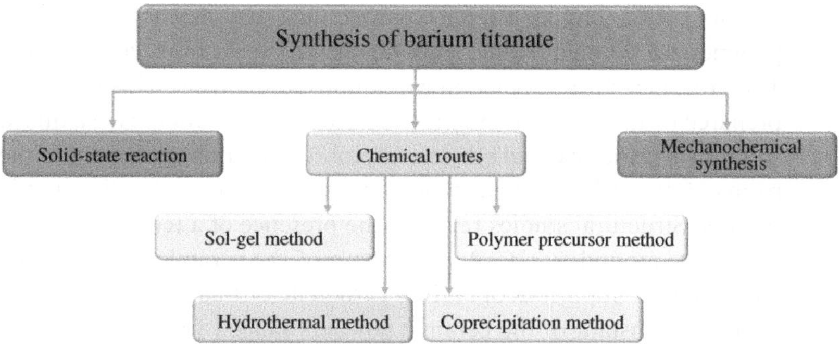

Figure 4.3 Synthesis routes of barium titanate.

The structural analysis revealed the presence of the expected tetragonal phase in the synthesized ceramic. However, impurities such as barium carbonate were also present, which decreased with the increase in reaction temperature. Submicron-sized particles were obtained through this synthesis route. However, agglomerated particles were observed when the reaction was carried out under low alkaline conditions [10]. Most of the synthesis routes of ferroelectric ceramic nanoparticles (NPs) require high-temperature conditions. This, in turn, calls for high energy requirements. Hence, now there is a need to shift toward low-energy requirement processes. Eutectic solvent-based methods for the synthesis of ferroelectric $BaTiO_3$ NPs have been carried out. A choline chloride/malonic acid eutectic solvent mixture was used to prepare $BaTiO_3$ NPs. This is a low-cost method that produces the ferroelectric ceramic in a pure phase. Initially, barium chloride is formed that progresses to form $BaTiO_3$ by reacting with anatase at relatively high temperatures. The nanomaterial showed properties comparable to those obtained from the solid-state synthesis method. This is a versatile method that could be explored for the synthesis of other functional oxides. Moreover, this technique could be used with biotemplates for implementation in emergent technologies [11]. The commonly used sol–gel method for $BaTiO_3$ preparation involves two basic steps: the partial hydrolysis of the metal alkoxide, followed by the evaporation and pyrolysis to yield an amorphous oxide. This approach has been used for the synthesis of doped $BaTiO_3$ ceramics. Yttrium-doped $BaTiO_3$ was prepared by this sol–gel technique that showed a positive temperature coefficient of resistivity. The precursor gel was calcined at an intermediate stage at around 600 °C, followed by a final calcination of the gel at 1300 °C. Structural studies indicated the presence of the tetragonal phase of the ceramic with a small impurity content of monoclinic

Ba_2TiO_4 [12]. The powdered ceramic has also been prepared by polymerized complex route. This was based on Pechini-type reaction. The precursor for this reaction was a transparent resin. This resin was prepared by the polymerization of barium and titanium ions in a mixed solution of citric acid and ethylene glycol. The thermal treatment of the resin at 500–900 °C resulted in the formation of the ferroelectric ceramic powder. Structural studies revealed the presence of a tetragonal phase of $BaTiO_3$ around 600 °C. A pure phase of the ceramic was obtained with no trace impurities of barium carbonate or titanium dioxide. The dielectric constant of the ceramic powders ranged from 1700 to 2840 that decreased with an increase in frequency [13]. Yet another route for the synthesis of $BaTiO_3$ is the coprecipitation process. This involves the mixing of ions at the molecular level. However, it is difficult to determine the pH conditions for the simultaneous precipitation of the Ba^{2+} and the Ti^{4+} ions, which poses a challenge during the synthesis of $BaTiO_3$ through this route. This route helps to obtain chemical homogeneity under optimized conditions of processing. Synthesis parameters such as pH, reagent concentration, presence of metal ions, and their stability can be maneuvered to obtain the simultaneous coprecipitation of the barium and titanate ions in their desired stoichiometric ratio.

(c) Mechanochemical synthesis: This is performed with the help of a planetary ball milling instrument in an air atmosphere and for a duration of up to four hours. Agglomerated particles were obtained with different morphologies and sizes. The grains were generally rounded in nature with the size of 20–50 nm. Mechanical treatment can efficiently reduce the particle size to the nanoscale, which is beneficial in the manufacturing of some products such as supercapacitor and ferroelectric materials [6].

4.4 Barium Titanate-Based Hybrid Piezoelectric Materials

Since the discovery of $BaTiO_3$, several research studies have been conducted in order to improve the properties of the materials containing it or to find its new derivatives. The main aim of developing such new materials is to cope with the limitations associated with pristine $BaTiO_3$. One route deals with the doping of different metal ions within the $BaTiO_3$ lattice so that effective distortion occurs. Another way of solving problems associated with $BaTiO_3$ is to make composite materials using $BaTiO_3$ as a piezoelectric filler. Finally, triboelectricity is often combined with piezoelectric property of $BaTiO_3$ in order to improve the electrical output. So, pristine $BaTiO_3$ is hybridized in various ways, targeting enhanced characteristics of the final product. In this section, several case studies have been discussed regarding hybrid $BaTiO_3$-based materials.

4.4.1 Modified $BaTiO_3$ Crystal Lattice

Ample number of works have been published based on electrical properties of doped $BaTiO_3$. $BaTiO_3$ was doped with Fe^{3+} and Nb^{5+} in order to study dependency of dielectric properties on the level of compressive stress. Fe^{3+} acts as an acceptor, whereas Nb^{5+} acts as a donor type of dopant. Both the dielectric constant and dielectric loss values were found to decrease with increasing stress levels for 1 mol% iron and 0.5 mol% niobium ions doped compound (BT–1Fe–0.5Nb). The reverse trends were observed when the compound (BT–1Fe–1.5Nb) was made using 1 mol% iron and 1.5 mol% niobium ions. These trends showed not much change for the compound (BT–1Fe–1Nb) with an equal molar ratio of iron and niobium ions. A closer look into this trend reveals that with increasing stress, the dielectric properties first enhance, followed by a reducing nature after a constant compressive stress value. Stress-induced domain reorientation and stress clamping of domain wall are the reasons for this type of dependency of dielectric properties on compressive stress. The influence of compressive stress on the ferroelectric hysteresis loop characteristics has also been explored. It has been found that hysteresis parameters such as dissipation energy, saturation polarization, remnant polarization, and coercive field decrease with increasing compressive force. The area within the polarization-electric field plot indicates the energy required to depolarize the ferroelectric material. This directly indicates the amount of domain involved in the switching process during the application of an electric field. The application of compressive force deteriorates the domain alignment in the direction of applied electric field, which requires a higher electric field to fully orient the domains. That is why saturation polarization reduces with increasing compressive force [14].

Su et al. prepared three-dimensional (3D) porous network of $0.82Ba(Ti_{0.89}Sn_{0.11})O_3–0.18(Ba_{0.7}Ca_{0.3})TiO_3$. Subsequently, polydimethylsiloxane (PDMS) was used to fill the micropores of the structure. Silver nanowires with PDMS were utilized to make the flexible electrode. The final structure was further poled with the help of a 10 kV/mm direct electric field under room temperature for a lapse of 24 hours. A 20% mass fraction of porous piezoelectric filler offered the optimum open-circuit voltage of 39 V and short-circuit current of 2.9 μA at an excitation of 35 N vertical force of 2 Hz frequency [15]. Dudem et al. fabricated a hybrid nanogenerator by incorporating Al-doped $BaTiO_3$ particles within PDMS [16]. Recently, $BaTiO_3$ was fluorinated by annealing the mixture of $BaTiO_3$ NPs and polyvinylidene difluoride (PVDF) powder. The fluorinated $BaTiO_3$ was then incorporated within the PVDF matrix prior to electrospinning. The electrospun nanofibrous membrane, thus formed, showed output voltage of 1.5 V under 2 N external force. This device was also tested for some practical mechanical forces coming from elbow flexion, walking, and slapping hands on a table. The output voltage was found to increase with an increase in elbow flexion angle, and 1.2 V output voltage was obtained for an elbow flexion angle of 120° [17].

4.4.2 Composite Containing $BaTiO_3$

Piezoelectric ceramics have very good electromechanical coupling coefficients; however, high density and brittleness are well-known disadvantages, restricting their application in various fields. On the other hand, piezoelectric polymers are flexible along with having lower density compared to ceramics. However, electromechanical coupling coefficients of polymers are quite less compared to those of piezoelectric ceramics. In order to realize the benefits of both materials and mitigate their associated limitations, both piezoelectric ceramics and polymers are often combined to make a composite material. In a typical composite structure, individual identities of the constituents remain unaltered, with an identifiable interface existing between the constituents. In the context of piezoelectric composites, the piezoelectric filler is generally a discrete phase dispersed in a continuous polymeric matrix. PDMS is widely explored as a polymeric matrix due to several interesting properties. Suo et al. incorporated $BaTiO_3$ NP within PDMS in order to make a composite film. A 20 wt% loading of NP offered the optimum condition for filler loading. A copper film was used as the tribo-positive material to boost up the electrical output arising from piezoelectric composite (Figure 4.4). The authors demonstrated the effect of poling direction on the mechanical-to-electrical conversion ability of the device. Forward poling exhibited highest electrical outputs followed by unpoled and reverse-poled samples. This phenomenon proves that electrical outputs can be manipulated by adjusting the directions of triboelectric charge transfer and piezoelectric polarization. In the triboelectric cum piezoelectric hybrid nanogenerator, the voltage was found to be approximately the sum of the voltages of individual triboelectric and piezoelectric effects [18].

Wang et al. studied the energy-harvesting ability of a hybrid nanogenerator comprising $BaTiO_3$ nanowire, carbon nanotube (CNT), and PDMS polymer matrix [19]. Han et al. hydrothermally grew $BaTiO_3$ NP onto

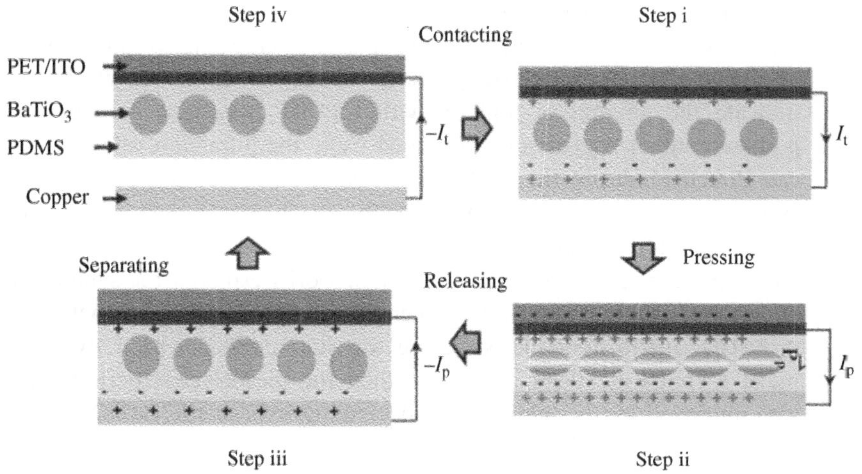

Figure 4.4 Combined mechanism of triboelectricity and piezoelectricity. Source: Reproduced with permission from Suo et al. [18]/American Chemical Society.

a graphene oxide (GO) two-dimensional structure. Pristine $BaTiO_3$ and $BaTiO_3$ hybridized with GO are separately incorporated within PDMS using spin coating technique. PDMS/active $BaTiO_3$ or $BaTiO_3$–GO/PDMS structure was made by spin coating different layers one by one. Aluminum and polyimide were used as electrode and encapsulating material, respectively, on both sides of the so-formed piezoelectric component. After hybridization, the size distribution of $BaTiO_3$ NPs was found to be more homogeneous. Improvement of GO crystallinity by reduction and fulfillment of $BaTiO_3$ oxygen vacancy was observed simultaneously as a result of hybridization. Uniform distribution of reduced GO after hybridization improves the interfacial contact between both species after the hydrothermal growth of $BaTiO_3$ onto GO. Hybridization also increases the local stress onto piezoelectric $BaTiO_3$ NPs by reinforcing the mechanical properties of the hybridized material. These four reasons can be attributed to the synergistic improvement of piezoelectric property of the hybridized material [20]. Wang et al. prepared $BaTiO_3$ nanofibers using the electrospinning technique. Further, these nanofibers, along with an electrode, were encapsulated within PDMS to make a piezoelectric device. Under bending excitation at 45 Hz, the device produced a peak-to-peak output voltage of 0.45 V under a resistance of 1 MΩ. With the help of piezoresponse force microscopy (PFM), the effective piezoelectric coefficient of the fibers was found to be 40 pm/V [21].

PVDF and its copolymers have the maximum piezoelectric properties among piezoelectric polymers. PVDF offers three different phases: α, β, γ, depending on the structure of the polymeric chain. The α-phase is thermodynamically stable, whereas the β-phase is electroactive and mainly responsible for piezoelectric property. Structural characterization can indicate the amount of β-phase within PVDF polymeric system. The β-phase content is generally quantified using the β-fraction value, which can directly indicate the piezoelectric performance of the PVDF-based systems. Processes such as electrical poling, mechanical stretching, electrospinning, and the addition of fillers can improve the β-fraction within the PVDF polymeric system. Several research works explored the synergistic effect of combining piezoelectric ceramics with PVDF-based materials. A $BaTiO_3$ micro-stone-like structure was incorporated within PVDF to make a composite flexible film. Further, the piezoelectric performance was improved by the addition of silver nanowires to the $BaTiO_3$/PVDF composite due to the incorporation of a conductive filler. This device offered a 14 V open-circuit output voltage and a 0.96 μA short-circuit current under 3 N force and 5 Hz excitation frequency, while the same values were 11 V and 0.78 μA for $BaTiO_3$/PVDF composite under the same conditions. The Ag/$BaTiO_3$-loaded device can harvest energy from vehicle movement and can be used in detecting human hand motion (Figure 4.5) [22].

Shi et al. studied the synergistic effect of incorporation of graphene nanosheets and $BaTiO_3$ NP within an electrospun PVDF nanocomposite web. It was found that with increasing graphene loading, the β-phase content of PVDF initially increased followed by a decreasing trend. The highest β-fraction value was found as 91.1% for 15 wt% $BaTiO_3$ loading and 0.15 wt% graphene loading. The so-formed nanocomposite membrane was able to charge capacitors, drive electronic watches, scavenge human motion, and light 15 LEDs [23]. Yaqoob et al. prepared multilayer piezoelectric structure with the help of the spin coating process. Graphene oxide (GO) was modified using amino compound and resultant substrate (n-Gr) made up a layer. PVDF and $BaTiO_3$ were mixed in dimethylformamide (DMF) prior to spin coating to make another layer of the piezoelectric nanogenerator (PENG). Three PENG structures were prepared: single layer of PVDF–$BaTiO_3$, double layer n-Gr/PVDF–$BaTiO_3$, and tri-layer PVDF–$BaTiO_3$/n-Gr/PVDF–$BaTiO_3$. These samples showed peak-to-peak open-circuit output voltages of 0.18, 0.8, and 1.5 V, respectively, when the samples were unpoled. All the testing was done under 2 N external force. As a result of poling process, the tri-layer sample offered a peak-to-peak open-circuit output voltage of 10 V, a short-circuit current of 2.5 μA, and a power of 5.8 μW. Under light finger tapping force, the device produced maximum output voltage of 2 V. The interface of the n-Gr and PVDF–$BaTiO_3$ acts as a charge blocker. In the bilayer structure, charge recombination is

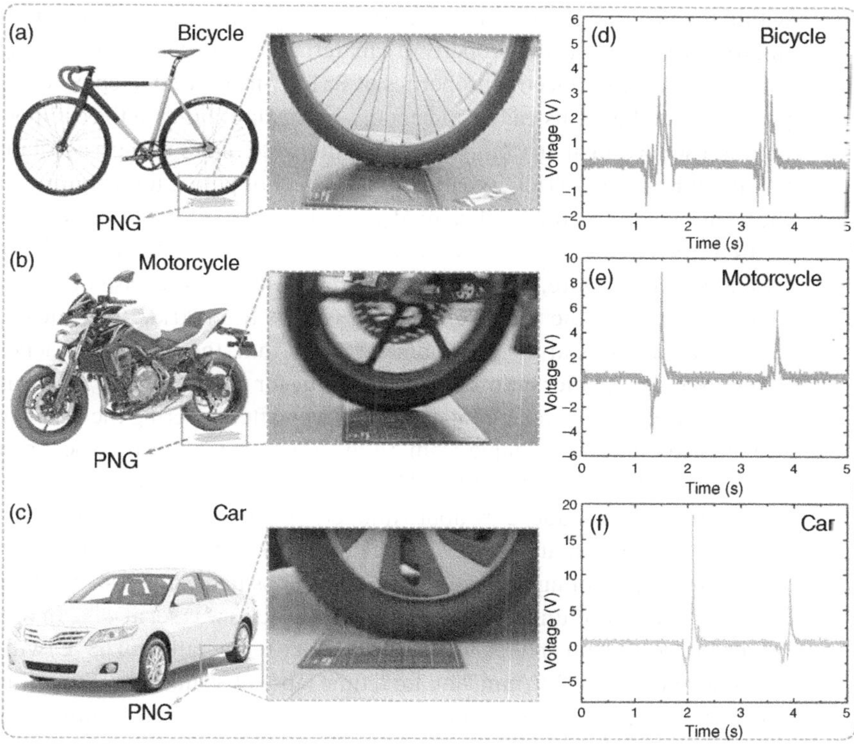

Figure 4.5 Application of piezoelectric nanogenerator (PENG) for mechanical energy harvesting from vehicle movement: (a–c) vehicles passing on the Ag/$BaTiO_3$ device and (d–f) corresponding electrical voltages. Source: Dudem et al. [22]/with permission from Elsevier.

possible between n-Gr layer and upper electrode, which can minimize the potential difference between the upper and lower electrodes. In the case of tri-layer PENG, charge recombination between r-Gr layer and electrodes is not possible due to presence of PVDF–$BaTiO_3$ layer in between. Ultimately, orientation of dipoles improves on both sides of n-Gr. That is why tri-layer structure exhibited maximum piezoelectric output voltage [24]. Kaur and Rana revealed the effect of $BaTiO_3$ and nickel oxide (NiO) incorporation within PVDF on the development of crystallinity and electroactive β-phase within $BaTiO_3$–NiO–PVDF composite film [25]. Yang et al. prepared 3D $BaTiO_3$–PVDF composite porous foam with the help of 3D printing and foaming technologies in order to overcome the limitations associated with 2D film deformation from piezoelectric energy-harvesting perspective [26].

In poling process, a material is subjected to an external electric field at a particular temperature for a fixed time interval. As a result, the dipoles within the material are forced to align in a particular direction. This process improves the piezoelectric property of a material. The effect of corona poling parameters on piezoelectric properties of $BaTiO_3$-incorporated epoxy composite was studied. Unsteady wind energy-harvesting ability of the composite was also investigated. Composites were prepared maintaining a constant volume fraction of $BaTiO_3$ as 32 vol%. The effect of poling temperature on the value of piezoelectric coefficient (d_{33}) was investigated and d_{33} was found as 0.4 pC/N at 25 °C, whereas the same was 2.5 pC/N at 65 °C. Beyond this particular temperature, d_{33} was found to decrease with increase in poling temperature. With the increase in poling voltage from 20 kV/mm, the d_{33} value also enhances up to 40 kV/mm. At 40 kV/mm, the d_{33} value reaches 2.3 pC/N and thereafter saturated. The d_{33} value for poling at 15 kV/mm was found to be zero. This value remained zero even after lowering the distance between corona tip to sample from 15 to 6 mm. The d_{33} of the specimen increased with rise in poling time. The d_{33} of poled sample decreased with time. The d_{33} was reduced by approximately 10% within a lapse of 7 days. The relative permittivity of the specimen was 16, which was found to be unaffected by the poling process. The poled composite specimen showed d_{33} of 4.3 pC/N when volume fraction of $BaTiO_3$ particles was set at 44%. In this particular case, corona poling was performed at a temperature of 65 °C, with a poling voltage of 40 kV/mm, and for a time lapse of 30 minutes. From this result, it can be concluded that d_{33} enhances with increase in volume fraction of $BaTiO_3$ particles, but at higher percentage of $BaTiO_3$, the composite becomes brittle in nature. Unsteady wind energy-harvesting tests revealed that lower velocity of air can generate higher output voltage compared to the same generated by higher velocity airflow. In this experiment, the lower frequency was associated with the source of airflow, which was rotating at a lower speed and vice versa. Higher frequency air flow produces very high strain rate on the specimen. This means that the new pressing signal may come to the material before complete recovery of the material from the compressed state, which reduces the variation of stress and ultimately leads to lowering in output voltage [27].

Surface modification of filler materials prior to composite manufacturing alters the interfacial bonding of the composite structure. Sundar et al. treated the surface of $BaTiO_3$ NP with the help of ethanol and a silane coupling agent, 3-glycidyloxypropyltrimethoxysilane. As ceramic piezoelectric filler and epoxy matrix are hydrophilic and hydrophobic in nature, respectively, suitable chemical coupling between these two constituents can improve the interfacial adhesion between them, resulting in enhanced interfacial polarization. Morphological study revealed that with increase in silane coupling agent concentration, the size of the nano-sized clusters

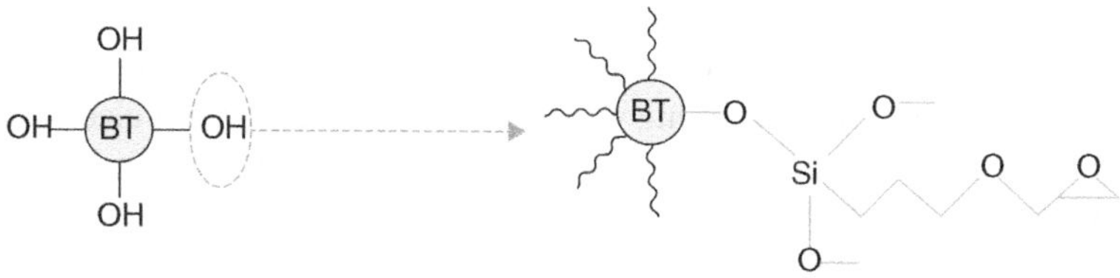

Figure 4.6 Substitution of hydroxyl group as a result of surface modification with silane. Source: Sundar et al. [28]/Public domain/CCBY 4.0/MDPI.

reduces. With increasing coupling agent concentration, the piezoelectric strain coefficient of the nanocomposite enhances due to better dispersion of the filler and improved interfacial polarization between the constituents. For silane-treated filler, with rising silane concentration, the resistivity becomes higher. While the silane concentration increases, the hydroxyl groups on the surface of $BaTiO_3$ NPs are substituted by those of silane coupling agent, thus reducing the formation of adsorbed water molecules on the $BaTiO_3$ surface (Figure 4.6). This is why conductivity decreases with enhancement in silane coupling agent concentration [28].

4.4.3 Piezoelectric Cum Triboelectric Energy Harvester

Another approach for preparing hybrid nanogenerator is to combine the output of another mechanism with that of piezoelectricity. Apart from piezoelectricity, triboelectric energy-harvesting technique can also convert wasted mechanical energy into electrical energy. When two material surfaces undergo contact-separation cycle, charge transfer happens between these two surfaces. This phenomenon is known as triboelectrification. Transferred charges can create potential differences between two electrodes attached to two different surfaces by electrostatic induction effect. The resultant potential difference can be used to power small electronics. As both the mechanisms, piezoelectricity and triboelectricity, can generate electrical output as a result of mechanical input force, these two physics effects are well utilized to improve the electrical outputs of nanogenerators. Zhu et al. designed a unique triboelectric cum PENG, which can allow the superposition of triboelectric and piezoelectric voltages without any coupling between them under point contact. The authors utilized piezoelectricity of $BaTiO_3$ sheet along with triboelectricity between polytetrafluoroethylene (PTFE) and copper. A hole was intentionally formed in the PTFE membrane to capture the piezoelectric polarization along with triboelectric charges. The device can offer a 45 V open-circuit output voltage, a $10.4\,\mu A/cm^2$ short-circuit current density, and a $32\,nC/cm^2$

transferred charge under a vertical force of 50 N at a frequency of 2 Hz. The individual peaks for triboelectrification and piezoelectricity had been identified. Forward and reverse poling were carried out and different electrical outputs were observed for these two cases. The peak for piezoelectricity has become reversed while the poling direction has been altered. This work vividly depicts how piezoelectricity and triboelectricity can be efficiently combined, and the direction of polarization can be manipulated to enhance the total electrical outputs [29]. Patnam et al. incorporated Ca-doped barium zirconate titanate microparticles within PDMS in order to study the piezoelectric and triboelectric energy-harvesting capability of this composite sample. Ca-doped barium zirconate titanate PDMS composite was used as negative triboelectric layer, whereas aluminum was taken for the positive layer. At 2 mol% loading of microparticles, the hybrid generator offered 550 V open-circuit voltage, a 34 μA short-circuit current, and a 23.6 W/m^2 power density [30]. Min et al. revealed the influence of the presence of ferroelectric materials within one surface of a triboelectric nanogenerator (TENG). P(VDF–TrFE) in pristine form, P(VDF–TrFE) with $BaTiO_3$ nanofiller in cubic and tetragonal forms were prepared via electrospinning technique, and these three membranes were tested for triboelectricity using PET as another tribo-pair. It was found that P(VDF–TrFE) with tetragonal $BaTiO_3$ offered maximum electrical outputs compared to the other two structures. The electrospun structure offered higher electrical outputs compared to spin-coated structure due to higher dipole orientation and greater amount of β-crystalline phase present within the electrospun membrane. Additionally, ferroelectric $BaTiO_3$ filler also boosts up the triboelectric output due to higher ferroelectric polarization, resulting in improved surface potential, in particular. The peak positive open-circuit voltage of P(VDF–TrFE)/tetragonal $BaTiO_3$ was found to be 315 V, which is 1.4 times increment compared to the pristine P(VDF–TrFE) sample. Peak short-circuit current density for P(VDF–TrFE)/tetragonal $BaTiO_3$ was 6.7 $\mu A/cm^2$, which was 3.4 times higher compared to the neat sample. With increase in nominal contact pressure at lower range (3.2–12.8 kPa), the voltage was found to be enhanced. Moreover, the sensitivity of the nanogenerator was highest for P(VDF–TrFE)/tetragonal $BaTiO_3$ membrane, followed by P(VDF–TrFE)/cubic $BaTiO_3$ membrane and pristine $BaTiO_3$/P(VDF–TrFE) membrane. Finally, ferroelectric-assisted TENG was used to charge a supercapacitor, which enables this device to be used in wearable electronic applications [31]. Zhang et al. for the first time, utilized triboelectricity, piezoelectricity, and pore dipole simultaneously in a hybrid device. In the context of TENG, under the action of electrostatic field, pores can act as pore dipoles, which enhances the surface charge density and ultimately increasing the charge density at the electrodes. Additionally, for PENG, presence of pore also generates pore dipoles under electrostatic induction by piezoelectric polarization. Pore dipoles, in turn, increase the change in dipole moment, which further raises the change in charge density at electrodes. The authors fabricated the device using $BaTiO_3$–PVDF composite fibrous structure as piezoelectric cum tribonegative layer, whereas natural rubber was utilized as tribopositive layer. The existence of nanopores on the surface, as well as within the $BaTiO_3$–PVDF composite fibers, can be attributed to the evaporation of high boiling point dimethyl sulfoxide (DMSO) during the electrospinning process. At optimum $BaTiO_3$ loading, nanoporous triboelectric cum PENG offered 2.12 times transfer charge density than the sum of charge densities arising from nanoporous piezoelectric layer and neat triboelectric layer ensuring synergistic coupling between three effects as mentioned earlier [32].

4.5 Applications

$BaTiO_3$-based materials have been extensively used in energy-harvesting applications [22, 33, 34]. Bowland et al. introduced $BaTiO_3$ film within carbon fiber-reinforced composite, which is useful in the context of energy-harvesting and passive sensing applications [33]. Zhang et al. prepared hybrid nanogenerator using $BaTiO_3$–PVDF composite fibrous structure as piezoelectric cum tribonegative layer, and natural rubber was utilized as tribopositive layer, which can effectively convert biomechanical energy into its electrical counterpart [32]. Su et al. prepared $0.82Ba(Ti_{0.89}Sn_{0.11})O_3$–$0.18(Ba_{0.7}Ca_{0.3})TiO_3$ piezoelectric porous network filled with PDMS. This device has a potential to harvest biomechanical energy, and along with it can realize tactile imitation [15]. Sahu et al. suggested $BaTiO_3$–PVDF composite film can be utilized to energize low-power electronics [35]. Choi and Jeong studied the piezoelectric properties of $BaTiO_3$-loaded nanocellulose films. Such a device can charge a microcapacitor [36]. Mahadeva et al. developed a paper-based piezoelectric material, which can act as a sensing element. They suggested that cellulosic wood fiber functionalized with nanostructured $BaTiO_3$ can be used for manufacturing economical microelectromechanical systems (MEMSs) [37]. Behera et al. proposed Ni-modified $BaTiO_3$ can be applied in the fields of microwave bandwidth and temperature sensors [38].

Jiang et al. revealed that photocatalytic degradation can be improved due to synergistic effect of $BaTiO_3$ and Ag. Ag was deposited onto $BaTiO_3$ perovskite to investigate the photocatalytic degradation of rhodamine B dye. As a result of pressure from sonication, electric field is generated within $BaTiO_3$, which in turn results in flow of electrons and holes in opposite directions. These electrons and holes react with the adsorbed oxygen and water, respectively, in order to form free radicals and ultimately improving the photocatalytic degradation activity. Ag acts as an electron sink for transportation of electrons outside the $BaTiO_3$ to separate the charge carriers. Ultimately, electrons are transported to Ag/dye solution interface. These factors improve the photocatalytic degradation activity [39]. Previously, Qian et al. also studied the piezo-catalytic effect of $BaTiO_3$–PDMS composite porous foam in degradation of rhodamine B dye solution [40].

Yuan et al. showed polyacrylonitrile (PAN)/$BaTiO_3$ electrospun membrane incorporated composite structure can be used in structural health monitoring applications. The authors have demonstrated that content of $BaTiO_3$ influences the piezoelectric property of the nanofibrous membrane. With increase in $BaTiO_3$ content up to 15 wt%, the piezoelectric property of the nanofibrous web enhances, followed by deterioration of piezoelectricity up to 25 wt% loading of $BaTiO_3$. This data corresponds to the increase in zigzag phase content of PAN up to a 15% loading of nanofiller. Further, the nanofibrous electrospun membrane was cured with carbon fiber. The piezoelectric PAN/$BaTiO_3$ electrospun nanofibrous web also retains its piezoelectric property within the carbon fiber composite structure, which is beneficial from structural health monitoring perspective [41].

Most of the vital activities in the human body are controlled by electrical signals. The electrical potential of 10–60 mV was measured in several locations of the human body. Human body parts such as skin, bone, cartilage, tendons, ligament, and DNA show piezoelectric properties. Dry bone shows piezoelectricity, whereas wet bone is not piezoelectric as presence of water molecules changes the crystal to more symmetric structure. External pressure deforms the hydrogen bonds present in polypeptide chains of collagen, which is responsible for piezoelectric nature of bone. Pressure-induced electrical signals act as a stimulus for bone growth and remodeling. Under the action of electrical field/electrical potential, cells can move and migrate directionally. As many tissues in the human body are piezoelectric in nature, it is natural to produce scaffolds having piezoelectric properties to regenerate tissues properly. For making piezoelectric scaffolds, PVDF has been vastly used as a piezoelectric polymer. However, its non-biodegradable nature hinders its use in this particular field [42]. A poled piezoelectric scaffold consists of positive charges on one side and negative charges on the other side. Negative charges influence the crystallization of inorganic ions, which in turn affects the behavior of deposited cells. Ehterami et al. reported that a piezoelectric $BaTiO_3$-based scaffold can be used in bone tissue regeneration and repair. They modified the scaffold with a Gel/HA nanocomposite coating, which offered significantly higher cell density compared to the pristine scaffold [43]. Shuai et al. further introduced Ag-decorated polydopamine functionalized $BaTiO_3$ NP into a PVDF scaffold (Figure 4.7). The authors also revealed the antibacterial activity of

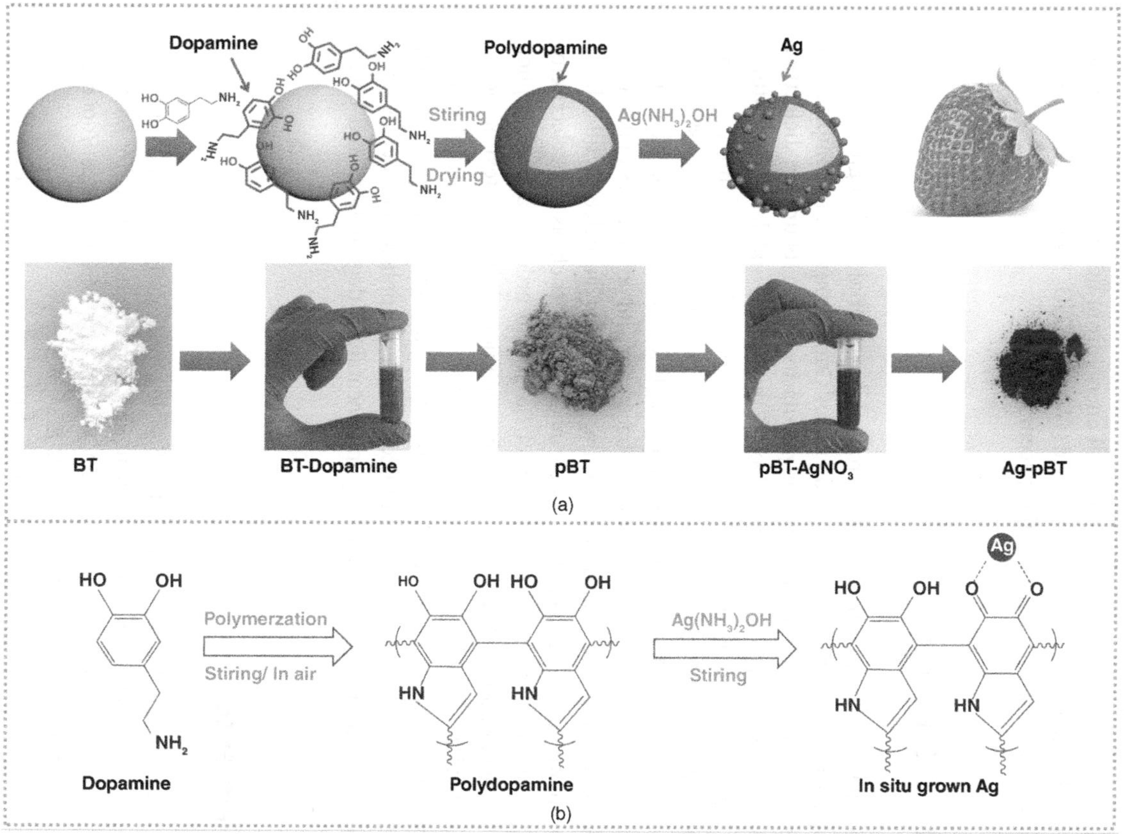

Figure 4.7 In situ growth of Ag NP onto polydopamine functionalized $BaTiO_3$: (a) Preparation process, (b) reaction mechanism. Source: Shuai et al. [44]/with permission from Elsevier.

such a $BaTiO_3$-based polymeric scaffold [44]. Conventional batteries used to power pacemakers have a limited lifespan. After discharging, the battery has to be replaced by a new one, which demands risky and painful surgery. This difficulty can be mitigated with the help of self-powered nanogenerators by converting tiny biomechanical energy from heartbeats to electrical signals. Kabir et al. suggested a hybrid (PVDF–$BaTiO_3$)/polyamide-11 could be used as a self-powered cardiac pacemaker [45].

4.6 Conclusion

$BaTiO_3$ is the first-discovered ferroelectric material. Although PZT has better piezoelectric properties and greater Curie temperature compared to $BaTiO_3$, the latter has prevailed its popularity due to its unique dielectric, piezoelectric, pyroelectric, ferroelectric properties, and lead-free nature. There are ample numbers of research works dealing with improvement of piezoelectric properties of this lead-free ceramic material. Most of the approaches involve hybridization of the pristine ceramic. First, crystal structure of $BaTiO_3$ is doped with different acceptor and donor types of dopants. Second, composite materials containing $BaTiO_3$ as piezoelectric filler have been manufactured. Finally, other mechanisms such as triboelectricity are combined with piezoelectricity of $BaTiO_3$ in order to achieve higher electrical outputs. Poling is an effective technique in the context of piezoelectricity. The direction of poling noticeably affects the electrical output of a piezoelectric cum triboelectric device. Apart from energy harvesting, $BaTiO_3$-based materials can be utilized in several applications such as sensing, structural health monitoring, photocatalytic degradation, and tissue engineering.

References

1 Seto, J. and Romero, P.A. (2016). Shaping it up: design and engineering of biominerals and crystalline materials from the bottom up. In: *Biomineralization and Biomaterials*, 3–50. Woodhead Publishing https://doi.org/10.1016/B978-1-78242-338-6.00002-8.

2 Holterman, J. and Groen. P. (2018). *An Introduction to Piezoelectric Materials and Components.*

3 Vijatović, M.M., Bobić, J.D., and Stojanović, B.D. (2008). History and challenges of barium titanate: part II. *Sci. Sintering* 40: 235–244. https://doi.org/10.2298/SOS0803235V.

4 Ertuğ, B. (2013). The overview of the electrical properties of barium titanate. *Am. J. Eng. Res.* 2: 1–7.

5 Stojanovic, B.D., Jovalekic, C., Vukotic, V. et al. (2005). Ferroelectric properties of mechanically synthesized nanosized barium titanate. *Ferroelectrics* 319: 65–73. https://doi.org/10.1080/00150190590965424.

6 Vijatović, M.M., Bobić, J.D., and Stojanović, B.D. (2008). History and challenges of barium titanate: part I. *Sci. Sintering* 40: 155–165. https://doi.org/10.2298/SOS0802155V.

7 Kong, L.B. (2008). Progress in synthesis of ferroelectric ceramic materials via high-energy mechanochemical technique. *Prog. Mater Sci.* 53: 207–322. https://doi.org/10.1016/j.pmatsci.2007.05.001.

8 Yako, K., Kakemoto, H., Tsurumi, T., and Wada, S. (2005). Domain size dependence of d_{33} piezoelectric properties for barium titanate single crystals with engineered domain configurations. *Mater. Sci. Eng., B* 120: 181–185. https://doi.org/10.1016/j.mseb.2005.02.031.

9 Ma, N., Zhang, B., Yang, W., and Guo, D. (2012). Phase structure and nano-domain in high performance of BaTiO 3 piezoelectric ceramics. *J. Eur. Ceram. Soc.* 32: 1059–1066. https://doi.org/10.1016/j.jeurceramsoc.2011.11.014.

10 Newalkar, B.L., Komarneni, S., and Katsuki, H. (2001). Microwave-hydrothermal synthesis and characterization of barium titanate powders. *Mater. Res. Bull.* 36: 2347–2355. https://doi.org/10.1016/S0025-5408(01)00729-2.

11 Boston, R., Foeller, P.Y., Sinclair, D.C., and Reaney, I.M. (2017). Synthesis of barium titanate using deep eutectic solvents. *Inorg. Chem.* 56: 542–547. https://doi.org/10.1021/acs.inorgchem.6b02432.

12 Kareiva, A., Tautkus, S., Rapalaviciute, R. et al. (1999). Sol-gel synthesis and characterization of barium titanate powders. *J. Mater. Sci.* 34: 4853–4857. https://doi.org/10.1023/A:1004615912473.

13 Vinothini, V., Singh, P., and Balasubramanian, M. (2006). Synthesis of barium titanate nanopowder using polymeric precursor method. *Ceram. Int.* 32: 99–103. https://doi.org/10.1016/j.ceramint.2004.12.012.

14 Phys JA. Changes in dielectric and ferroelectric properties of hybrid-doped barium titanate ceramics under compressive stress 2017;124102. https://doi.org/10.1063/1.3042228.

15 Su, H., Wang, X., Li, C. et al. (2021). Enhanced energy harvesting ability of polydimethylsiloxane-BaTiO3-based flexible piezoelectric nanogenerator for tactile imitation application. *Nano Energy* 83: 105809. https://doi.org/10.1016/j.nanoen.2021.105809.

16 Dudem, B., Bharat, L.K., Patnam, H. et al. (2018). Enhancing the output performance of hybrid nanogenerators based on Al-doped $BaTiO_3$

composite films: a self-powered utility system for portable electronics. *J. Mater. Chem. A* 16101–16110. https://doi.org/10.1039/c8ta04612c.
17 Wang, S., Zhang, L., Wang, L. et al. (2022). Fluorinated barium titanate nanoparticles for wearable piezoelectric power generation. *ACS Appl. NANO Mater.* https://doi.org/10.1021/acsanm.1c03777.
18 Suo, G., Yu, Y., Zhang, Z. et al. (2016). Piezoelectric and triboelectric dual effects in mechanical-energy harvesting using $BaTiO_3$/polydimethylsiloxane composite film. *ACS Appl. Mater. Interfaces* 6–12. https://doi.org/10.1021/acsami.6b11108.
19 Wang, Y., Zhang, X., Guo, X. et al. (2018). Hybrid nanogenerator of $BaTiO_3$ nanowires and CNTs for harvesting energy. *J. Mater. Sci.* 13081–13089. https://doi.org/10.1007/s10853-018-2540-9.
20 Kyu, J., Som, D., Rang, Y. et al. (2021). In-depth exploration of the synergistic interplay between perovskite barium titanate nanoparticles and two-dimensional graphene oxide for flexible piezoelectric nanogenerators. *Appl. Surf. Sci.* 538: 147962. https://doi.org/10.1016/j.apsusc.2020.147962.
21 Wang, F., Mai, Y., Wang, D. et al. (2015). High quality barium titanate nanofibers for flexible piezoelectric device applications. *Sens. Actuators, A* 233: 195–201. https://doi.org/10.1016/j.sna.2015.07.002.
22 Dudem, B., Kim, D.H., Bharat, L.K., and Yu, J.S. (2018). Highly-flexible piezoelectric nanogenerators with silver nanowires and barium titanate embedded composite films for mechanical energy harvesting. *Appl. Energy* 230: 865–874. https://doi.org/10.1016/j.apenergy.2018.09.009.
23 Shi, K., Sun, B., Huang, X., and Jiang, P. (2018). Synergistic effect of graphene nanosheet and $BaTiO_3$ nanoparticles on performance enhancement of electrospun PVDF nanofiber mat for flexible piezoelectric nanogenerators. *Nano Energy* 52: 153–162. https://doi.org/10.1016/j.nanoen.2018.07.053.
24 Yaqoob, U., Uddin, A.S.M.I., and Chung, G.S. (2017). A novel tri-layer flexible piezoelectric nanogenerator based on surface-modified graphene and PVDF-$BaTiO_3$ nanocomposites. *Appl. Surf. Sci.* 405: 420–426. https://doi.org/10.1016/j.apsusc.2017.01.314.
25 Kaur, G. and Rana, D.S. (2020). Synthesis and comprehensive study of polyvinylidene fluoride–nickel oxide–barium titanate (PVDF–NiO–$BaTiO_3$) hybrid nanocomposite films for enhancement of the electroactive beta phase. *J. Mater. Sci. - Mater. Electron.* 31: 18464–18476. https://doi.org/10.1007/s10854-020-04390-8.

26 Yang, C., Chen, F., Sun, J., and Chen, N. (2021). Boosted mechanical piezoelectric energy harvesting of polyvinylidene fluoride/barium titanate composite porous foam based on three-dimensional printing and foaming technology. *ACS Omega* 6: 30769–30778. https://doi.org/10.1021/acsomega.1c04998.

27 Wang, Z. and Narita, F. (2019). Corona poling conditions for barium titanate/epoxy composites and their unsteady wind energy harvesting potential. *Adv. Eng. Mater.* 21: 2–9. https://doi.org/10.1002/adem.201900169.

28 Sundar, U., Lao, Z., and Cook-Chennault, K. (2019). Investigation of piezoelectricity and resistivity of surface modified barium titanate nanocomposites. *Polymers (Basel)* 11. https://doi.org/10.3390/polym11122123.

29 Zhu, Q., Dong, L., Zhang, J. et al. (2020). All-in-one hybrid tribo/piezoelectric nanogenerator with the point contact and its adjustable charge transfer by ferroelectric polarization. *Ceram. Int.* 46: 28277–28284. https://doi.org/10.1016/j.ceramint.2020.07.329.

30 Patnam, H., Dudem, B., Alluri, N.R. et al. (2020). Piezo/triboelectric hybrid nanogenerators based on Ca-doped barium zirconate titanate embedded composite polymers for wearable electronics. *Compos. Sci. Technol.* 188: 107963. https://doi.org/10.1016/j.compscitech.2019.107963.

31 Min, G., Pullanchiyodan, A., Dahiya, A.S. et al. (2021). Ferroelectric-assisted high-performance triboelectric nanogenerators based on electrospun P(VDF-TrFE) composite nanofibers with barium titanate nanofillers. *Nano Energy* 90: 106600. https://doi.org/10.1016/j.nanoen.2021.106600.

32 Zhang, J.H., Zhou, Z., Li, J. et al. (2022). Coupling enhanced performance of triboelectric-piezoelectric hybrid nanogenerator based on nanoporous film of poly(vinylidene fluoride)/$BaTiO_3$ composite electrospun fibers. *ACS Mater. Lett.* 4: 847–852. https://doi.org/10.1021/acsmaterialslett.1c00819.

33 Bowland, C.C., Malakooti, M.H., and Sodano, H.A. (2017). Barium titanate film interfaces for hybrid composite energy harvesters. *ACS Appl. Mater. Interfaces* 9: 4057–4065. https://doi.org/10.1021/acsami.6b15011.

34 Villa, S.M., Mazzola, V.M., Santaniello, T. et al. (2019). Soft piezoionic/piezoelectric nanocomposites based on ionogel/$BaTiO_3$ nanoparticles for low frequency and directional discriminative pressure sensing. *ACS Macro Lett.* 8: 414–420. https://doi.org/10.1021/acsmacrolett.8b01011.

35 Sahu, M., Hajra, S., Lee, K. et al. (2021). Piezoelectric nanogenerator based on lead-free flexible pvdf-barium titanate composite films for driving low power electronics. *Crystals* 11: 1–10. https://doi.org/10.3390/cryst11020085.

36 Choi, H.Y. and Jeong, Y.G. (2019). Microstructures and piezoelectric performance of eco-friendly composite films based on nanocellulose and barium titanate nanoparticle. *Composites, Part B* 168: 58–65. https://doi.org/10.1016/j.compositesb.2018.12.072.

37 Mahadeva, S.K., Walus, K., and Stoeber, B. (2014). Fabrication and testing of piezoelectric hybrid paper for MEMS applications. In: *2014 IEEE 27th International Conference on Micro Electro Mechanical Systems*, 620–623. IEEE https://doi.org/10.1109/MEMSYS.2014.6765717.

38 Behera, C., Patel, P., Pradhan, N., and Choudhary, R.N.P. (2022). Studies of structural, dielectric and electrical characteristics of nickel-modified barium titanate for device applications. *J. Mater. Sci. - Mater. Electron.* 33: 1657–1669. https://doi.org/10.1007/s10854-022-07709-9.

39 Jiang, X., Wang, H., Wang, X., and Yuan, G. (2021). Synergetic effect of piezoelectricity and Ag deposition on photocatalytic performance of barium titanate perovskite. *Sol. Energy* 224: 455–461. https://doi.org/10.1016/j.solener.2021.06.032.

40 Qian, W., Zhao, K., Zhang, D. et al. (2019). Piezoelectric material-polymer composite porous foam for efficient dye degradation via the piezo-catalytic effect. *ACS Appl. Mater. Interfaces* 11: 27862–27869. https://doi.org/10.1021/acsami.9b07857.

41 Yuan, L., Fan, W., Yang, X. et al. (2021). Piezoelectric PAN/$BaTiO_3$ nanofiber membranes sensor for structural health monitoring of real-time damage detection in composite. *Compos. Commun.* 25: 100680. https://doi.org/10.1016/j.coco.2021.100680.

42 Ribeiro, C., Sencadas, V., Correia, D.M., and Lanceros-Méndez, S. (2015). Piezoelectric polymers as biomaterials for tissue engineering applications. *Colloids Surf., B* 136: 46–55. https://doi.org/10.1016/j.colsurfb.2015.08.043.

43 Ehterami, A., Kazemi, M., Nazari, B. et al. (2018). Fabrication and characterization of highly porous barium titanate based scaffold coated by Gel/HA nanocomposite with high piezoelectric coefficient for bone tissue engineering applications. *J. Mech. Behav. Biomed. Mater.* 79: 195–202. https://doi.org/10.1016/j.jmbbm.2017.12.034.

44 Shuai, C., Liu, G., Yang, Y. et al. (2020). A strawberry-like Ag-decorated barium titanate enhances piezoelectric and antibacterial activities of polymer scaffold. *Nano Energy* 74: 104825. https://doi.org/10.1016/j.nanoen.2020.104825.

45 Kabir, H., Kamali Dehghan, H., Mashayekhan, S. et al. (2022). Hybrid fibrous (PVDF-$BaTiO_3$)/ PA-11 piezoelectric patch as an energy harvester for pacemakers. *J. Ind. Text.* 51: 4698S–4719S. https://doi.org/10.1177/15280837211057575.

5

BNT-Based Hybrid Piezoelectric Materials

Abhishek Sasmal[1,2], *Shrabanee Sen*[1], *and Arunachalakasi Arockiarajan*[2,3]

[1] *Functional Materials and Devices Division, CSIR-Central Glass & Ceramic Research Institute, Kolkata, West Bengal, India*
[2] *Department of Applied Mechanics, Indian Institute of Technology Madras (IIT Madras), Chennai, Tamil Nadu, India*
[3] *Ceramic Technology Group-Center of Excellence in Materials and Manufacturing Futuristic Mobility, Indian Institute of Technology Madras (IIT Madras), Chennai, Tamil Nadu, India*

5.1 Introduction

In present days, mechanical energy harvesting using inorganic–organic hybrid flexible piezoelectric devices has attracted numerous attentions due to their multiple advantages, including superior applicability for flexible and wearable purposes and high sensitivity toward ambient mechanical vibrations. For the development of this type of device, different polymer materials such as polyvinyl chloride (PVC) [1], polydimethylsiloxane (PDMS) [2], poly(vinylidene fluoride) (PVDF), and its copolymers and terpolymers [3–5] have been explored widely and are under explicit research in present days. Among the different polymer materials used for this purpose, some exhibit inherent piezoelectricity, while some do not. Therefore, in order to fabricate inorganic–organic hybrid piezoelectric energy harvesters, the polymers which do not exhibit piezoelectricity (such as PDMS), must require piezoelectric inorganic materials dispersed within their matrix [2]. Whereas, the polymers which are inherently piezoelectric in nature (such as PVDF) do not strictly require piezoelectric inorganic materials dispersed within their matrix for the development of hybrid piezoelectric nanogenerators. In these cases, the incorporation of a small

Hybrid Materials for Piezoelectric Energy Harvesting and Conversion, First Edition.
Edited by S. Wazed Ali, Satyaranjan Bairagi, and Shahid Ul Islam.
© 2024 John Wiley & Sons, Inc. Published 2024 by John Wiley & Sons, Inc.

amount of non-piezoelectric materials within their matrix can also improve their piezoelectric performance by improving their polar chain conformation through interfacial interaction [3, 6, 7]. However, if the incorporated particles exhibit inherent piezoelectricity, the effects of interaction and the piezoelectric properties of fillers can act simultaneously, which in turn, effectively improves the output mechanical energy-harvesting activity of the final device. Therefore, it is obvious that the piezoelectric fillers, either incorporated in a non-piezoelectric or a piezoelectric polymer matrix, are advantageous in all aspects for the development of hybrid inorganic–organic flexible piezoelectric devices. Hence, inorganic piezoelectric materials are most often used for this purpose by the scientific community all over the world.

There are several piezoelectric materials, including $Pb(ZrTi)O_3$ [8], $BaTiO_3$ [9], ZnO [10], $BiFeO_3$ [11], $(KNa)NbO_3$ [12], $Bi_4Ti_3O_{12}$ [13], $(BaCa)TiO_3$–$Ba(TiZr)O_3$ [14], $ZnSnO_3$ [15], and $CsPbX_3$ (X = I, Br) [16, 17], which have been widely explored for this purpose. However, despite being an excellent lead-free piezoelectric and ferroelectric material with high spontaneous polarization (38 μC/cm^2) and Curie temperature (320 °C) [18], $(BiNa)TiO_3$ (BNT) has been explored very rarely to date for the development of hybrid piezoelectric energy harvesters. Therefore, there is a large scope for future research in widely exploring this material for the development of flexible inorganic–organic piezoelectric energy harvesters. Thus, this chapter focuses on describing the limitations of BNT, the ways to overcome the limitations, the widespread application of this material, and the current trend of research and future scopes of using it for the development of inorganic–organic hybrid piezoelectric energy-harvesting devices.

5.2 Key Limitations of BNT and Ways to Overcome the Limitations

One of the key limitations of BNT is its high conductivity resulting in a very large leakage current [18, 19]. In the BNT lattice, the highly polarizable Bi^{3+} ions with their 6s lone pair electrons and the weak interaction between Bi and O (weak Bi—O bond) renders a very low diffusion barrier for the migration of oxygen ions [19, 20]. As a result of this fact, the material poses very high conductivity. On the other hand, the coercive field of BNT is very high due to the doubling of the unit cell within the crystal structure and its intrinsic high spontaneous polarization [19, 21]. As a result of these facts, the poling of BNT ceramics becomes very difficult, and as a result a very feeble piezoelectric coefficient (d_{33} ~ 80 pC/N) [22] is achieved. Moreover, this material poses a very low depolarization temperature, which places it on the rear side of real-life applications [18, 19].

In order to overcome these issues, many attempts including phase structure tuning, domain and/or domain wall engineering, and making composite samples through the addition of other materials to BNT have been taken by the scientific community all over the world. Success in overcoming the issues related to BNT ceramics has also been achieved by following the mentioned techniques [18, 19, 23]. Thus, BNT ceramics that can withstand high field could be designed now with an improved relaxor ferroelectric nature, which, in turn, improves the energy storage activity of the resulting BNT-based composite ceramics [24–29]. In most cases, the piezoelectric properties and polarization of BNT ceramics are increased [19, 30–34] by phase structure tuning and domain and/or domain wall engineering through the preparation of complex ceramics by mixing other suitable materials with BNT.

5.3 Applicability of BNT-Based Materials for Piezoelectric Energy Harvesting

As researchers succeed in improving the piezoelectric properties of BNT, it is obvious that piezoelectric energy harvesters with improved performance by using this material can be developed. Manotham et al. [35] showed the improvement of piezoelectric voltage constant (g_{33}) of sintered $0.94Bi_{0.5}Na_{0.5}TiO_3$–$0.06BaTiO_3$ pellets by the addition of B_2O_3 as an interstitial dopant. This result proved the energy-harvesting activity of the system. In a similar procedure, Jarupoom et al. [36] also explored the energy-harvesting activity of $(1-x)(Bi_{0.5}Na_{0.5})TiO_3$–$xBi(Mg_{0.5}Ti_{0.5})O_3$ (x = 0–0.20)-based ceramic systems. Zhang et al. [37] also developed a real-life energy harvester (Figure 5.1) with excellent energy-harvesting performance by using Mn^{2+} substituted $0.92(Bi_{0.5}Na_{0.5})TiO_3$–$0.04(Bi_{0.5}K_{0.5})TiO_3$–$0.04BaTiO_3$ (0.92BNT–0.04BKT–0.04BT) ceramic pellets.

These reports prove the superior applicability of the BNT-based ceramic system for vibration energy harvesting. However, the main focus of this chapter is to enlighten the applicability of BNT for the development of organic–inorganic hybrid flexible systems for mechanical energy harvesting. It is true that the augmentation of polarization and piezoelectric properties of any ceramic-based materials can improve the piezoelectric energy conversion (mechanical to electrical) performance of the resulting hybrid polymer-based devices when incorporated within their matrix. However, in organic polymer systems, fillers are most often incorporated in

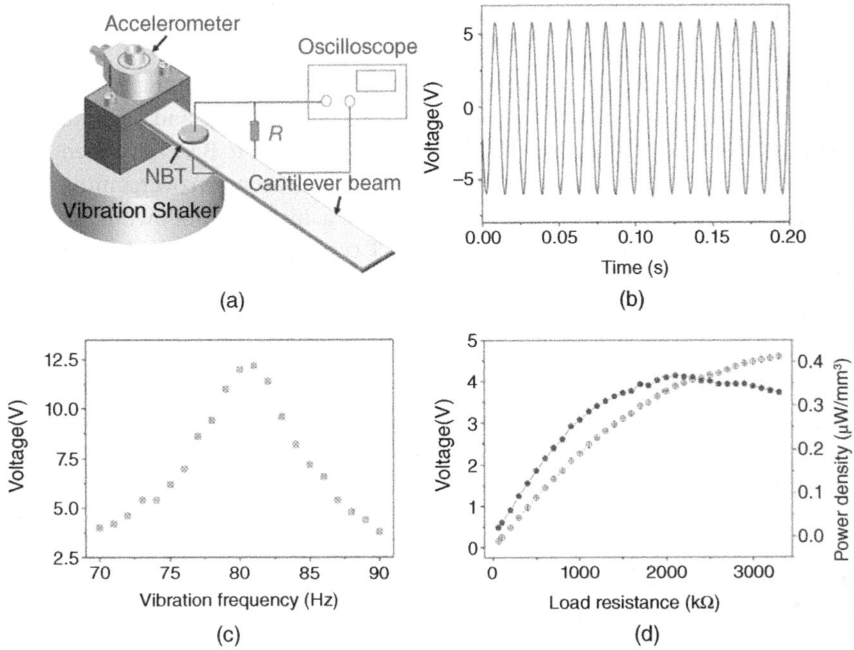

Figure 5.1 Piezoelectric energy harvesters using Mn^{2+} substituted $0.92(Bi_{0.5}Na_{0.5})TiO_3$–$0.04(Bi_{0.5}K_{0.5})TiO_3$–$0.04BaTiO_3$ ceramic pellets as reported by Zhang et al. [37] (a) schematic illustration of the measurement of the output energy-harvesting performance, (b) time-dependent, (c) frequency dependent, and (d) load resistance-dependent output voltage of the fabricated device. Source: Reproduced with permission from Zhang et al. [37]. Copyright (2020) Elsevier B.V.

particle form. Thus, the factors which determine the output piezoelectric energy-harvesting characteristics of BNT-based bulk ceramics may not play a role in the same way when BNT particles are incorporated into a polymer matrix. In fact, in some of the recent reports, it has been shown that the improvement of conducting nature of fillers can effectively augment the output mechanical energy-harvesting activity of the resulting polymer-based composites through the enhancement of charge transport within the hybrid material and material to the electrode [10, 38–40]. Hence, the highly conducting nature of BNT may not be a big obstacle to the improvement of piezoelectric energy-harvesting performance of BNT-based polymer composites. Therefore, it is expected that the BNT, either modified or unmodified, will effectively augment the energy-harvesting activity of their polymer-based composites. If it is modified in such a manner that its conducting nature is reduced and piezoelectric property is improved, it will improve the desired energy-harvesting activity of the resulting polymer-based system by the effect of its inherent piezoelectricity and the interfacial interaction (wherever applicable). On the other hand, if it is not modified (i.e., exhibits low piezoelectric performance and highly conducting nature), the output energy conversion activity of the resulting polymer-based system will be improved by the effect of the improved charge transport by means of its highly conducting nature and the interfacial interaction (wherever applicable). Thus, BNT and BNT-based materials appear to be very good candidates for piezoelectric energy harvesting when incorporated into an organic polymer matrix. The succeeding sections describe the present state-of-the-art research in this field and the future prospects.

5.4 BNT/Piezoelectric Polymer-Based Piezoelectric Energy Harvesters

Due to the limitations of BNT, as described earlier, the research works on the BNT-based polymer composites were found to be very rarely reported in the early stage of the past decade. However, as the research community succeeded in overcoming the limitations of BNT, the experimental research work in the field of BNT-based polymer composites is increasing day by day, especially in the past few years. The main modification of BNT, which was done by the research community, was the reduction of its conducting nature and hence the leakage current. Therefore, the preliminary focus of researchers was to develop low-loss dielectric and ferroelectric polymer composites for energy storage applications by using modified BNT fillers. There are many reports in present days where the BNT-based fillers have been incorporated into PVDF-based ferroelectric polymers for effective dielectric, ferroelectric, and capacitive energy storage applications [41–46]. However, as the present chapter is focused only on piezoelectric energy harvesting, the application of BNT for the same is mainly discussed here.

Though the application of BNT in polymer matrix for piezoelectric energy harvesting is very rarely explored till date, this system is attaining potential and gradually increased attention to the researchers in present days for being applied in the mentioned purpose. There are some reports where the BNT has been used as a filler in piezoelectric PVDF-based composites for enhanced piezoelectric and pyroelectric applications. Mahdi et al. [47] fabricated PVDF-trifluoroethylene (PVDF–TrFE)-based composite films by incorporating BNT particles and have shown their excellent pyroelectric response driven by the tuning of polarization. They have also shown the piezoelectric and pyroelectric response of $0.88(Bi_{0.5}Na_{0.5})TiO_3$–$0.084(Bi_{0.5}K_{0.5})TiO_3$–$0.063BaTiO_3$/PVDF–TrFE-based composite systems prepared via solvent-casting technique followed by hot pressing [48]. The improvement of ferroelectric polarization and pyroelectric activity of P(VDF–TrFE) induced by this filler material has also been investigated by them [49]. Ji et al. [50] fabricated $0.78Bi_{0.5}Na_{0.5}TiO_3$–$0.22SrTiO_3$ (BNT–ST) incorporated PVDF–TrFE-based composite membrane by electrospinning technique and have designed the energy harvesting module by sandwiching the membrane of certain size between two interdigitated copper electrodes. The final module has shown excellent frequency sensing activity. Petrovic et al. [51] prepared $0.94[(Bi_{0.5}Na_{0.5})TiO_3]$–$0.06BaTiO_3$/PVDF-based composite systems by hot-pressing technique and have shown their capacitive energy storage and mechanical energy harvesting applications. Recently, Raj et al. [52] have incorporated the BNT filler into the PVDF matrix, and have shown a beautiful relationship between the filler content and phase formation within the fabricated composite films via the tape casting technique. They have also fabricated an efficient piezoelectric nanogenerator device by using the film with optimum polar phase and sandwiching it between two aluminum electrodes with the help of Kapton encapsulation. The device has shown excellent power output (power density $\sim 11.8\,mW/m^2$) and has been used for effective real-life applications. Thus, it has been observed that the BNT-based fillers can effectively tune the polarity of PVDF-based matrix and can improve the output of piezoelectric energy conversion ability by the combined effect of improved polarity and inherent piezoelectric properties of BNT. Therefore, the applicability of BNT in inorganic–organic hybrid piezoelectric energy-harvesting devices has been proven, which has attracted researchers all over the world for the development and investigation of this system for the mentioned purpose.

Not only the simple incorporation of BNT-based fillers into piezoelectric polymer matrices, but also other complex device fabrication strategies and materials' modification techniques have been employed in order to improve the performance of BNT-based hybrid piezoelectric materials. Bouad et al. [53] have developed a bilayered device based on BNT and P(VDF–TrFE) in order to achieve improved piezoelectric performance. For this purpose, $(Bi_{0.5}Na_{0.5})TiO_3$ thin film was first grown on $LaNiO_3$ electrode-coated Si/SiO_2 substrate via a sputtering technique. After H_2O_2 treatment, this thin film was coated with dopamine-functionalized polymethyl methacrylate (PMMA) followed by the coating of the P(VDF–TrFE) layer via the spin coating technique. This complex technique helped in improved interfacial coupling resulting in enhanced piezoelectric behavior. Dai et al. [54] fabricated thermoplastic polyurethane (TPU)-BNT-based composite co-blended with PVDF. The improvement of piezoelectricity of the resulting composite system supported by mechanical elasticity was obtained in this system [54, 55]. Yu et al. [56] fabricated a self-healable multilayered piezoelectric sensor by using a similar kind of blended composite. For this purpose, PVDF–TPU–PCL (polycaprolactone)–BNT and PVDF–TPU–PCL–ATO (Sb-doped SnO_2)-based composite films were developed. The first one was used as a piezoelectric layer and the last one served as the electrode. The combined multilayered structure of these two films was developed via the hot-pressing technique. The resulting optimized device showed excellent piezoelectric energy harvesting, biomechanical vibration sensing, and pressure sensing activities.

Apart from the direct piezoelectric energy conversion and vibration sensing applications, these kinds of hybrid piezoelectric materials can be utilized in other diverse fields too. For example, the excellent piezo-photocatalytic performance of BNT [57–59] can be incorporated into these types of hybrid materials and devices in order to achieve excellent multifunctionality. In this regard, Zhou et al. [60] fabricated a porous three-dimensional composite foam based on BNT–PVDF for efficient piezo-photocatalytic application. The incorporation of BNT into PVDF helped in improving the piezoelectricity of the composite system by β-phase formation. On the other hand, the unique porous structure made the composite system sufficiently flexible and deformable under ultrasonication, which, in turn, helped in generating a built-in alternating electric field through the bi-piezoelectric integration of PVDF and BNT. Moreover, due to this porous structure, the availability of active sites for the desired photocatalytic reaction increased noticeably. As a result of all these facts, the system showed excellent piezo-photocatalytic activity. The performances of PVDF and BNT–PVDF-based three-dimensional porous foam are reported by Zhou et al. [60] under the dark and light illumination conditions during the stirring (ST) and ultrasonication (US), as presented in Figure 5.2a,b. The corresponding piezo-photocatalytic mechanism is schematically illustrated in Figure 5.2c,d. On the basis of the discussion about BNT-based hybrid piezoelectric materials, it can be considered that the BNT-based systems are emerging as very unique and efficient multifunctional materials for real-life applications.

5.5 BNT/Non-Piezoelectric Polymer-Based Piezoelectric Energy Harvesters

Being a lead-free piezoelectric material with an excellent piezoelectric response, BNT-based materials have drawn potential attraction in

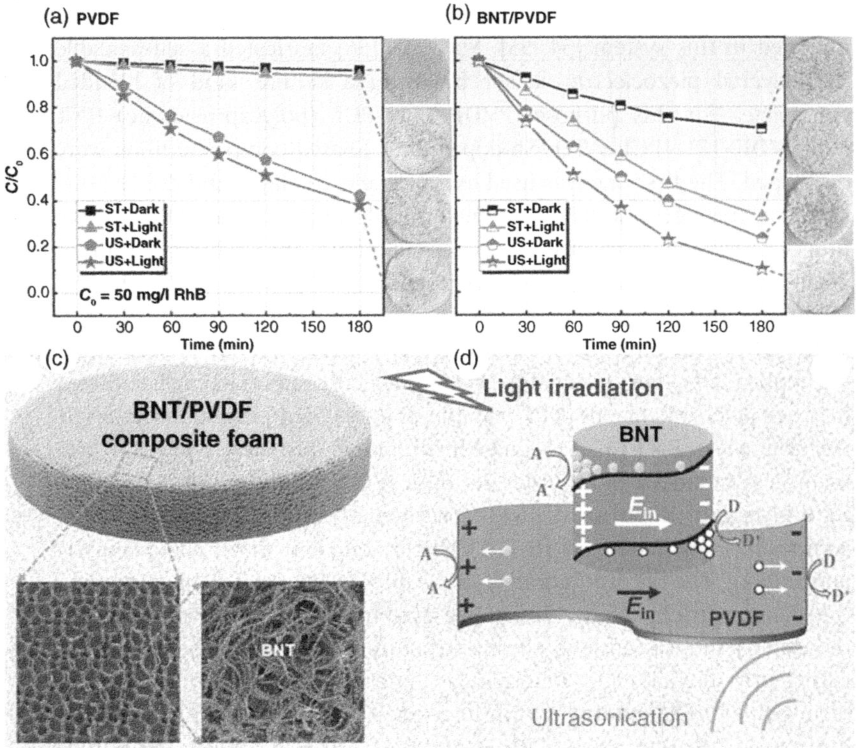

Figure 5.2 Relative variation of the RhB solution concentration ($C_0 = 50$ mg/l) over time under ST + dark, ST + light, US + dark, and US + light (degradation conditions) in the presence of (a) PVDF and (b) BNT–PVDF composite foams as reported by Zhou et al. [60] along with the schematic illustration of (c) the BNT–PVDF-based porous composite foam and (d) the associated piezo-photocatalytic mechanism. Source: Reproduced with permission from Zhou et al. [60]. © 2022/Elsevier.

present days. As the BNT exhibits inherent piezoelectricity, it can be used as filler in different non-piezoelectric polymers too for the development of inorganic–organic hybrid materials for piezoelectric energy harvesting. Unlike PVDF, other polymers such as PDMS and ecoflex show better performance in terms of stretchability, resistance toward mechanical deformation, etc., which make them more useful in different real-life applications. Therefore, researchers are focusing on such types of devices nowadays. For the fabrication of these kinds of devices, piezoelectric filler material is a very

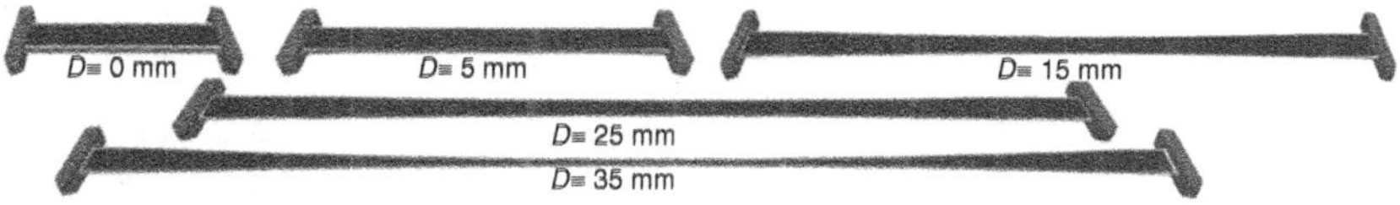

Figure 5.3 Schematic presentation of BNKT/ecoflex-based flexible and stretchable piezoelectric nanogenerator reported by Raj et al. [63] at various stretching displacement (*D*) values. Source: Reproduced with permission from Raj et al. [63]. Copyright (2021) Elsevier Ltd.

important requirement. Recently, Raj et al. [61] have developed BNT/polycaprolactone-based lightweight flexible nanogenerator with significant energy-harvesting activity (power density ~ 4 mW/m^2) for real-life motion monitoring applications. Batra et al. [62] have developed BNT/PDMS-based flexible energy harvester by spin coating the composite solution on an ITO-coated flexible polyethylene terephthalate (PET) substrate. Raj et al. [63] have also developed a super-stretchable (Figure 5.3) and flexible piezoelectric energy harvesters by using $0.8Bi_{0.5}Na_{0.5}TiO_3$–$0.2Bi_{0.5}K_{0.5}TiO_3$ (BNKT) filler in ecoflex matrix and electroding the composite films with carbon grease. Yang et al. [64] have used the nonwoven electrospun $Bi_{0.5}(Na_{0.83}K_{0.17})_{0.5}TiO_3$ fabric for developing flexible piezoelectric nanogenerators. In an almost similar strategy, Cao et al. [65] also developed piezoelectric nanogenerators by using 0.94BNT–0.06BT electrospun fibers on a flexible interdigitated electrode, and encapsulating with PDMS. Yu et al. [66] followed a unique and complex strategy to develop a high-performance piezoelectric/piezoresistive sensor integrated with thermochromic functionality. For this purpose, they developed a color-changing piezoelectric film and a piezoresistive layer and then laminated them together. The schematic structure of the final device is shown in Figure 5.4. Along with good thermochromic performance, the device has shown excellent vibration sensing, pressure sensing, and human motion sensing applications, which are associated with their good piezoelectric/piezoresistive properties In all of these cases, the piezoelectric energy-harvesting activities of the final devices have been found to be mainly governed by the inherent piezoelectricity of the BNT-based materials. Thus, if the BNT-based materials can be tuned in such a way that their piezoelectricity can be improved, they can be effectively used for the development of flexible inorganic–organic piezoelectric hybrid energy-harvesting and sensing devices.

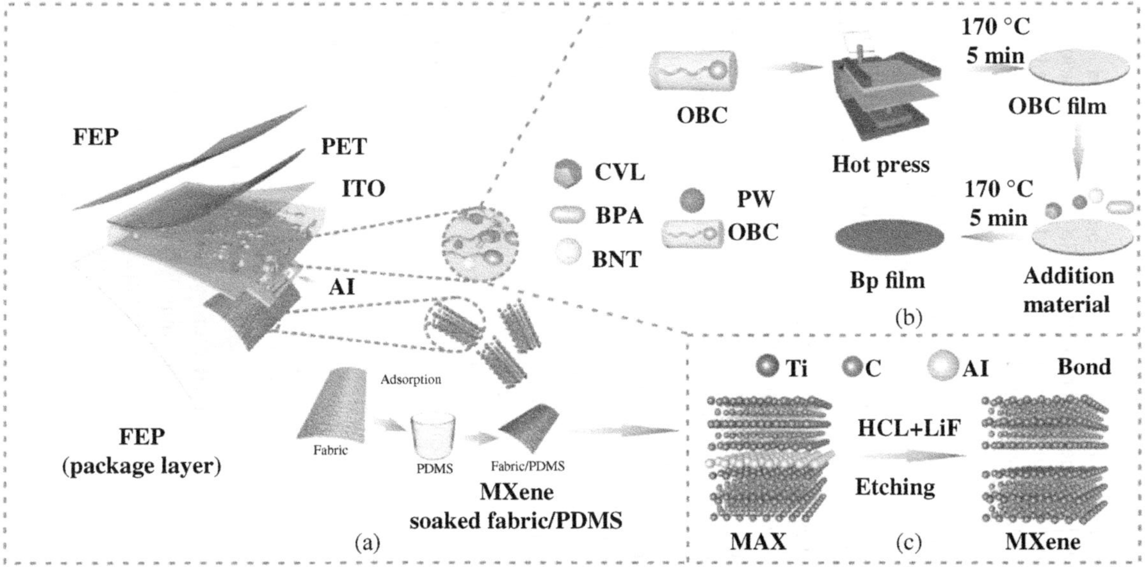

Figure 5.4 (a) Schematic illustration of the piezoelectric/piezoresistive device structure, as reported by Yu et al. [66] along with the associated schematic presentation of (b) the fabrication process of the color-changing piezoelectric film (BP) and (c) the preparation procedure of MXene materials (FEP: fluorinated ethylene propylene [external encapsulation layer], OBC: olefin block copolymer, CVL: crystal violet lactone, BPA: bisphenol, PW: paraffin wax). Source: Reproduced with permission from Yu et al. [66]. Copyright (2023) Elsevier B.V.

5.6 BNT-Based Other Mechanical Energy Harvesters

Not only piezoelectric energy harvesting but also BNT-based materials incorporated into different other polymer matrices can also be effectively used for triboelectric and piezo-tribo hybrid energy harvesting. Due to very high output energy-harvesting activity, hybrid energy harvesting is of immense interest in recent times. Therefore, BNT-based materials, exhibiting superior piezoelectricity and a certain degree of conducting nature can be very useful to enhance the output current and voltage of the hybrid devices simultaneously. The BNT-based materials can effectively augment the dielectric properties of polymer-based composite systems [67], which, in turn, can improve the desired triboelectric performance [68]. The piezoelectricity of BNT improves the output electricity of the hybrid devices. Whereas its high value of conductivity, compared to other piezoelectric ceramics, can improve the output current of the resulting hybrid device. Due to these advantages, BNT-based materials are being extensively studied in recent times for hybrid energy-harvesting applications [69–72]. Both the piezoelectric and non-piezoelectric polymer matrices have been used for this purpose. For example, He et al. [71] achieved a very high output open-circuit voltage of 2170 V with 47 W/m^2 power density by fabricating electrospun PVDF–HFP/Ag nanowires/Mn doped BNT–BT-based hybrid nanogenerators. Zhang et al. [70] also achieved high output voltage and current density (1020 V and 32.5 $\mu A/cm^2$) from 85BNT–15$SrTiO_3$-based hybrid energy harvesters. Li et al. [73] developed PDMS–BNT-based hybrid nanogenerators with high output open-circuit voltage (~370 V). Similarly, Yu et al. [74] succeeded in achieving a high value of output voltage (~120 V) by using BNT–BT incorporated PDMS-based hybrid piezoelectric material as a triboelectric layer. Thus, the BNT-based materials have shown their superior applicability in piezo-tribo hybrid energy harvesting. Hence, though the literature available in this regard is very less in number, the superiority of BNT-based materials is attracting the research community in very recent times for application in mechanical energy harvesting.

5.7 Challenges and Future Scopes

As described earlier, BNT possesses high conductivity and a large coercive field, which limits their piezoelectric applications. Therefore, the main challenge of this material is to tune these two properties simultaneously through proper materials' engineering techniques, which researchers are trying to resolve. In many cases, researchers have succeeded in resolving these issues. As a result of this fact, high piezoelectricity in BNT-based materials has now been achieved, which has made it suitable for the development of inorganic–organic hybrid materials for piezoelectric energy harvesting. On the other hand, as already described in this chapter, the existence of certain conducting nature of BNT is also helpful in improving the energy-harvesting performance of its polymer-based hybrid devices. Thus, with the discovery of these facts and the improvement of desired properties of BNT through proper techniques, the scope of using this material in piezoelectric and hybrid energy-harvesting devices has increased. Due to the known limitations of BNT, its use in inorganic–organic energy harvesters has been limited so far. Therefore, the research work available in this field is very less in number. However, with the discovery of the abovementioned facts, this material is gaining immense interest in the research community for the fabrication and development of mechanical energy harvesters. As the available number of research in this field is very less, there is a large scope for future research in this field through suitable tuning of different parameters required to enhance the performance of inorganic–organic composite-based piezoelectric and hybrid energy harvesters. Furthermore, along with the materials' optimization, other external techniques which have been utilized to augment the energy-harvesting activity of other material-based energy harvesters may also be used for BNT-polymer-based hybrid materials for achieving effective performance enhancement.

5.8 Conclusion

This chapter describes the limitations, challenges, and future scopes of research for the development of inorganic–organic composite-based piezoelectric and hybrid mechanical energy harvesters using BNT. The present state-of-the-art of research in this field is also elaborately discussed here. The unique properties of BNT which can effectively improve the output properties of the resulting nanogenerator devices are also explained here. Finally, based on the present state-of-the-art, the large future scopes of research for the development of BNT-based hybrid inorganic–organic piezoelectric and piezo-tribo hybrid devices have been emphasized here.

Acknowledgments

A. Sasmal and S. Sen thank the Director, CSIR-CGCRI. A. Sasmal thanks the Institute Post-Doctoral Fellowship Scheme of IIT Madras for financial assistance. A. Arockiarajan acknowledges the Institute of Eminence Research Initiative Project on Materials and Manufacturing for Futuristic Mobility (Project no. SB22231272MMMOEX008702).

References

1 Paria, S., Karan, S.K., Bera, R. et al. (2016). *Ind. Eng. Chem. Res.* 55: 10671.

2 Su, H., Wang, X., Li, C. et al. (2021). *Nano Energy* 83: 105809.

3 Sasmal, A., Devi, P.S., and Sen, S. (2022). *Macromol. Mater. Eng.* 307: 2200071.

4 Parangusan, H., Ponnamma, D., and Almaadeed, M.A.A. (2017). *RSC Adv.* 7: 50156.

5 Bhavanasi, V., Kumar, V., Parida, K. et al. (2016). *ACS Appl. Mater. Interfaces* 8: 521.

6 Chinya, I., Sasmal, A., and Sen, S. (2020). *J. Alloys Compd.* 815: 152312.

7 Dutta, B., Kar, E., Bose, N., and Mukherjee, S. (2018). *ACS Sustainable Chem. Eng.* 6: 10505.

8 Pal, A., Sasmal, A., Manoj, B. et al. (2020). *Mater. Chem. Phys.* 244: 122639.

9 Jian, G., Jiao, Y., Meng, Q. et al. (2020). *Adv. Mater. Interfaces* 7: 2000484.

10 Pratihar, S., Patra, A., Sasmal, A. et al. (2021). *Soft Matter* 17: 8483.

11 Sasmal, A., Sen, S., and Devi, P.S. (2019). *Phys. Chem. Chem. Phys.* 21: 5974.

12 Bairagi, S. and Ali, S.W. (2019). *Eur. Polym. J.* 116: 554.

13 Hajra, S., Padhan, A.M., Sahu, M. et al. (2021). *Nano Energy* 89: 106316.

14 Gao, X., Zheng, M., Yan, X. et al. (2019). *J. Mater. Chem. C* 7: 961.

15 Sasmal, A., Patra, A., Devi, P.S., and Sen, S. (2021). *Compos. Sci. Technol.* 213: 108916.

16 Chen, H., Zhou, L., Fang, Z. et al. (2021). *Adv. Funct. Mater.* 31: 2011073.

17 Maity, K., Pal, U., Mishra, H.K. et al. (2022). *Nano Energy* 92: 106743.

18 Zhang, Y., Kim, H., Wang, Q. et al. (2020). *Nanoscale Adv.* 2: 3131.

19 Zhou, X., Xue, G., Luo, H. et al. (2021). *Prog. Mater Sci.* 122: 100836.

20 Islam, M.S. (2000). *J. Mater. Chem.* 10: 1027.

21 Jones, G.O. and Thomas, P.A. (2002). *Acta Crystallogr., Sect. B: Struct. Sci.* 58: 168.

22 Hiruma, Y., Nagata, H., and Takenaka, T. (2009). *J. Appl. Phys.* 105: 084112.

23 Bai, W., Zheng, P., Wen, F. et al. (2017). *Dalton Trans.* 46: 15340.

24 He, J., Liu, X., Zhao, Y. et al. (2022). *ACS Appl. Electron. Mater.* 4: 735.
25 Zhu, C., Cai, Z., Luo, B. et al. (2021). *ACS Appl. Mater. Interfaces* 13: 28484.
26 Yang, H., Cai, Z., Zhu, C. et al. (2022). *ACS Sustainable Chem. Eng.* 10: 9176.
27 Ma, W., Zhu, Y., Marwat, M.A. et al. (2019). *J. Mater. Chem. C* 7: 281.
28 Zhu, C., Cai, Z., Luo, B. et al. (2020). *J. Mater. Chem. A* 8: 683.
29 Zhao, M., Xia, W., Liang, Y. et al. (2022). *J. Mater. Sci. - Mater. Electron.* 33: 21702–21712.
30 Shang, K., Shi, W., Yang, Y. et al. (2022). *Ceram. Int.* 1.
31 Zhang, Y.R., Li, J.F., Zhang, B.P., and Peng, C.E. (2008). *J. Appl. Phys.* 103: 074109.
32 Shen, Y., Xie, H., Sun, Y. et al. (2021). *Bull. Mater. Sci.* 44: 100.
33 Pattipaka, S., James, A.R., and Dobbidi, P. (2018). *J. Alloys Compd.* 765: 1195.
34 Fernandez-Benavides, D.A., Gutierrez-Perez, A.I., Benitez-Castro, A.M. et al. (2018). *Materials (Basel)* 11: 361.
35 Manotham, S., Jaita, P., Butnoi, P. et al. (2022). *J. Alloys Compd.* 897: 163021.
36 Jarupoom, P. and Jaita, P. (2022). *J. Asian Ceram. Soc.* 10: 498.
37 Zhang, Y., Liu, X., Wang, G. et al. (2020). *J. Alloys Compd.* 825: 154020.
38 Ghosh, S.K., Kim, J., Kim, M.P. et al. (2022). *ACS Nano* 16: 11415.
39 Ghosh, S.K., Kim, M.P., Na, S. et al. (2022). *Nano Energy* 100: 107438.
40 Sasmal, A., Patra, A., and Arockiarajan, A. (2022). *Appl. Phys. Lett.* 121: 133902.
41 Yi, Z., Wang, Z., Nian, W. et al. (2021). *ACS Appl. Energy Mater.* 4: 13528.
42 Lin, Y., Zhang, Y., Zhan, S. et al. (2020). *J. Mater. Chem. A* 8: 23427.
43 Li, J., Chen, G., Lin, X. et al. (2020). *J. Mater. Sci.* 55: 2503.
44 Wang, Z., Nian, W., Wang, T. et al. (2018). *J. Mater. Sci. - Mater. Electron.* 29: 9129.
45 Xu, Z., Weng, L., Guan, L. et al. (2021). *J. Mater. Sci. - Mater. Electron.* 32: 28129.
46 Moharana, S., Sai, S., and Mahaling, R.N. (2018). *J. Adv. Dielectr.* 8: 19.
47 Mahdi, R.I., Gan, W.C., Halim, N.A. et al. (2015). *Ceram. Int.* 41: 13836.
48 Mahdi, R.I. and Abd Majid, W.H. (2016). *RSC Adv.* 6: 81296.
49 Mahdi, R.I., Gan, W.C., Abd Majid, W.H. et al. (2018). *Polymer (Guildf)* 141: 184.
50 Ji, S.H., Cho, J.H., Jeong, Y.H. et al. (2016). *Sens. Actuators, A* 247: 316.
51 Vijatovic Petrovic, M., Cordero, F., Mercadelli, E. et al. (2021). *J. Alloys Compd.* 884: 161071.

52 Raj, N.P.M.J., Abisegapriyan, K.S., Khandelwal, G., and Kim, S.J. (2022). *Sustainable Energy Fuels* 6: 674.

53 Bouad, V., Fadel, A., Mohan, S. et al. (2022). *ACS Appl. Polym. Mater.* 4: 7258.

54 Dai, Z., Feng, Z., Feng, C. et al. (2021). *J. Appl. Polym. Sci.* 138: e49818.

55 Shehata, N., Nair, R., Boualayan, R. et al. (2022). *Sci. Rep.* 12: 8335.

56 Yu, C., Xu, J., Zhang, Y. et al. (2023). *J. Alloys Compd.* 958: 170509.

57 Zhao, Z., Wei, L., Li, S. et al. (2020). *J. Mater. Chem. A* 8: 16238.

58 Zhou, X., Sun, Q., Zhai, D. et al. (2021). *Nano Energy* 84: 105936.

59 Xu, X., Lin, X., Yang, F. et al. (2020). *J. Phys. Chem. C* 124: 24126.

60 Zhou, X., Sun, Q., Xiao, Z. et al. (2022). *J. Environ. Chem. Eng.* 10: 108399.

61 Raj, N.P.M.J., Abisegapriyan, K.S., Khandelwal, G. et al. (2020). *Sustainable Energy Fuels* 4: 5636.

62 Batra, K., Sinha, N., and Kumar, B. (2019). *J. Mater. Sci. - Mater. Electron.* 30: 6157.

63 Raj, N.P.M.J., Khandelwal, G., and Kim, S.J. (2021). *Nano Energy* 83: 105848.

64 Yang, Q., Wang, D., Zhang, M. et al. (2016). *J. Alloys Compd.* 688: 1066.

65 Cao, Z., Wang, J., Zhang, C. et al. (2022). *J. Alloys Compd.* 918: 165766.

66 Yu, C., Liu, K., Xu, J. et al. (2023). *Chem. Eng. J.* 459: 141648.

67 Rashmiranjan, P., Mohanty, D., Nayak, P.K. et al. (2022). *Polym. Sci. Ser. B Polym. Chem.* 64: 539.

68 Sasmal, A., Patra, A., Maity, S. et al. (2022). *Sustainable Energy Fuels* 6: 4652.

69 Zhao, K., Meng, J., Zhong, M. et al. (2022). *Small* 94: 2202792.

70 Zhang, J.H. and Hao, X. (2020). *Nano Energy* 76: 105074.

71 He, Y., Wang, H., Sha, Z. et al. (2022). *Nano Energy* 98: 107343.

72 Ji, S.H., Lee, W., and Yun, J.S. (2022). *J. Eur. Ceram. Soc.* 42: 1414.

73 Li, B., Liu, H., Sun, Y. et al. (2022). *J. Mater. Sci. - Mater. Electron.* 33: 5335.

74 Yu, C., Xu, J., Yang, T. et al. (2022). *J. Mater. Sci. - Mater. Electron.* 33: 22588.

6

$ZnSnO_3$-Based Hybrid Piezoelectric Materials

Anupam Chowdhury[1], Srijan Das[1], Mukta Nitin Mirlekar[1], and S. Wazed Ali[1,2]

[1] *Department of Textile and Fibre Engineering, Indian Institute of Technology Delhi, Hauz Khas, New Delhi, India*
[2] *School of Interdisciplinary Research (SIRe), Indian Institute of Technology Delhi, New Delhi, India*

6.1 Introduction

The energy crisis existing in the present day and age calls for new types of energy generators. One such very promising group of energy harvesters is piezoelectric energy harvesters that can convert ubiquitous mechanical energy to usable electrical energy. Piezoelectric nanogenerators (PENG) can again be composed of piezoelectric polymers such as PVDF, piezoelectric ceramics such as $BaTiO_3$, or a combination of both. Usually, composite designs are aimed at the present day.

Lead-free perovskites are in high demand because of their potential applications in the case of sensors, nanogenerators, actuators, optoelectronics, and so on [1–4]. Among these, the ones that show appreciable piezoelectricity are highly sought after because of their applications in PENGs and actuators. Zinc stannate (ZTO) is one such lead-free piezoelectric ceramic that has gained wide interest because of its attractive properties such as electrical and structural ordering upto 700 °C, excellent polarization of 59 $\mu C/cm^2$ along *c*-axis, and ability to modulate its properties based on synthesis conditions [5]. Also, it reportedly has high electron mobility, negligible visible light absorption, and electrical conductivity, making it attractive for various applications [6, 7]. Moreover, other than these, ZTO has also been reported to be antimicrobial [8] and can be used as a photocatalyst for dye degradation [9].

A lot of emphasis has been placed recently on studying the effect of various synthesis parameters on the properties of ZTO. The different nanostructures

Hybrid Materials for Piezoelectric Energy Harvesting and Conversion, First Edition.
Edited by S. Wazed Ali, Satyaranjan Bairagi, and Shahid Ul Islam.
© 2024 John Wiley & Sons, Inc. Published 2024 by John Wiley & Sons, Inc.

possess their own properties, and these are dependent on the defects in the crystal or the bulk [10]. Different synthesis methods have so far been used for developing ZTO nanostructures. Also, in applications regarding piezoelectricity, ZTO, along with other fillers, has been used in order to improve piezoelectric output. ZTO has been mainly used as a filler for improving or conferring piezoelectric properties to specific polymer matrix. This allows for composite designs that allow flexibility combined with high piezoelectric output.

This chapter primarily discusses the various routes for the synthesis of ZTO nanostructures, followed by the properties and morphological structures of the produced nanostructures. Then the hybrids of ZTO have been discussed, and finally, instances of usage of ZTO as PENGs are provided.

6.2 Synthesis of Zinc Stannate

ZTO has multiple forms, of which one occurs at temperatures between 300 and 500 °C and another at over 600 °C during crystallization via solid-state synthesis pathway. The former one is the metastable $ZnSnO_3$ form having a face-centered perovskite structure, while the latter is the more stable zinc orthostannate (Zn_2SnO_4) form having cubic spinel structure (Figure 6.1). The typical synthesis methods of ZTO are as follows: thermal evaporation, chemical vapor deposition, sol–gel synthesis, coprecipitation, hydrothermal reaction, ion-exchange reaction, solid-state reaction, and electrospinning [6].

It is not easy to obtain a single pure phase of ZTO. In the majority of the cases using high temperatures, a mixture of Zn_2SnO_4, $ZnSnO_3$, and SnO_2 is

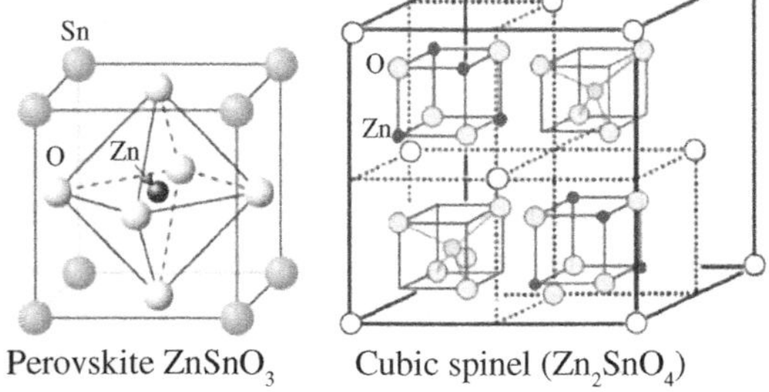

Figure 6.1 Structure of $ZnSnO_3$ and Zn_2SnO_4. Source: Copyright Taylor and Francis 2011. Reproduced with permission from Baruah and Dutta [6].

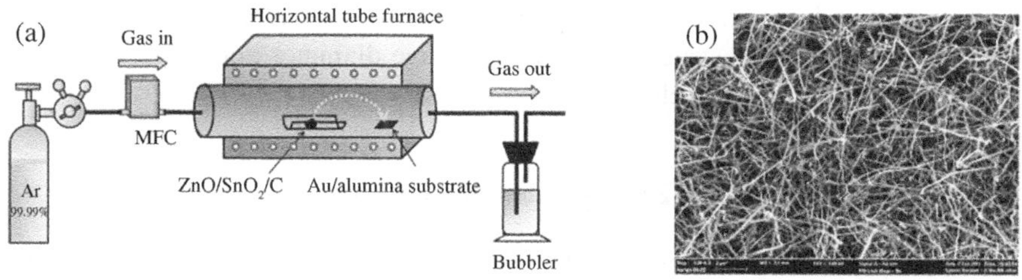

Figure 6.2 Schematic of synthesis of ZTO nanowires using thermal evaporation (a) and field-emission scanning electron micrograph (FE-SEM) image of the nanowires (b). Source: Reproduced with permission from Tharsika et al. [11], © 2015/Elsevier.

obtained [6]. The different synthesis methods of ZTO are described in subsequent sections.

6.2.1 Synthesis of ZTO by Thermal Evaporation

Thermal evaporation is a synthesis method, where the evaporative materials are heated off. Zn_2SnO_4 nanowires have been previously developed following this method with the aid of a gold catalyst. The method includes the use of ZnO, SnO_2, and activated carbon in the molar ratio of 9 : 1 : 10, milling them to get a homogeneous mixture, and then placing it in a furnace. High temperatures varying from 800 to 950 °C were used for different amounts of time. A schematic of the process is shown in Figure 6.2. A furnace temperature of 900 °C was found to be optimum for the growth of nanowires. The SEM image of the nanowires is shown in Figure 6.2 [11].

In another study, cubic spinel ZTO was synthesized via the thermal evaporation method [12]. The nanowires obtained had a length of about 10 μm. The emission band of the nanowires was centered around 637 and 710 nm, which was attributed to the presence of an excess of Zn^{2+} in the structure and point defects, respectively.

6.2.2 Synthesis of ZTO by Chemical Vapor Deposition (CVD)

CVD is a well-known technique to produce high-quality materials under vacuum. This has been used to grow a variety of materials [13, 14]. Lim et al. used CVD to grow ZTO on gold nanoparticles-coated Si-based substrates [15]. ZnO and SnO were used as starting materials, and a mixture of argon and oxygen was used as a gas. The synthesized ZTO was found to have cubic crystalline structure and had average diameter and length of 50 nm and 15 μm, respectively.

6.2.3 Synthesis of ZTO by Sol–Gel Synthesis

Ayesha et al. synthesized ZTO via sol–gel synthesis method. Zinc acetate and tin chloride were used as starting materials in approximately 2 : 1 molar ratio [16]. Thereafter, NaOH and HCl were sequentially added to obtain ZTO nanoparticles. The average particle size of the developed ZTO was found to be 13.5 nm and had a cube-like morphology. The SEM images of the produced ZTO nanoparticles are shown in Figure 6.3.

6.2.4 Synthesis of ZTO by Coprecipitation

Mekprasart et al. synthesized cubic spinel-type ZTO using coprecipitation method, using zinc chloride and tin chloride pentahydrate as starting materials, where ZTO was precipitated by adding NaOH dropwise [17]. The powder obtained was then calcinated at 900–1100 °C for obtaining the required structure. The particle size varied from 45 to 109 nm for different calcination temperatures.

6.2.5 Synthesis of ZTO by Hydrothermal Reaction

Jeronisa et al. synthesized ZTO nanoparticles via the hydrothermal route, where NaOH was used as a mineralizer, taking Zn : Sn : NaOH in the molar ratio of 2 : 1 : 8 [8]. The slurry obtained after the addition of NaOH was transferred to a reactor and was heated at 200 °C for 24 hours to obtain ZTO

Figure 6.3 FE-SEM images of ZTO nanoparticles synthesized via sol–gel method at different magnifications. Source: Reproduced with permission from Ayesha et al. [16], © 2020/Elsevier.

nanoparticles. The crystallite size was found to be 20.95 nm, and the band gap was reported to be 3.9 eV.

In another study, ZTO nanoparticles were synthesized from zinc acetate and tin chloride using KOH as the mineralizer [18]. The reaction was carried out at 180 °C for 12 hours, and irregular cubic-shaped nanoparticles were obtained with an average size between 20 and 50 nm.

6.2.6 Synthesis of ZTO by Ion-Exchange Reaction

Kovacheva et al. obtained ZTO nanostructures using ion-exchange method between solid $\alpha\text{-}Li_2SnO_3$ and $ZnCl_2 \cdot KCl$ [19]. The reaction proceeds as follows:

$$(Li_2SnO_3)_{solid} + (ZnCl_2)_{melt} = (ZnSnO_3)_{solid} + (2LiCl)_{melt} \tag{6.1}$$

The precursor Li_2SnO_3 was placed in a reaction vessel containing 6 g of $ZnCl_2 \cdot KCl$ which further melts at 350 °C. The reaction time varied between 24 and 196 hours. Thereafter, ZTO was obtained by purification, washing, and filtering. The obtained ZTO after the reaction when underwent thermal treatment above 600 °C produced a mixture of ZTO and SnO_2.

6.2.7 Synthesis of ZTO by Solid-State Reaction

The solid-state process of synthesis of ZTO involves mechanical grinding followed by high-temperature calcination. Das and Sujatha Devi produced cubic spinel ZTO using a mixture of ZnO nanorods and SnO_2 nanoparticles in 2 : 1 ratio [20]. First, the mixture was prepared by grinding and then calcined at 1000 °C for 8 hours. The nanoparticles formed exhibited a length of 300 nm and width of 50–100 nm.

Another study reported the formation of single-phase polycrystals of Zn_2SnO_4 via solid-state synthesis [21]. The process involved ball milling of ZnO and SnO_2 at first followed by sintering at 1300 °C for 2 hours.

6.2.8 Synthesis of ZTO by Electrospinning

Electrospinning is a technique for producing nanofibers [22]. Here, the polymer solution is subjected to high voltage, which results in the formation of a Taylor cone. The polymer solution is subsequently stretched and elongated to form fibers [23]. The formed fibers are then collected on a grounded plate or roller. Choi et al. produced electrospun mats of ZTO by using $Zn(CH_3COO)_2 \cdot 2H_2O$ ($Zn(OAc)_2$), $Sn(CH_3COO)_4$ ($Sn(OAc)_4$), and poly(vinylacetate) (PVAc) as precursors in a DMF solution [24]. The formed electrospun webs were subsequently pressed between preheated plates at 120 °C for 5 minutes.

6.3 Morphologies and Properties of Zinc Stannate

As discussed earlier, ZTO shows different morphologies depending on the synthesis conditions, the two main phases being the metastable $ZnSnO_3$ phase and the thermodynamically stable orthostannate phase. The transformation of ZTO from the former to the latter was studied by Bora et al. using Raman spectroscopy [25]. According to the study, at around 250 °C, $ZnSnO_3$ phase was dissolved into an amorphous phase. It started to recrystallize at 500 °C as a result of Ostwald ripening and then formed stable orthostannate crystals at around 750 °C.

6.4 Uses of Zinc Stannate and Hybrids in Piezoelectric Nanogenerators

ZTO has been widely used for developing PENGs taking account of its inherent piezoelectric properties. Lee et al. [5] developed hybrid PENG made up of perovskite ZTO nanocubes and polydimethylsiloxane (PDMS) composites without the use of any poling treatment. The nanogenerator showed an output voltage of 20 V and a current density of 1 $\mu A/cm^2$. A 40 wt% concentration of $ZnSnO_3$ has been found to be the optimum concentration value after which there is no increase in voltage value even if the concentration of $ZnSnO_3$ is increased. The piezoelectricity of $ZnSnO_3$ is mainly due to the large and spontaneous polarization generated along its *c*-axis. This is due to the greater displacement of Zn ions than that of the Sn ions. Apart from this, the excellent robustness, durability, and sturdy nature of nanogenerators, could be used in automobiles and biomedical applications for energy harvesting under vertical compression.

To further improve the performance of ZTO-based nanogenerators, hybrid materials have been used with the addition of another component. Alam et al. fabricated a PENG based on $ZnSnO_3$, PDMS, along with multi-wall carbon nanotubes (MWCNTs) as a filler [26]. The filler of MWCNTs has contributed to enhance the output voltage up to 40 V and the short-circuit current of 0.4 μA. The reasons for enhanced voltage due to the filler were mainly because of the better dispersion of $ZnSnO_3$ in the PDMS matrix. Another reason could be the parallelization of c-axis of $ZnSnO_3$ along the thickness of the hybrid nanogenerator, which made it easier to polarize when an external pressure is applied. MWCNTs being conductive in nature could reduce the internal resistance of nanogenerators giving high output characteristics. The output of the fabricated device is sufficient to power up the LEDs and can charge the capacitor too.

The presence of high additional electrons in the $ZnSnO_3$ nanoparticles makes it less efficient to generate voltage as a PENG [27]. In order to overcome this issue, the PENG based on p–n junction of copper oxide and ZTO nanorods was fabricated. The use of CuO is because of its presence of holes, which nullifies the free electrons of $ZnSnO_3$ by the formation of depletion layer. In order to enhance the output voltage of nanogenerator, it was further spin coated with PDMS–RGO (reduced graphene oxide). The role of RGO is charge trapping and it can also act as a charge transportation layer. 27 V and 0.9 μA were the voltage and current values of the fabricated device, respectively. The fabricated device could be used in charging portable electronic devices and can also harvest biomechanical energy from human motions.

1–5 wt% aluminum nanoparticles were doped in ZTO PENG by Shawon et al. (2022). The 2 wt% of aluminum nanoparticles (Al NP) showed the optimum values of electrical output voltage of 110 V and current of 13 μA [28]. If more wt% of Al NPs added, then it causes weak insulation of nanocomposite causing electrical breakdown and hence, 2 wt% concentration is required (Figure 6.4). The presence of Al NPs changes the crystal structure of ZTO by its cationic presence and reduces the size of ZTO nanocubes and gives a higher dielectric constant. When external force is applied to the nanocomposite, polarization produces the piezo potential, which creates the dipole moment by shifting cations and anions on different sides, creating potential between the electrodes. The augmented voltage makes the device to be used as an energy harvester and in sensory applications to detect human movements by mounting it in a helmet.

Using solution-casting method, a nanogenerator was developed by the use of ZTO nanoparticles (35 wt%) and plasticized PVC as a matrix, giving a voltage and current of 40 V and 1.4 μA, respectively [29]. The fabricated device can power up LEDs and can charge capacitors.

PENG was fabricated using electrospun nanofibers of PVDF/$ZnSnO_3$/MoS_2 by the hydrothermal route [30]. The enhanced output is mainly because of the contributing interface of both MoS_2 and ZTO with the PVDF matrix. The MoS_2 and ZTO both interact with the –CF and –CH groups of PVDF and hence, enhance the electroactive β phase by mobilization of the bonds (Figure 6.5). The device produced a voltage and current of 26 V and

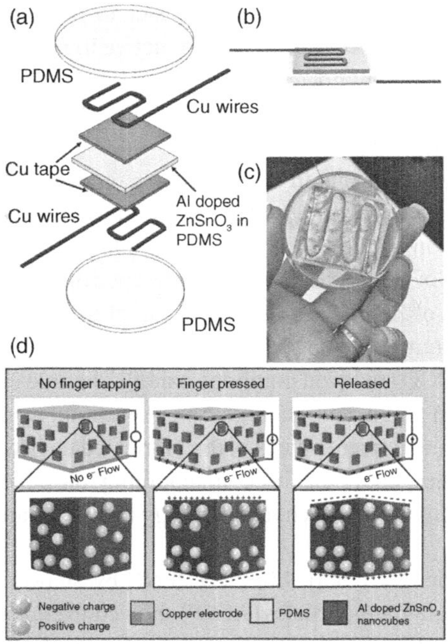

Figure 6.4 Schematic of (a) developed Al-doped $ZnSnO_3$ piezoelectric film and (b) film with electrodes, (c) demonstration of the size of the film, (d) schematic of the working mechanism of nanogenerator. Source: Reproduced with permission from Shawon et al. [28], © 2022/Elsevier.

0.5 μA, respectively. The dipoles arrange when the external force is applied, giving voltage across the material. Biomechanical energy harvesting and body movement sensing can be done with the help of the fabricated device and it is durable for 3000 cycles.

Ferroelectricity can be tuned by choosing the appropriate metal doping of $ZnSnO_3$. In one of the studies by Barman et al., $ZnSnO_3$ was doped by Ca and Ba. A lattice strain of 0.8–3% was reported for these synthesized materials. Saturated polarization (Ps) was 2.19 $\mu C/cm^2$ for Ca-doped samples and 2.66 $\mu C/cm^2$ for Ba-doped samples. Coercive electric field (E_c) was 20 kV/cm, much lower than that of E_c = kV/cm for the undoped sample. Selective doping also led to engineered band gap of 2.8 eV for these materials [31].

Veeralingam et al., studied n–n heterojunction in coaxial core–shell $ZnSnO_3@In_2O_3$ nanofibers. The d_{33} value was reported to be 254 pm/V by piezoresponse force microscopy (PFM). $ZnSnO_3$ belonged to rhombohedral crystal class with the R_3C space group and In_2O_3 belonged to cubic crystal class with $I2_13$ space group. Both of these are non-centrosymmetric in nature. The non-centrosymmetric nature of $ZnSnO_3$ and In_2O_3 led to a local negative potential at their interface resulting in the upward alignment

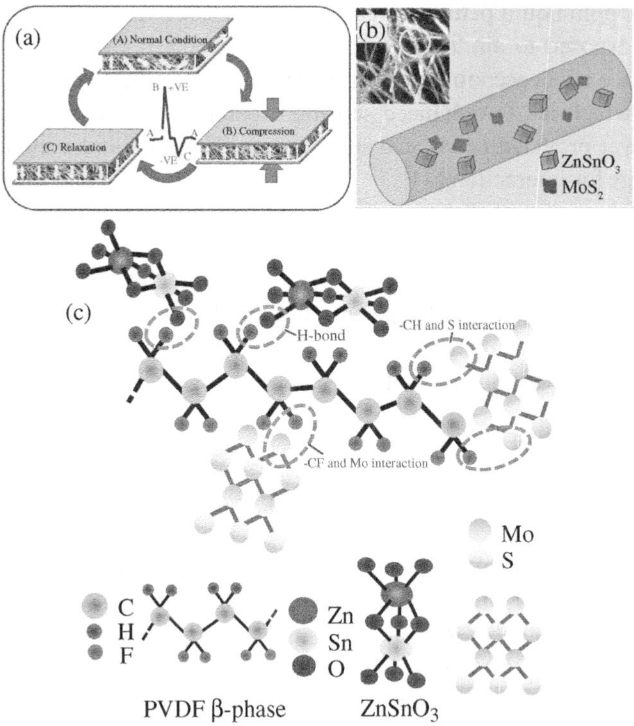

Figure 6.5 Schematic representation of the mechanism voltage generation in ZTO/MoS_2-based nanogenerator, (b) structural representation of the nanofibers developed, (c) molecular interaction of β-PVDF, ZSO, and MoS_2. Source: Reproduced with permission from Muduli et al. [30], © 2021/American Chemical Society.

of the conduction band and valence band of In_2O_3, and increasing the Schottky barrier height at the interface of both the materials [32].

The improvement of dielectric permittivity in ceramic fillers is of paramount importance for improved piezoelectric performance of the nanogenerator. Sasmal et al. have reported an improved dielectric permittivity of Ag-doped $ZnSnO_3$. The space charge polarization has improved to a great extent due to the Ag doping of $ZnSnO_3$. The nanocomposite films showed low dielectric loss of <0.05 at 10 kHz. The output peak-to-peak open-circuit voltage increased to 20 V for Ag-doped $ZnSnO_3$ as compared to 11 V for non-doped $ZnSnO_3$-loaded PVDF. The rectified DC voltage was able to charge a 10 μF capacitor in 270 s [33].

ZnO is a promising piezoelectric material, and when used along with $ZnSnO_3$, it can perform a surface coupling effect. The output voltage when exposed to 8000 ppm liquid petroleum gas (LPG) was 0.089 V, smaller than 0.533 V, when exposed to air. The sensitivity against 8000 ppm LPG was 83.23. The surface effect accounted for the morphology-tuned properties of $ZnO/ZnSnO_3$, which led to enhanced sensing capabilities. Moreover, the nanoparticle layer provided Lewis acid sites for gas adsorption and a small activation energy for interaction [34].

The surface decoration of ceramics can lead to enhanced dielectric permittivity and reduced dielectric loss. Sasmal et al. has studied the enhanced dielectric permittivity of ZnO-decorated $ZnSnO_3$.The rectified output AC voltage was used to charge a 10 μF capacitor and used for powering LEDs. The nanogenerator showed open-circuit voltage of 23 V and generated 40 V when combined with triboelectric material [35].

6.5 Conclusion

In summary, in this chapter, different synthesis methods for $ZnSnO_3$ are discussed. It is generally difficult to control morphology in solid-state synthesis; however, it usually results in less unreacted products as impurity phases. High-temperature methods are predominant in solid-state synthesis of $ZnSnO_3$. In hydrothermal synthesis method, it is difficult to obtain a single-phase $ZnSnO_3$, as it is composed of a mixture of SnO_2, Zn_2SnO_4, ZnO, and $ZnSn(OH)_6$, depending on reactant molar concentrations and hydrothermal reaction conditions. It is imperative to tune the crystal structure by introducing lattice strain in the crystal structure. Doping by different metal ions can be promising toward this aspect. Doping also results in band gap engineering and changed crystal structure. The excess ions present in $ZnSnO_3$ can be suppressed by incorporation of n–p heterojunction, thereby resulting in reduced dielectric loss. Different approaches of n–n heterojunctions of core–shell morphologies can be an interesting area to work upon. Different metal ion decorated $ZnSnO_3$ can increase the dielectric permittivity and augment space charge polarization, thereby resulting in enhanced piezoelectric activity.

Acknowledgments

The authors AC and WA would like to acknowledge the generous funding by DST under Device Development Programme (DDP) (Project no: DST/TDT/DDP-05/2018 [G]).

References

1 Pandey, R., Sb, G., Grover, S. et al. (2019). Microscopic origin of piezoelectricity in lead-free halide perovskite: application in nanogenerator design. *ACS Energy Lett.* 4: 1004–1011. https://doi.org/10.1021/ACSENERGYLETT.9B00323/SUPPL_FILE/NZ9B00323_SI_001.AVI.

2 Wang, X., Zhang, T., Lou, Y., and Zhao, Y. (2019). All-inorganic lead-free perovskites for optoelectronic applications. *Mater. Chem. Front.* 3: 365–375. https://doi.org/10.1039/C8QM00611C.

3 Ali, R., Hou, G.J., Zhu, Z.G. et al. (2018). Predicted lead-free perovskites for solar cells. *Chem. Mater.* 30: 718–728. https://doi.org/10.1021/ACS.CHEMMATER.7B04036/ASSET/IMAGES/LARGE/CM-2017-04036G_0007.JPEG.

4 Zhang, Y., Kim, H., Wang, Q. et al. (2020). Progress in lead-free piezoelectric nanofiller materials and related composite nanogenerator devices. *Nanoscale Adv.* 2: 3131–3149. https://doi.org/10.1039/C9NA00809H.

5 Lee, K.Y., Kim, D., Lee, J.H. et al. (2014). Unidirectional high-power generation via stress-induced dipole alignment from $ZnSnO_3$ nanocubes/polymer hybrid piezoelectric nanogenerator. *Adv. Funct. Mater.* 24: 37–43. https://doi.org/10.1002/ADFM.201301379.

6 Baruah, S. and Dutta, J. (2011). *Zinc Stannate Nanostructures: Hydrothermal Synthesis.* http://www.tandfonline.com/action/journalInformation?show=aimsScope&journalCode=tsta20#.VmBmuzZFCUk 12. https://doi.org/10.1088/1468-6996/12/1/013004

7 Sun, S. and Liang, S. (2017). Morphological zinc stannate: synthesis, fundamental properties and applications. *J. Mater. Chem. A* 5: 20534–20560. https://doi.org/10.1039/C7TA06221D.

8 Jeronsia, J.E., Joseph, L.A., Jaculine, M.M. et al. (2016). Hydrothermal synthesis of zinc stannate nanoparticles for antibacterial applications. *J. Taibah Univ. Sci.* 10: 601–606. https://doi.org/10.1016/J.JTUSCI.2015.12.003.

9 Jain, S., Shah, A.P., and Shimpi, N.G. (2020). An efficient photocatalytic degradation of organic dyes under visible light using zinc stannate (Zn_2SnO_4) nanorods prepared by microwave irradiation. *Nano-Struct. Nano-Objects* 21: 100410. https://doi.org/10.1016/j.nanoso.2019.100410.

10 Dou, J. and Chen, Q. (2021). Zinc stannate nanostructures for energy conversion†. *Chin. J. Chem.* 39: 367–380. https://doi.org/10.1002/CJOC.202000369.

11 Tharsika, T., Haseeb, A.S.M.A., Akbar, S.A. et al. (2015). Gas sensing properties of zinc stannate (Zn_2SnO_4) nanowires prepared by carbon

assisted thermal evaporation process. *J. Alloys Compd.* 618: 455–462. https://doi.org/10.1016/J.JALLCOM.2014.08.192.

12 Du, Y. and Ding, P. (2010). Thermal evaporation route to zinc stannate nanowires and the cathodoluminescence of the individual nanowires. *J. Alloys Compd.* 502: L14–L16. https://doi.org/10.1016/J.JALLCOM.2010.04.121.

13 Xia, L. (2021). Importance of nanostructured surfaces. *Bioceramics* 5–24. https://doi.org/10.1016/B978-0-08-102999-2.00002-8.

14 Choy, K.L. (2003). Chemical vapour deposition of coatings. *Prog. Mater. Sci.* 48: 57–170. https://doi.org/10.1016/S0079-6425(01)00009-3.

15 Lim, T., Kim, H., Meyyappan, M., and Ju, S. (2012). Photostable Zn_2SnO_4 nanowire transistors for transparent displays. *ACS Nano* 6: 4912–4920. https://doi.org/10.1021/NN300401W/ASSET/IMAGES/LARGE/NN-2012-00401W_0007.JPEG.

16 Ayesha, B., Jabeen, U., Naeem, A. et al. (2020). Synthesis of zinc stannate nanoparticles by sol-gel method for photocatalysis of commercial dyes. *Results Chem.* 2: 100023. https://doi.org/10.1016/J.RECHEM.2020.100023.

17 Mekprasart, W., Tangcharoen, T., Nakhanivej, P., and Pecharapa, W. (2016). Synthesis and characterization of cubic-like zinc stannate powders prepared by co-precipitation method. *Chiang Mai J. Sci.* 43: 345–350.

18 Dinesh, S., Barathan, S., Premkumar, V.K. et al. (2016). Hydrothermal synthesis of zinc stannate (Zn_2SnO_4) nanoparticles and its application towards photocatalytic and antibacterial activity. *J. Mater. Sci. Mater. Electron.* 27: 9668–9675. https://doi.org/10.1007/S10854-016-5027-Y/FIGURES/12.

19 Kovacheva, D. and Petrov, K. (1998). Preparation of crystalline $ZnSnO_3$ from Li_2SnO_3 by low-temperature ion exchange. *Solid State Ionics* 109: 327–332. https://doi.org/10.1016/S0167-2738(97)00507-9.

20 Das, P.P. and Sujatha Devi, P. (2014). Formation of self-assembled defect-free Zn2SnO4 nanostructures from binary oxides without the Kirkendall effect. *Inorg. Chem.* 53: 10797–10799. https://doi.org/10.1021/ic501598h.

21 Nikolić, M.V., Ivetić, T., Paraskevopoulos, K.M. et al. (2007). Far infrared reflection spectroscopy of Zn_2SnO_4 ceramics obtained by sintering mechanically activated ZnO–SnO_2 powder mixtures. *J. Eur. Ceram. Soc.* 27: 3727–3730. https://doi.org/10.1016/J.JEURCERAMSOC.2007.02.040.

22 Jose Varghese, R., Sakho, E.H.M., Parani, S. et al. (2019). Introduction to nanomaterials: synthesis and applications. In: *Nanomaterials for Solar*

Cell Applications, 75–95. Elsevier https://doi.org/10.1016/B978-0-12-813337-8.00003-5.

23 Davoodi, P., Gill, E.L., Wang, W., and Shery Huang, Y.Y. (2021). Advances and innovations in electrospinning technology. In: *Biomedical Applications of Electrospinning and Electrospraying*, 45–81. Elsevier https://doi.org/10.1016/B978-0-12-822476-2.00004-2.

24 Choi, S.H., Hwang, D., Kim, D.Y. et al. (2013). Amorphous zinc stannate (Zn_2SnO_4) nanofibers networks as photoelectrodes for organic dye-sensitized solar cells. *Adv. Funct. Mater.* 23: 3146–3155. https://doi.org/10.1002/adfm.201203278.

25 Bora, T., Al-Hinai, M.H., Al-Hinai, A.T., and Dutta, J. (2015). Phase transformation of metastable $ZnSnO_3$ upon thermal decomposition by in-situ temperature-dependent raman spectroscopy. *J. Am. Ceram. Soc.* 98: 4044–4049. https://doi.org/10.1111/jace.13791.

26 Alam, M.M., Ghosh, S.K., Sultana, A., and Mandal, D. (2015). Lead-free ZnSnO3/MWCNTs-based self-poled flexible hybrid nanogenerator for piezoelectric power generation. *Nanotechnology* 26: 165403. https://doi.org/10.1088/0957-4484/26/16/165403.

27 Paria, S., Ojha, S., Karan, S.K. et al. (2020). Approach for enhancement in output performance of randomly oriented $ZnSnO_3$ nanorod-based piezoelectric nanogenerator via p-n heterojunction and surface passivation layer. *ACS Appl. Electron. Mater.* 2: 2565–2578. https://doi.org/10.1021/ACSAELM.0C00467/SUPPL_FILE/EL0C00467_SI_002.AVI.

28 Shawon, S.M.A.Z., Carballo, Z.D., Vega, V.S. et al. (2022). Surface modified hybrid $ZnSnO_3$ nanocubes for enhanced piezoelectric power generation and wireless sensory application. *Nano Energy* 92: 106653. https://doi.org/10.1016/J.NANOEN.2021.106653.

29 Paria, S., Karan, S.K., Bera, R. et al. (2016). A facile approach to develop a highly stretchable PVC/$ZnSnO_3$ piezoelectric nanogenerator with high output power generation for powering portable electronic devices. *Ind. Eng. Chem. Res.* 55: 10671–10680. https://doi.org/10.1021/ACS.IECR.6B02172/ASSET/IMAGES/LARGE/IE-2016-02172N_0009.JPEG.

30 Muduli, S.P., Veeralingam, S., and Badhulika, S. (2021). Interface induced high-performance piezoelectric nanogenerator based on a electrospun three-phase composite nanofiber for wearable applications. *ACS Appl. Energy Mater.* 4: 12593–12603. https://doi.org/10.1021/ACSAEM.1C02371/ASSET/IMAGES/LARGE/AE1C02371_0006.JPEG.

31 Barman, S. and Datta, A. (2022). Coexisting ferroelectricity and photoconductivity in doped LN-type ZnSnO3 nanospikes. *Mater. Chem. Phys.* 290: 126507. https://doi.org/10.1016/J.MATCHEMPHYS.2022.126507.

32 Veeralingam, S. and Badhulika, S. (2022). Enhanced carrier separation assisted high-performance piezo-phototronic self-powered photodetector based on core-shell $ZnSnO_3@In_2O_3$ heterojunction. *Nano Energy* 98: 107354. https://doi.org/10.1016/J.NANOEN.2022.107354.

33 Sasmal, A., Patra, A., Devi, P.S., and Sen, S. (2021). Space charge induced augmented dielectric permittivity and improved energy harvesting ability of nano-Ag decorated $ZnSnO_3$ filled PVDF based flexible nanogenerator. *Compos. Sci. Technol.* 213: 108916. https://doi.org/10.1016/J.COMPSCITECH.2021.108916.

34 Fu, Y., Nie, Y., Zhao, Y. et al. (2015). Detecting liquefied petroleum gas (LPG) at room temperature using $ZnSnO_3$/ZnO nanowire piezo-nanogenerator as self-powered gas sensor. *ACS Appl. Mater. Interfaces* 7: 10482–10490. https://doi.org/10.1021/ACSAMI.5B01822/ASSET/IMAGES/MEDIUM/AM-2015-01822X_0012.GIF.

35 Sasmal, A., Medda, S.K., Devi, P.S., and Sen, S. (2020). Nano-ZnO decorated $ZnSnO_3$ as efficient fillers in PVDF matrixes: toward simultaneous enhancement of energy storage density and efficiency and improved energy harvesting activity. *Nanoscale* 12: 20908–20921. https://doi.org/10.1039/D0NR02057E.

7

$ZnFe_2O_4$-Based Hybrid Piezoelectric Materials

Ipsita Chinya[1,2,3] *and Shrabanee Sen*[3]

[1] *Department of Chemical Engineering, Dr. V R Godhania College of Engineering and Technology, Porbandar, Gujarat, India*
[2] *Academy of Scientific and Innovative Research, CSIR-CGCRI, Kolkata, West Bengal, India*
[3] *Functional Materials and Devices Division, CSIR-Central Glass & Ceramic Research Institute, Kolkata, West Bengal, India*

7.1 Introduction

The most indispensable resource is the energy that influences the quality of modern human day-to-day lifestyle. Considering the intimidation of global warming and the enormous consumption of traditional fossil fuels, the search for alternative green and renewable energy sources with low carbon emissions is essential to carry on sustainable advances in human civilization [1–3]. The recent expansion on the Internet of Things (IoT), micro-electromechanical systems (MEMS), and wireless sensor networks for health monitoring, military and security surveillance, infrastructure intensive care [1], and so on, requires a large amount of small-scale energy in portable and wearable form. However, traditional power sources such as external batteries have apparent boundaries in terms of a short lifetime, bulkiness, periodical maintenance, and e-pollution problems along with environmental menace [1–4]. In this regard, the development of ubiquitously available nanogenerator (NG) self-powered autonomous electronics devices has been highly attractive in the past decades.

The electroactive polymer Polyvinylidene fluoride (PVDF) and its copolymers (Poly vinylidene fluoride-co-hexafluoro propylene (PVDF–HFP), Polyvinylidene fluoride- co- trifluoroethylene (PVDF–TrFE))-based two electrodes-sandwiched structure NG are the best choice as an active material. The significantly higher piezoelectric strain constant ($|d_{31}/d_{33}| > 1$) of the polymer with their good mechanical strength, excellent film-forming

Hybrid Materials for Piezoelectric Energy Harvesting and Conversion, First Edition.
Edited by S. Wazed Ali, Satyaranjan Bairagi, and Shahid Ul Islam.
© 2024 John Wiley & Sons, Inc. Published 2024 by John Wiley & Sons, Inc.

ability, cost-effectiveness, chemical inertness, and weather-resilient characteristics are much improved compared to that of others [5–7], making them the most suitable materials for NG applications. To enrich polar phase content in PVDF and its copolymer, commonly, an external electrical poling treatment followed by subsequent mechanical stretching with a high stretching ratio is applied. Otherwise, applying high temperature or pressure, annealing, and so on are also very frequently used. However, all these processes cause surface inhomogeneity, leading to degradation of the mechanical property and early breakdown, resulting in a decrease in energy storage characteristics of the film [8]. These all are unfavorable for final device fabrication. Additionally, these processes are very tedious, non-scalable, pricey, and not at all an industrial-friendly method [5, 6] for large-scale production. Thus, enriching polar phase content in PVDF through the incorporation of an external nucleating agent such as nanoceramics to implement self-polarization behavior [9] in the matrix is the most commonly tried by researchers. Different well-known ferroelectric and non-ferroelectric ceramics such as doped and undoped Ba, Ca and Pb titanate, $LiTaO_3$, ZnS, InN, and GaN [10] with different shapes and sizes have been investigated as nucleating agents. However, the extreme dielectric field contrast between the well-known doped or undoped ceramics with the polymer matrix leads to deterioration of the electrical properties of the composite from theoretically predicted values [11–13]. On the other hand, non-oxide nitride materials are very expensive, and their synthesis in phase-pure state requires sophisticated techniques. Although other non-ferroelectric single metal oxides such as TiO_2, ZnO, WO_3, V_2O_5, and SiO_2 [13–17] are very promising candidates as nucleating agents, during their synthesis, some crystal defects arise which may alter the piezoelectric properties of the composite from sample to sample. Henceforth, iron oxide and various ferrite materials with moderate permittivity have been immensely investigated in recent times [18, 19]. Their easy synthesis procedure with different shapes and sizes, and easy variation of tetrahedral and octahedral sites with different metal concentrations for synthesis of tailor-made multifunctional material may add an extra advantage for use of nanocomposites for different applications. Moreover, the piezoelectricity of the composite film can be simply altered by proper selection of the morphology of the ceramic fillers or adjusting their concentration. Another important feature is the surface compatibility of the nanofiller with the polymer matrix. Appropriate selection of surface functionalizer of ceramic fillers may help to enhance this incompatibility. The modifier layer, attached through chemical bonding or interacting by physical forces with interfacial layer, leading to encourages growth of β-phase with all trans-PVDF on the surface of the nanoparticle [20–22].

0–3 piezo-composites by dispersing zero-dimensional ceramic particles in a continuous PVDF matrix (three-dimensional) have been rigorously studied by previous researchers. Several advantages in terms of high-volume production with low cost, customizability in shape of the final product (such as film, belt, and textile) intacting adequate flexibility [23–25] are the primary intentions for their selection. However, at lower filler loading causing inferior dielectric property and poor remnant polarization are the main limitations for their real field applications. Hence, to cope with this limitation, high volume fractions (>50 vol%) of fillers loading with high flexibility and good mechanical performance of the composites [26–28] are desired. The value of the percolation threshold for the composite with high aspect ratio (AR) fillers is less compared to the low AR fillers due to easier formation of interconnecting pathways in the system [29, 30]. Additionally, as high AR fillers with low Brunauer–Emmer–Teller (BET) surface area show low surface energy with lower agglomeration and improve the overall performance of the composites [30]. On other hand, for cubic fillers, a spontaneous polarization along the *Z*-axis exists, which is a source of piezoelectricity. This axis tries to align in parallel to the thickness of the composite. Upon imparting pressure to the composite, the degree of alignment changes, and thus gives a higher piezo response value [31, 32].

Lead-free, spinel -Zinc Ferrite (ZF) has exhibited a widespread range of applications in microelectronics [33], pseudocapacitors [34], anode materials for lithium-ion batteries [35], semiconductor photocatalysis [36], high-density data storage [37], ferrofluid technology [38], magnetic resonance imaging [39], and so on. However, combining ZF with PVDF in the field of NG applications was started to explore from 2016 onwards. A high remnant polarization value ($1\ \mu C/cm^2$) of ZF [39, 40] may help to increase the synergistic remnant polarization and energy storage behavior of the composite.

In this chapter, a summarization of the detailed fabrication strategy of ZF/PVDF composite as flexible, lightweight, and self-powered bio-safe NG was presented. In this regard, all the associated parameters affecting polar phase formation in PVDF were investigated followed by an evaluation of respective dielectric, ferroelectric, and piezoelectric properties.

7.2 Current Scenario, Challenges in This Field and Scope of the Chapter

For the effective utilization of NG in wearable electronics, the proposed generator should have enough flexiblilty, be bio-safe, and free from any hazardous materials. β-PVDF is the most responsive piezoelectric polymer used as self-powered flexible NG application. Its strong piezoelectric response, chemical and mechanical durability, mark it as a valuable material for this application. Diverse stratagems have been established to obtain electroactive β phase of PVDF, primarily concentrating on advance of specific process technology or the inclusion of specific fillers. PVDF-metal oxide-based piezoelectric composites are very promising materials in this field. However, shape, size, and concentration of metal oxide are three major factors for the nucleation and stabilization of electroactive β-PVDF. On the other hand, agglomeration of ceramic fillers and surface incompatibility between ceramic filler and polymer matrix are also very crucial for fabrication of homogeneous composite film. Understanding all of these parameters and optimizing them to obtain a bio-safe, flexible, and wearable polymer composite-based NG for self-powered sensor applications are still a very inspiring and challenging field of research. Thus, there is ample scope and opportunity to develop a new, economically viable, and flexible material for piezoelectric NG to empower milli to micro-watt range electronic devices.

7.3 Fabrication Strategy of the Nanocomposites

ZF with different architectures was fabricated through various wet chemical methods, such as sol–gel auto combustion, hydrothermal, and solvothermal using different precursor materials and chelating agents. Here, all the precursors and chelating agents used for the synthesis of different nanostructure ZF with different wet chemical techniques were summarized (Table 7.1).

The nanocomposite can be prepared by different techniques, such as injection molding [41–43], tape casting [44, 45], co-extrusion [46, 47], hot pressing [48, 49], solvent casting [50, 51], and spin-coating method [52]. Here, all discussed composites are fabricated by a simple solvent-casting technique. The two phases of mixing are done using polar solvents, such as N, N-Dimethylacetamide (DMAc) and N-Dimethylformamide (DMF). The complete protocol was described through a diagram (Figure 7.1).

Different characterization techniques such as structural, morphological, thermal, dielectric, ferroelectric, and piezoelectric have been employed for both ZF (to investigate shape, size, phase purity, crystallinity, and electrical properties) and the polymer and their composites (for phase identification, morphology, and electrical properties).

Table 7.1 Different synthesis techniques with their precursors and chelating agents.

Method	Precursors	Chelating agent	Shape/code used
Sol–gel auto combustion	Zinc -Nitrate, -Ferric -Nitrate	PEG (400)	Uniform spherical particle shape/ZF-P
	Zinc -Nitrate, -Ferric Nitrate	Citric acid	Spherical particle shape/ZF(C)
Solvothermal	Zinc -Nitrate, -Ferric Nitrate	Oleic acid	Cube/ZF-c
Electrospinning	Zinc -Nitrate, -Ferric -Nitrate	Polyvinyl pyrrolidone/ Ethanol	Rod-shaped
Hydrothermal	Zinc Chloride, Ferric -Nitrate	Hydrazine	Rod-shaped/(ZF-r) (Lower aspect ratio)
	Zinc Chloride, Ferric -Nitrate	Potassium-Hydroxide	Rod-shaped/(ZF-R) (Higher aspect ratio)
	Zinc -Nitrate, Ferric -Nitrate	Sodium Acetate	Uniform spherical particle shape/ZF-s

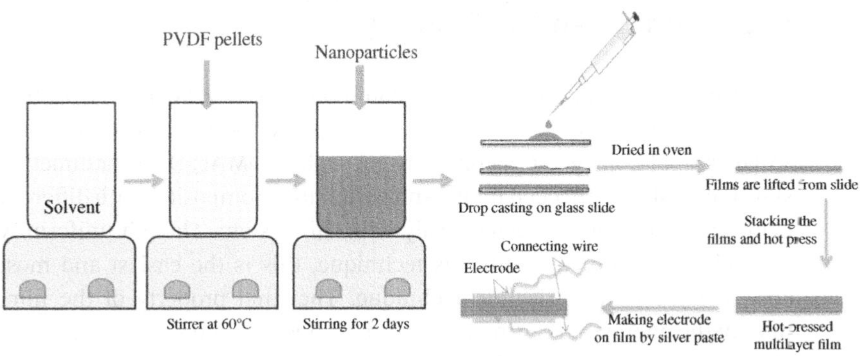

Figure 7.1 Schematic representation of fabrication protocol of the nanocomposite.

All the characterization tools and techniques performed here are enlisted here:

High-temperature furnace (Okay Electrical Furnace, Bysakh and Co), oven (Spac and Service), autoclave, Magnetic Stirrer (Remi, 5MLH Plus), Centrifuged machine (Remi PR-24), Lamination Hot-pressing Instrument (Carver Washbash Indiana), Electrical Poling unit (Neo-Teletronix Pvt. Lmt), Zeta Potential Study (Horiba, SZ-100 analyzer), X-ray diffractometer

(XRD, X'pert Pro MPD XRD system, PAN analytical), Fourier-transform infrared (FTIR, FTIR-Spectrum 2, Perkin Elmer) spectroscopy, FTIR with ATR mode (FTIR spectrophotometer, Shimadzu, 8400s), Ultraviolet–visible (UV) spectroscopy (UV-Vis Spectrophotometer, UV-2450, Shimadzu), X-ray photoelectron spectroscopy (XPS, ULVAC-PHI, Inc.), Field Emission Scanning Electron Microscope (FESEM, SEM Supra 35VP, Carl Zeiss) with EDAX facility, high-resolution transmission electron microscope (HRTEM, Tecnai G2 30ST, FEI), Atomic Force Microscopy (AFM, Park Systems NX10), Piezo-response force microscopy (PFM), Ferroelectric measurement system (Ceramic Multilayer Actuator Test Bench, Aixact System GmbH, TF Analyzer 2000), TGA/DSC (Shimadzu thermal analyzer TA-501), LCR Meter (6500 B Wayne Kerr), Oscilloscope (Agilent DSO2014A), Corona poling unit (50 kV), Piezoelectric coefficient (d33) meter (PiezoMeter PM300), Sputter coater (Leica EM ACE200), Source meter (Keysight B2911A), Multimeter (Fluke True RMS 115), Picoammeter, External force, pressure application system with different frequency.

7.4 The Controlling Factors of β-phase Formation in Composites and its Property

Selection of the solvent: The most common and versatile film fabrication with adjustable thickness and area is the solvent-casting technique. As PVDF is a polar molecule, a polar solvent, such as DMAc, DMF, or dimethyl siloxane (DMSO) is needed for two-phase mixing. Composites with different connectivity can be fabricated easily with this system. Though uniformity of thickness is difficult with this technique, this is the easiest and most cost-effective film fabrication technique. The final property of the film, mainly the energy-harvesting property, is strongly dependent on the extent of polarity present in the solvent.

The concentration of the filler: The most important controlling factor is the concentration of fillers in the matrix. -For a lower percolation threshold, even a small fraction of filler loading results in an alteration in dielectric property, attributable to its transition from an insulator to a conductor state [53]. However, comparatively high percolation threshold systems are not too sensitive compared to their lower counterpart. The piezoelectric coefficient of the composite is proportional to the remnant polarization of the system, which is further linearly related to the effective dielectric constant of the system. To increase the effective dielectric constant, a higher amount of filler loading is required, which increases the

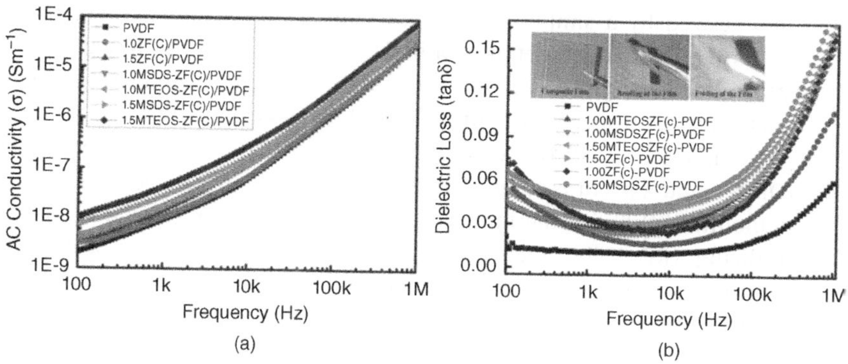

Figure 7.2 Variation in AC conductivity and dielectric loss with frequency of PVDF, ZF(C)–PVDF, and surface-modified ZF(C)–PVDF composites with different fractions of nanoparticle loading.

conductive pathway in composite increasing dielectric loss (Figure 7.2) and premature breakdown of nanocomposite.

However, a large number of porosities are observed in the microstructure of large-volume fraction filler-loaded nanocomposites. This porosity is beneficial for piezoelectric polarization [54], as the micropore can trap a large amount of charge at dielectric boundaries called space charge. So optimization of filler-to-polymer ratio is the crucial factor for attaining maximum dielectric, ferroelectric, and piezoelectric properties of composite for NG fabrication. So, the reports [21, 22, 32, 41, 55, 56] are mainly focused to discuss this particular effect. However, the optimum threshold concentration is also a function of shape, size, and interfacial interactions. So this effect will be discussed along with the other controlling factors.

The radius of the nanofiller: The radius of the filler plays a very significant role in β-phase nucleation in PVDF. Mackay et al. [57] reported that dispersed nanoparticles get swelled up within a polymer matrix if the radius of nanoparticles was shorter than that of the radius of gyration (R_g) of the polymer [58]. The calculated value of R_g - of PVDF is 27.5 nm, based on a freely joint chain model [59]. With the addition of fillers into PVDF, it swells, thereby increasing its R_g. This increment in R_g of PVDF is manifested by altering it to a more extended β structure [59]. Thus, below the critical nanoparticle size, the nucleation efficiency of the filler is reduced. On the other hand, if the nanoparticle radius is much higher than R_g, then mixed-phase conformation is achieved. An attempt was made to establish this fact very systematically by Chinya and Sen [60]. They observed a drastic change in the electrical properties of the composite below or above the critical diameter, rather than a threshold diameter of the filler. ZF was synthesized by a simple sol–gel auto-combustion technique using citric acid [61]. The diameter of the synthesized ZF was fine-tuned by calcination at different temperatures, and the variation in size was investigated by analyzing the respective bright-field TEM images (Figure 7.3).

Composites of PVDF with different-sized ZF(C) were prepared by DMF-based solvent-casting method. The correlation between the size of nanofiller (ZF) with the induced polar PVDF phase fraction was studied utilizing XRD, FTIR, and FESEM analyses. Further, the effect of this size restriction on the dielectric and energy storage behavior of the ZF/PVDF composite was studied. The obtained results were further explicated by analyzing a proposed theoretical model (Figure 7.4). The ion–dipole interaction between ZF (with different sizes) and participated dipoles of PVDF was considered here. Calcined ZF was considered with three different diameters; for example, 20, 30, and 50 nm. The nucleated polar phase in PVDF was established by the number of polymer dipoles that participated in direct interaction with the negative charge of the ZF surface. The average monomer unit ($\overline{X_n}$) in the PVDF chain was calculated from the following equation (Eq. (7.1)) [62]:

$$\overline{M_n} = M_o \times \overline{X_n} \tag{7.1}$$

where $\overline{M_n}$ and M_o represent number average molecular weight of PVDF and the molecular weight of one monomer unit, respectively. Using the abovementioned equation, the calculated value of the PVDF polymer chain is composed of ~1671 monomer units. Positively charged polymer dipoles around H atoms are attached with ZF, leading to the induction of dipolar orientation of the monomer units [21, 22, 63]. Now, considering the size of individual PVDF monomers around 1.28 Å [63, 64], 30 nm in diameter ZF, 44% of the total PVDF monomers can be in contact with the surface of the nanoparticle by direct participation through strong electrostatic interaction. However, when the diameter of ZF increases to 50 nm, all monomers can accommodate the surface, and the remaining space can accommodate another polymer chain, resulting in a repulsion with the existing PVDF chain and leading to the formation of mixed conformations. In the case of 20 nm size, 29% of the total monomer is only possible to adjust on the surface of ZF, i.e. the efficiency is again reduced. The probable mechanism is schematically presented in Figure 7.4. This exhibited that the nucleation efficiency of ZF is abridged when the diameter of ZF is above or below the critical diameter, i.e. 30 nm.

The geometry of the filler: The geometry of fillers is also a critical aspect for determining the ultimate properties such as energy density and the piezo response of polymer nanocomposites. Interfacial interactions

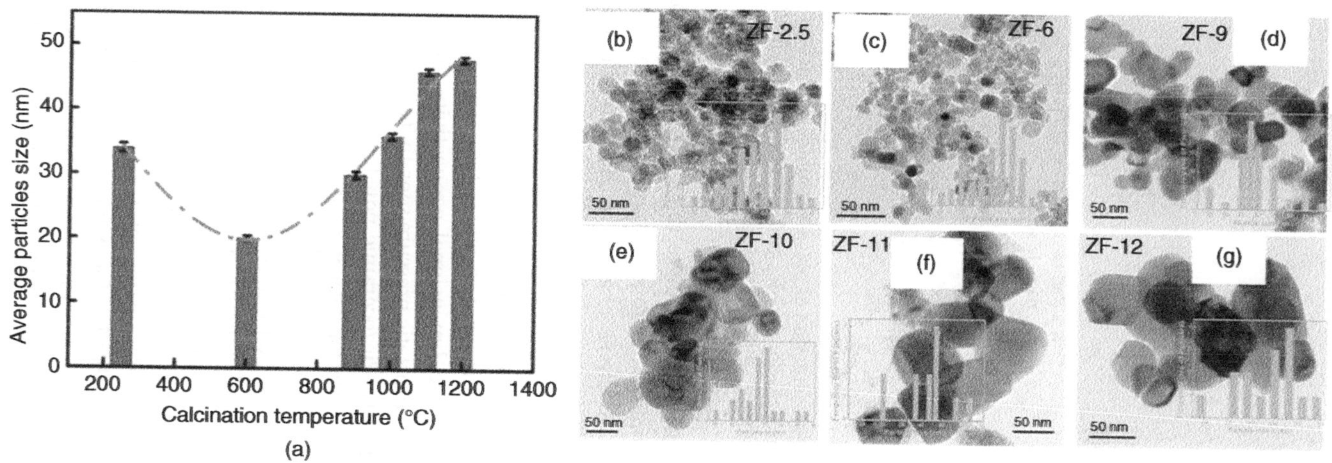

Figure 7.3 (a) Variation of their calculated average size with calcination temperature, (b–g) corresponding TEM images.

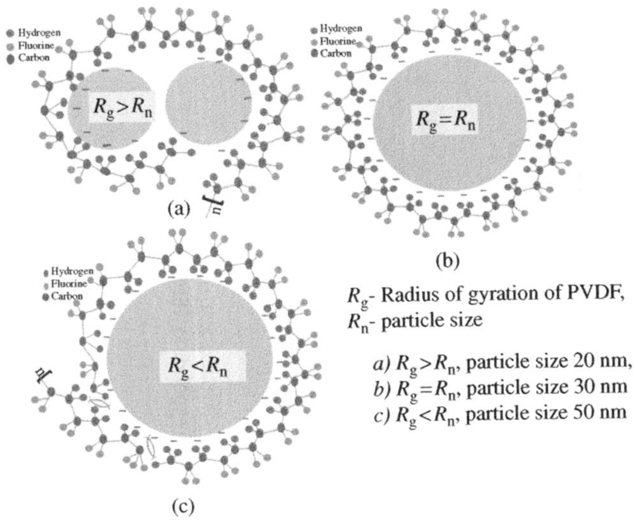

Figure 7.4 Schematic representation of polymer dipoles interacting with ZF nanoparticles of different sizes: (a) $R_g > R_n$, particle size ~20 nm, (b) $R_g = R_n$, particle size ~30 nm, and (c) $R_g < R_n$, particle size ~50 nm.

vary with the shape of the filler due to their different surface areas and thus, create different interaction zones with polymer dipoles. Accordingly, the orientation of PVDF dipole changes, leading to different conformations. The connectivity in nanocomposite depends on the shape of fillers [65]. Among the different connectivity in piezoelectric ceramic–polymer composites, 3-3, 1-3, and 0-3 are most extensively studied because of their easy fabrication process compared to others [66]. The ceramic filler and polymer matrix with three-dimensional connectivity are known as 3-3 composites with high hydrostatic pressure sensitivity, low permittivity, low density, and more rigidity. Piezoceramic powders dispersed in a polymer matrix with 0-3 connectivity have several advantages such as high-volume production, low cost, and product customizability (such as film, belt, and textile) with satisfactory flexibility [66–68]. This type of composite can be produced on large scale. 1-3 composites have one-dimensionally connected ceramic phase in polymer matrix with superior piezoelectric properties. Rod-like nanofillers having higher AR can orient in matrix as "*in-plane*" direction during the solution-casting process. When this composite is subjected to an applied mechanical force, nanorods try to orient themselves more in "in-plane" direction and increase anisotropy in the susceptibility of the composite [68, 69]. In consistence with literature survey and prior discussion, it can be concluded that ion–dipole interaction between the negatively charged ZF surface with —CH_2 dipole of PVDF is the most usual and commonly acknowledged mechanism for direct induction of polar phase in the PVDF matrix. Here, the geometry of the filler plays a foremost role in this mechanism [21, 22, 70–72]. The electric field vector around the surface of ZF varies with their geometry [73]. Accordingly, the respective interaction behavior is changed. Computational and investigational studies exhibited [74–76] that fillers with tuned morphology, such as nanowires and nanofibers, are more efficient for catalyzing the polar phase transition compared to spherical or cubic geometries. Furthermore, fillers of different shapes have different percolation thresholds, which play an important role in the determination of electrical properties of ferrite–polymer composites. Chinya et al. [32] investigated the effect of the geometry of ZF nanostructure on the polar phase nucleation and electrical performance of PVDF-based nanocomposites. Three different nanostructures of ZF, i.e. spherical (ZF-s), cubic (ZF-c), and rod (ZF-r), were synthesized by hydrothermal and solvothermal techniques. Details of precursors and chelating aids used are described in Table 7.1. Their PVDF-based nanocomposites were fabricated using DMAc as a solvent, with varying filler-to-polymer ratios to determine the electrical percolation threshold for each system. The electroactive polar phase nucleation in PVDF through the addition of ZFs was confirmed from detailed XRD and FTIR studies.

The evaluated polar phase fractions, with increasing filler content, are graphically represented in Figure 7.5a. The percolation threshold of ZF in PVDF is lowest for rod-shaped fillers (3 wt% concerning PVDF) compared to cubic (5 wt%) and spherical (6 wt%). This signifies that the incorporation of nanorods, i.e. one-dimensional fillers, can more efficiently overcome the drawback of high-weight fraction filler loading to obtain a superior dielectric/ferroelectric composite [32]. The individual amount of β and γ phase was calculated for three composites (3ZF-r/PVDF, 6ZF-s/PVDF, and 5ZF-c/PVDF representing maximum polar phase fraction from each set) systems, from the intensity ratio of 1276 and 1233 cm^{-1} ($I_{1276/1233}$) in FTIR spectra were plotted in Figure 7.5b. Respective FTIR absorption spectra are presented in the inset of Figure 7.5b.

To justify further this interfacial interaction, the damping constant ($2r_{dc}$) allied with the frequency of –CH_2 vibration (ν-CH_2) was calculated by considering the frequency shift of this vibration band in the composite compared to neat PVDF [77]. A clear frequency shifting of –CH_2 asymmetric (ν_{as} at 3020 cm^{-1}) and symmetric (ν_s at 2980 cm^{-1}) stretching vibrational modes was observed (inset of Figure 7.5c) in the composites compared to neat PVDF, which indicates strong interfacial interaction between the filler surface and the polymer dipole. Assuming the change in the effective mass

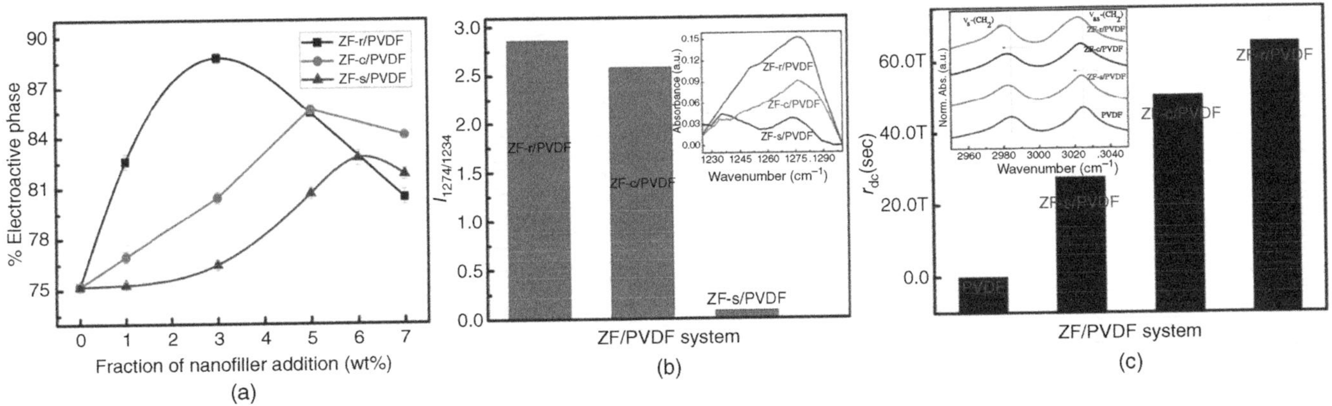

Figure 7.5 (a) Increment in polar phase fraction, (b) quantitative β/γ ratio, (c) variation of damping coefficient, r_{dc}, of different morphology ZF–PVDF composites; FTIR spectra of PVDF and nanocomposites in 1200–1300 cm region (inset b) and 3100–2900 cm region, (inset c).

of $-CH_2$ dipoles owing to the interfacial interaction, a damping (parameter, $2r_{dc}$) may arise causing red shifting in the frequency (i.e. wavenumber) of CH_2 stretching vibration [9, 78–80]. Figure 7.5c shows that the damping constant $2r_{dc}$ of composites changes as a function of the geometry of the filler. The value was maximized for the ZF–r/PVDF composite.

Comparative dielectric constant study at room temperature also indicates the superiority of a one-dimensional nanofiller (Figure 7.6a). The dielectric threshold too aroused earlier for them compared to other systems due to the easier formation of an intra-particle connecting network, ensuing increment in conductivity pathway development in composite with reduced internal resistance in the matrix [72]. They also calculated relaxation time as the reciprocal of the frequency at which the composite exhibited the lowest dielectric loss. The value abruptly reduced from its initial value (for pure PVDF = 166 μS), passing through a minimum and further

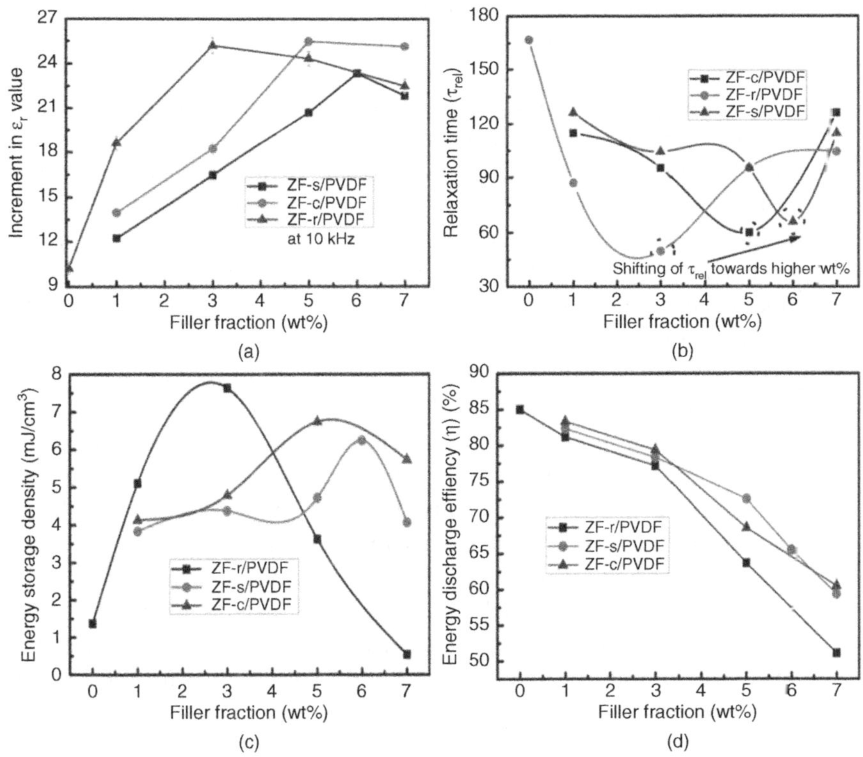

Figure 7.6 (a) Variation of dielectric constant, (b) relaxation time, (c) energy storage density value, and (d) energy discharge efficiency of different filler (rod, cube, and spherical shaped) fraction loaded ZF-PVDF nanocomposites.

increasing sharply (Figure 7.6b). These outcomes implied that sideways with dominant dipolar polarization, interfacial polarization has also played an important role [81]. Dipolar polarization upsurges because of a rise in dipole density (metal-oxide dipoles), whereas the latter rises due to enhancement in interfacial interaction. The large dielectric gap between the two elements develops a strong Maxwell–Wagner–Sillars polarization (MWS) polarization [82–84]. The increase in τ_{real} value afterward reaching minima was sharpest for ZF–r/PVDF composites representing the presence of higher interfacial interaction in the system.

Comparative energy storage density of neat PVDF and ZF/PVDF composite films at 10 Hz was measured from room temperature P–E loops (electrical polarization versus electric field). Trusting similarity with tradition in polar phase formation and dielectric constant, the corresponding ferroelectric threshold concentrations were 3 wt% (for rod ZF/PVDF composite), 5 wt% (for cubic ZF/PVDF composite), and 6 wt% (for spherical ZF/PVDF composite). After reaching the threshold concentration, more lossy loops were formed. The corresponding calculated (by integrating the area between the closed loop of the *P–E* curve and positive *Y*-axis) storage energy density (U′) (Figure 7.6c) and energy discharge efficiency (ratio of storage energy density with the total energy density) (Figure 7.6d) are also plotted. The highest energy storage performance was shown by 3ZF–r/PVDF composite (7.64 mJ/cm^3) at 1 kV applied field and 10 Hz frequency. Comparing the results, the rod-shaped filler-loaded composite (3ZF–r/PVDF) exhibited maximum η value (77.2%) with a much lower filler concentration. These results show the reliability of the material for capacitor applications.

The sensitivity of the composite toward human finger tapping was validated by piezo output characteristics, such as open-circuit AC voltage (V_{oc}), DC voltage, and short-circuit current (I_{sc}) measurement of three optimized films from three systems (rod, cubic, and spherical ZF nanostructure-loaded/PVDF composites) (results shown in Table 7.2). The rod-shaped filler-loaded composite exhibited a shorter output voltage lifetime (SVLT) (full width at half maximum [FWHM] of the voltage peak), compared to cubic- and spherical-shaped ZF-incorporated systems (Table 7.2). This leads to an increment of output voltage [85, 86]. Higher power density was obtained for 2.96 $\mu W/mm^3$ (3ZF–r/PVDF) comparative to 5ZF–c/PVDF (1.83 $\mu W/mm^3$) and 6ZF-s/PVDF (~0.73 $\mu W/mm^3$) (Table 7.2). The superior performance of 3ZF–r/PVDF composite may be utilized for various practical applications such as scavenging vibration of the human body movements, glowing of commercially available light emitting diodes (LEDs), and charging of different capacitors.

Table 7.2 Comparative output AC and DC voltage, short-circuit (SC) current, short voltage lifetime, and power density value of rod, cubic, and spherical shape nanoparticle-incorporated PVDF nanocomposites (threshold concentration is only considered here).

Systems	AC (V)	DC(V)	SC current (μA)	SVLT (ms)	PD (μW/mm³)
3ZFr/PVDF	39.10	3.02	51.34	3.15	2.96
5ZF–c/PVDF	29.86	2.43	17.67	5.39	1.83
6ZF–s/PVDF	19.69	1.56	1.9	7.04	0.73

Further, P. Durga Prasad et al. [55] addressed the issues of uniformity of fillers (average diameter of fiber confirmed by AFM study is 341 nm) by synthesized ZF via the electrospinning technique using polyvinyl pyrrolidone (PVP) and then incorporated in PVDF using DMF as the solvent. This uniformity leads to an enhanced filler-to-polymer ratio (15 wt% obtained a maximum polar phase of 88%) with an enhanced output of 3.2 V at a 50 MΩ load resistance under a 1.5 N mechanical force application.

Interfacial interaction: Another problem of polymer–ceramic composite fabrication is the interfacial area between the two major constituents, which plays a significant part in the determination of dielectric, piezoelectric, and ferroelectric properties of composite [87]. One of the promising ways to raise interfacial interaction between filler and polymer matrix is surface modification of ceramic fillers. Surface modification improves the dispersion of ceramic filler in the polymer matrix and increases compatibility by forming bonding through chemical means with an interfacial layer, which encourages the growth of all-trans β-phase in PVDF on the nanoparticle's surface [88, 89]. Surface functionalization employing different groups was investigated previously by many researchers to increase interaction between the polymer and nanoparticle through dipolar interaction [21, 22], electrostatic interaction [90], or hydrogen bonding [91, 92]. The modifier can alter surface charge [88] and thus affect the polarization of the adjacent polymer moiety. On the other hand, there is also a controversy that surface treatment leads to a decrease in the electrical property [93]. According to the predicted hypothetical multilayer core model, the intermediate layer which binds both inorganic filler and organic matrix can enhance/decrease interface compatibility [89]. Hence, the selection of a suitable modifier is a crucial factor. Martins et al. [94] used three different surfactants such as anionic, cationic, and nonionic surfactants, and compared their effects. They have mentioned that polar phase formation as well as properties of the composite can be tuned by selecting the proper surface functional group.

Table 7.3 Summary of surface functionalized Iron oxide/-Ferrite–polymer composites.

Ferrite	Polymer	Modifier group	References
Fe_3O_4	PVDF	TEOS	[95]
Fe_3O_4	PVDF	Oleic acid	[96]
$CoFe_2O_4$	PVDF	Anionic (Sodium dodecyl sulfate, SDS), nonionic (Triton X-100), and cationic (Cetyltrimethylammonium bromide, CTAB) surfactants	[94]
$CoFe_2O_4$, $NiFe_2O_4$	PVDF	An aqueous solution of citric acid	[97]
$CoFe_2O_4$	PVDF	Metallic Co	[98]
$NiFe_2O_4$	PVDF	Metallic Ni	[98]
$NiFe_2O_4$	PVDF	Polypyrrole	[99]
$GaFeO_3$	PVDF	SDS	[100]
$BiFeO_3$	PVDF	SDS	[101]
$BiFeO_3$	PVDF–HFP	H_2O_2	[102]
$Ni_{0.5}Zn_{0.5}Fe_2O_4$	PVDF	PEO–silane	[103]
$ZnFe_2O_4$	PVDF	Tetraethyl orthosilicate(TEOS), SDS	[21]
$ZnFe_2O_4$	PVDF	Polyethylene glycol (PEG)	[22]

Researchers have functionalized the surface of iron oxide and different ferrites [95] to improve compatibility with polymer matrix and enhance the polar phase fraction, as well as the electrical properties of the composite, employing different modifiers. Some of them are summarized in Table 7.3.

In general, surface treatment may be possible in three distinguished ways. Accordingly, three different modifiers, such as TEOS (small molecule attachment such as SiO_2), SDS (anionic type surfactant), and PEG-6000 (compatible heteropolymer) were selected. Anionic surfactant SDS builds a soft layer on the ZF surface, resulting in the formation of a bipolar assembly that can interrelate with both organic PVDF and inorganic ZF through hydrophobic and hydrophilic interactions simultaneously [21]. This also increased the effective interaction area by disrupting aggregated ferrite morphology. On the other hand, hydrolyzed TEOS can form an inorganic robust hard layer of silica on the ZF surface, which further interacts with polymer dipoles and forms a six-membered ring complex. This stable complex

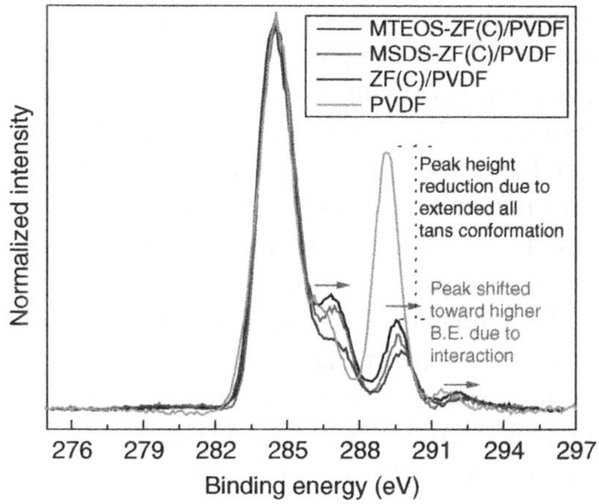

Figure 7.7 Comparative XPS-spectra of modified and unmodified ZF(C)–PVDF with neat PVDF.

structure not only increases compatibility between the filler and polymer surface but also behaves as an intermediate insulating layer to avert the migration of negative charges from the ZF surface. Thus, ion–dipole interaction density gets reduced between the polymer and the filler. The increased number of residual ions on the ZF surface leads to a reduction in free ion mobility and other defects in the nanocomposite structure [21]. Comparing XPS spectra (Figure 7.7) of composites (with optimum concentration) with neat PVDF, it was observed that the peak appeared (at 284.4 eV) for the sp^3 hybridized —CH— group existed in all spectra, which indicated no alteration has taken place in the skeleton of PVDF after composite formation [21, 104]. The presence of other three carbon moieties, i.e. —CH_2 bond (side chain), —CF covalent bond (main chain/skeleton), and —CF_2 bond (side chain) with binding energy at 286.3, 289.13, and 291.61 eV, were observed in neat PVDF film. After the addition of ZF(C) nanoparticles in the PVDF matrix, the binding energy corresponding to both side chains and main chain shifted toward higher binding energy (286.9, 289.6, and 292.2 eV, respectively) due to the change in the electronic environment of CH_2 and CF_2 dipoles of PVDF, as well as attainment of extended all-trans chain conformation (confirmed from FTIR and XRD study) [21]. Theoretically, the peak area corresponding to 286.3 and 291.6 eV must be equal to neat PVDF, but it was not observed in Figure 7.7. This might be due to the interaction of the PVDF chain with the polar DMF solvent [21, 105]. A considerable peak area enhancement was observed in the normalized —CH_2 peak of modified composites compared to unmodified composites showing a clear indication of increased interfacial interaction between modification layers and —CH_2 dipoles of PVDF leading to the formation of extended all-trans (trans-trans-trans, TTT) conformation in modified composites [21, 22, 106].

The peak area corresponding to the —CF_2 bond remains unchanged in surface-treated composites, which indirectly indicates that the —CF_2 bond does not contribute to the ion–dipole interaction phenomenon. Another important observation was an increment in the peak area for the -CH_2 bond and —CF bond of the main chain, which was comparatively higher for TEOS-modified composite compared to SDS-modified composite. These results indicated that more extended β-phase formation was attained in TEOS-modified composite.

The probable mechanism for this assumed interaction is schematically presented in Figure 7.8. After being modified with SDS, negative surface-charged ZF(C) powders (evident from zeta potential value) become more negatively charged due to the presence of hydrophilic SO_4^{2-} group in the surface of ZF(C). This highly negatively charged SDS-modified surface interacts with the positively charged H atom in the —CH_2 dipole of PVDF through ion–dipole interaction. Thus, the tendency of the F atom in the —CF_2 bond of PVDF forms an extended TTT conformation corresponding to electroactive β PVDF. Similarly, in the case of modification with TEOS, TEOS forms $Si(OH)_4$ in ethanol. This further reacts with absorbed -OH ion on ZF(C) surface and during condensation, forming one —O—Si—O— bond on the ZF(C) surface, releasing one water molecule for each reaction. This —O—Si—O— bond is further coordinately associated (lone pair–dipole interaction) with the H atom of the —CH_2 dipole of PVDF, leading to the formation of a corresponding polar phase of PVDF. Thus in both cases, the formation of extended TTT conformation is possible leading to enhancement in electroactive phases [21].

The effective dielectric permittivity (ε_{eff}) of composite is influenced by several factors, such as the dielectric constant of the individual components (ε_1 and ε_2 are dielectric constant polymer and nanofiller, respectively), volume fraction (φ_1 and φ_2 are the respective volume fractions of polymer and the filler, $\varphi_1 + \varphi_2 = 1$), the morphology of filler, dispersion and interfacial interaction between the two constituent phases. Various theoretical studies with models were proposed by researchers to predict the effective dielectric constant of a nanocomposite, which were developed and founded on some assumptions [107–119].

The maximum similarity was observed for Furukuwa Model, with deviation at lower frequency region (Figure 7.9a). Other important findings were during the energy storage density study and voltage generation capability study (Figure 7.9b,c) of the fabricated composite. Almost similar electrical properties were shown for surface-modified and electrically poled unmodified nanocomposites. This is a clear indication of the induction of self-polarization characteristics in surface-modified nanocomposite, which can minimize the ungainliness of the external poling process.

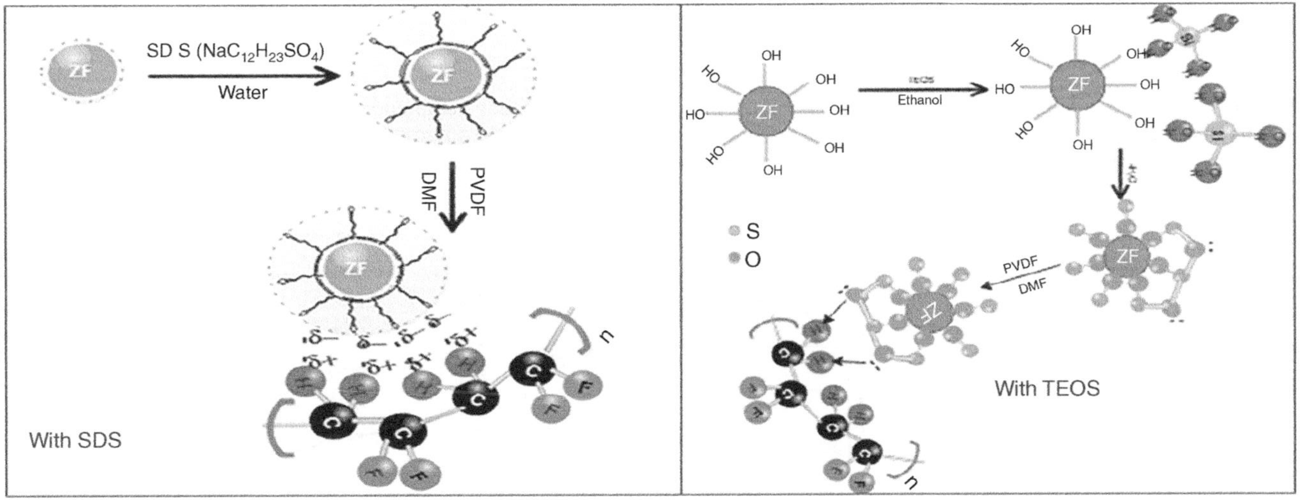

Figure 7.8 Schematic representation of the suggested interaction of the modified surface layer of Zinc Ferrite with the dipole of PVDF.

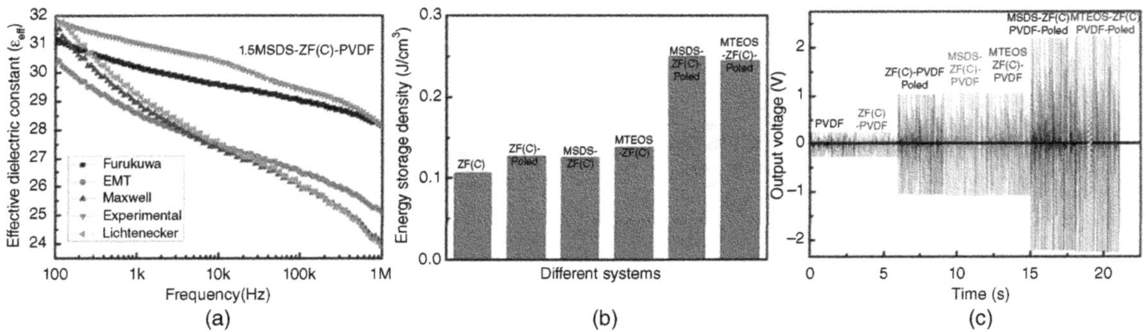

Figure 7.9 (a) Comparison of the frequency-dependent effective dielectric constant of 1.5 MSDS–ZF(C)–PVDF composite with different theoretical models, comparative (b) energy storage density values, and (c) output AC voltage obtained (with single finger tapping on the surface of the composite) from 1.5 wt% unmodified and surface-modified ZF(C)–PVDF composites (both poled and unpoled conditions) with neat PVDF.

Surface treatment of ZF nanofiller (ZF-P), synthesized using PEG-400 as a chelating agent through sol–gel auto-combustion route) with a heteropolymer (PEG-6000) over ZF surface was also performed, where the soft coating of PEG-6000 layer easily covers the surface of ZF by the small size —OH groups [120]. It led to an increment of interfacial interaction between the ZF surface and $—CH_2$ dipoles of PVDF inside the composite, leading to enhanced orientation of the dipoles of PVDF more effectively. Introducing the soft boundary layer of PEG, the aggregation of fillers is minimized (TEM image, Figure 7.10).

From the comparative FESEM study of ZF–P/PVDF and ZF–PEG/PVDF nanocomposites (shown in Figure 7.11a, b), uniform distribution of PEG-coated ZF was noticed over the PVDF matrix.

Additionally, diminished spherulite morphology was observed (shown in the inserted Figure 7.11a) in a neat PVDF matrix, corresponding to a nonpolar α-phase structure [21, 22, 121]. The α-spherulitic structure was a kinetic control product that immediately grew during solvent evaporation at the time of composite preparation. Composites exhibited a mixture of spherulitic and lamellar morphology, revealing the co-presence of polar and nonpolar structures, respectively [122]. The presence of very few numbers of spherulitic structures observed in PEG-coated composite indicated superior load-bearing capability compared to unmodified composite. This was also a sign that the modified composite exhibited a high Young modulus value [121, 123]. Young modulus is directly related to the piezo-characteristics of the material. So, the morphological study gives a modified composite that would exhibit better piezo-response behavior.

A significant observation was noticed that the filler-to-polymer ratio enhanced the determination of the polar phase (Figure 7.12a), and thus helped to enhance the dielectric constant. Figure 7.12b shows that for ZF–P/PVDF, the threshold concentration is 6 wt%, and after that, a plateau nature was observed. Whereas, in polyglycolated composite, the critical concentration is 10 wt%. The robust PEG layer behaves as a coupling agent in the matrix between organic and inorganic parts, thus increasing the filler-to-polymer ratio. These layers also assist to improve MWS polarization of the interface by creating an interaction zone with the Gouy–Chapman diffuse layer of polyglycolated ZF (ZF–PEG) and PVDF dipoles [22, 124]. This effect was cleared after the study of temperature-dependent dielectric properties (Figure 7.12c,d). A growing trend was noticed in permittivity values for all composites with enhancing temperatures up to 120 °C. Beyond 120 °C, the β phase cannot exist [22, 125] and thus dipolar orientation decreases leading to a decrease in permittivity value. An increasing trend was noticed in dielectric permittivity over the temperature range and it was

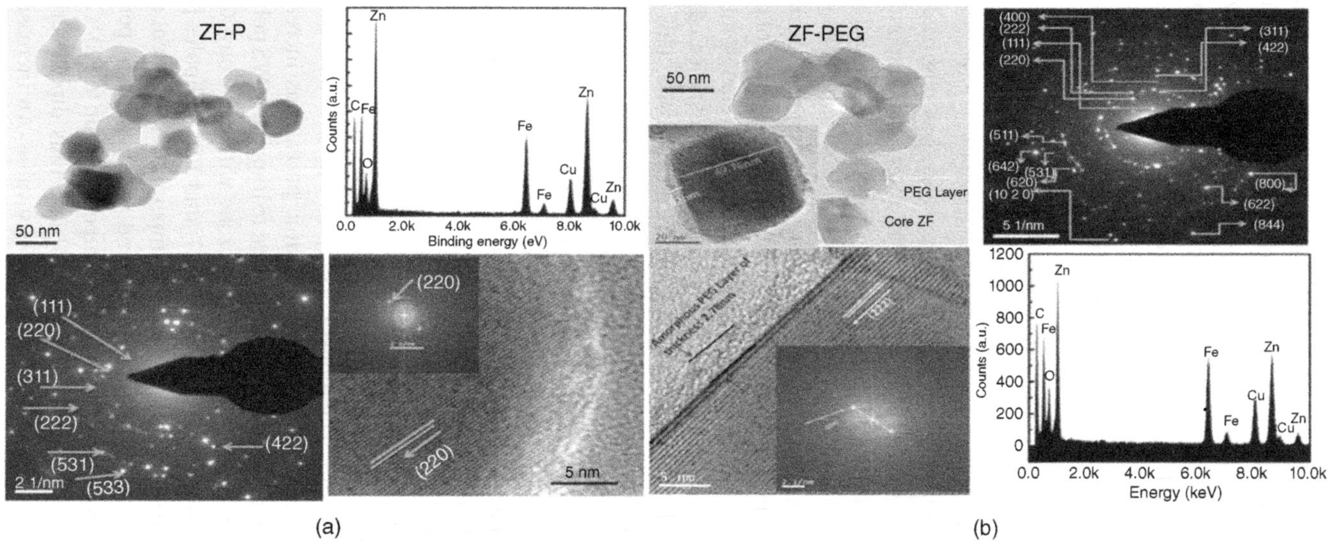

Figure 7.10 Comparative TEM images of (a) ZF and (b) ZF–PEG powders.

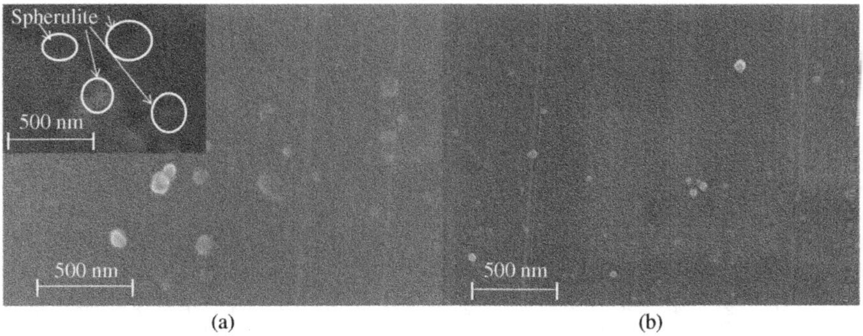

Figure 7.11 Comparative FESEM micrograph of (a) ZF–P/PVDF and PVDF (inserted) and (b) ZF/PEG–PVDF.

more stripper for surface-modified composites. This can be interpreted by higher interfacial polarization occurring due to entrapping more charges [126]. As the charge entrapment occurred in the surficial polymer layer, the effect of temperature on loss characteristics was insignificant. No loss peak related to the glass transition of the polymer chain looked into the loss curve in the experimental temperature range, as this might occur at −35 °C [127].

Piezo-voltage generation performance was also investigated based on the ferroelectric properties of ZF–PEG/PVDF composites. Considering both dielectric constant value, energy storage density, and discharge efficiency, 12 wt% ZF–PEG/PVDF composite was selected as the optimized material for further piezo-output measurement. The maximum 18 V open-circuit piezo output was obtained.

7.4.1 Effect of Conducting Filler Addition

In addition to functionalization of the surface, incorporating a third-phase conductive supplementary nanofiller, such as graphite, reduced graphene oxide (rGO), carbon black (CB), single/multi-wall carbon nanotube (SWCNT/MWCNT), silicon carbide (SiC) nanowire, poly(hedral oligomeric silsesquioxane) (POSS) fiber, in a binary composite matrix is a recent approach to reduce the local effective electric field mismatch. Alongside this, the third phase aids in the role of the scattering agent and helps to float nanofillers throughout the surface of the nanocomposite by preventing local sedimentation of nanofillers on the lowermost surface of the nanocomposite [128, 129]. Furthermore, a low volume fraction conducting filler reinforces the composite and hence attributes to an improved Young modulus, Y, leading to enhancing piezo-response of nanocomposite utilizing short-circuit current (I_{sc}) (Eq. 7.2) (A, d_{33}, and $\dot{\varepsilon}$ are the effective

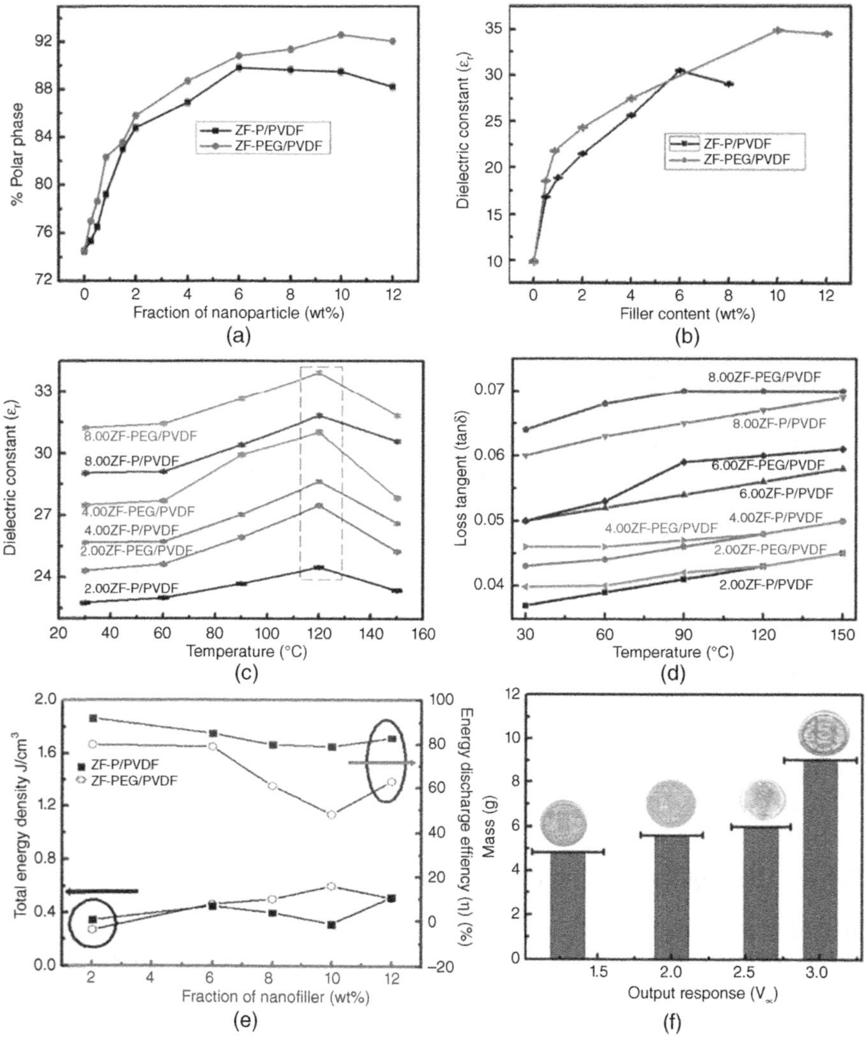

Figure 7.12 Comparative variation (a) in polar phase fraction, (b) frequency-dependent dielectric constant, (c) temperature-dependent dielectric constant, (d) dielectric loss, (e) total energy density and η value of ZF/PVDF and ZF–PEG/PVDF composite with increment in nanofiller loading, (f) impact sensing performance of the 12 wt% nanocomposite.

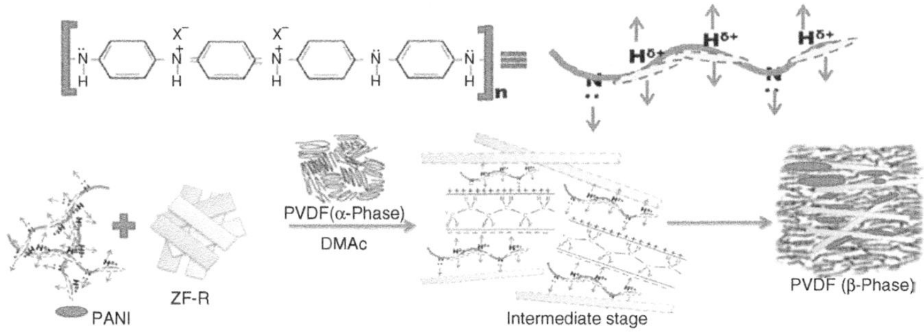

Figure 7.13 Schematic representation of the interaction between ZF surface, PANI chain, and $-CH_2$ dipole of PVDF.

contact area, piezoelectric coefficient, and Poisson's ratio of the composite, respectively) [130, 131].

$$I_{sc} = d_{33} \times Y \times A \times \dot{\varepsilon} \tag{7.2}$$

This third phase easily generates conduction pathways by its network formation, and minimizes the internal resistance of the devices [131–133], yielding a short voltage lifetime. These further lead to a rise in the output piezo-voltage. To investigate this fact in further detail, higher AR filler, such as rod-shaped ZF (ZF-R) (synthesized via a two-step hydrothermal route, for detail, refer to Table 7.1) was selected for introduction into the PVDF matrix and polyaniline (PANI) (electrochemically synthesized) as a conducting supplementary filler. PANI is a lucrative alternative to MWCNT or rGO for the fabrication of a piezoelectric NG. Being a semiconducting polymeric material, PANI may disperse easily in other polymer matrices. Besides, its ease of synthesis, nontoxic nature, and thermal stability at higher temperatures have added extra advantages. So PANI can efficiently decrease the heterogeneity of two different dielectric fields and promote the *in situ* polar phase growth in PVDF. Accordingly, enhanced electrical properties of the composite were obtained depriving of any additional treatment.

The plausible interaction between negatively charged ZF-Rs with $—CH_2$ dipoles of PVDF contributes to β-phase nucleation [134]. Moreover, π electron clouds over N in PANI may react with $—CH_2$ dipoles and stabilized β-phase. The schematic of the interaction is presented in Figure 7.13.

Integration of ZF-Rs in PVDF increased both dipolar and interfacial polarization, executing in an increment in the dielectric constant value. Further, the incorporation of PANI as a ternary filler in the matrix contributes to enhancing interfacial polarization only. This led to a sharp increase in the permittivity value in PANI-decorated composites at lower frequency regions (above >10 kHz, interfacial polarization cannot put

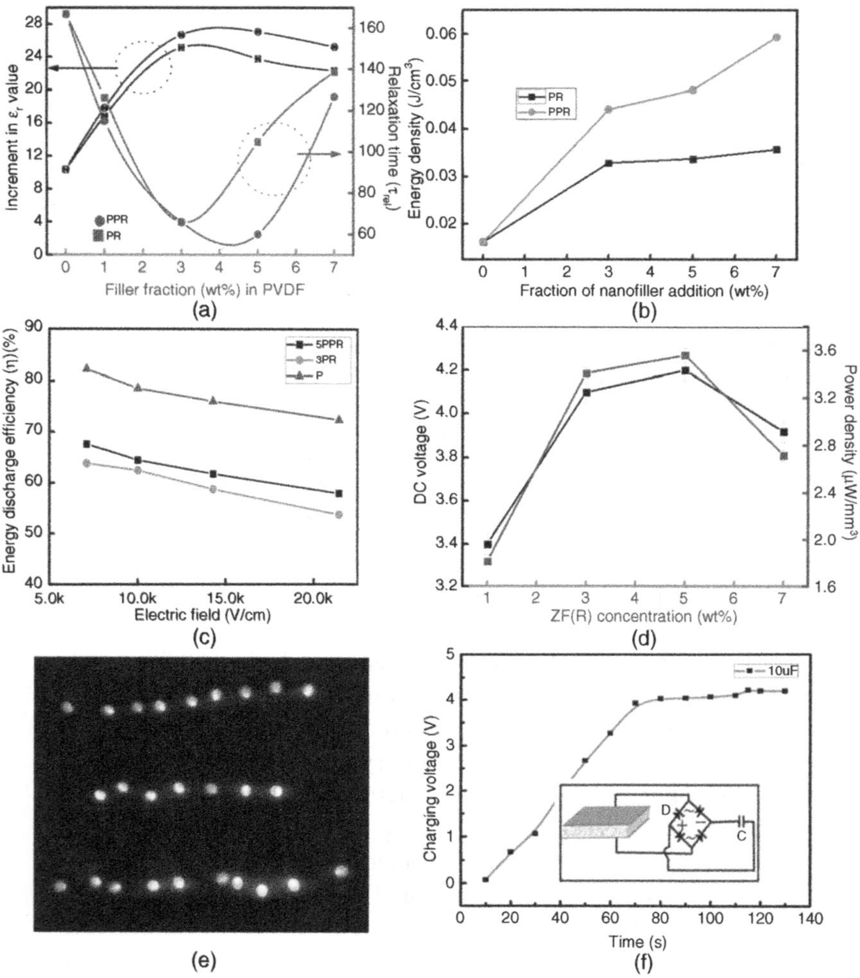

Figure 7.14 Variation in (a) dielectric constant value at 10 kHz and relaxation time of both composite films with increasing ZF-R concentration, (b) energy density, (c) energy discharge efficiency of both composites (3PR and 5PPR) and neat PVDF with a variation of the electric field, (d) output DC voltage and produced power density with a variation of ZF-Rs concentration.

through this sudden frequency change) [135–137]. The maximum dielectric constant value was offered by 5wt% PANI modified ZF-R/PVDF (5PPR) composite (Figure 7.14a). The concept for the increment in interfacial polarization owing to PANI incorporation was further verified by relaxation time (τ_{rel}) calculation. Higher interfacial polarization minimized the τ_{rel} value. The trend in the value of τ_{rel} (Figure 7.14a) (for pure PVDF, the value was 166.81 μs) with an increase in filler concentration, passed through minima for 3wt% ZF-R/PVDF (3PR) (65.79 μs) composites was observed. The addition of PANI further depressed the τ_{rel} value (5PPR, the minimum value was 59.95 μs) (Figure 7.14b). A relatively slim P–E hysteresis loop with low remnant polarization was found with a higher η value (Figure 7.14c). Obtained piezo DC output voltage and power density (Figure 7.14d) from both composites indicate that PANI-decorated composites possessed higher amounts of inductive charges, with much faster charging/discharging speed compared to unmodified composite. The maximum 4.2 V DC output voltage with a power density of 3.56 μw/mm^3 was obtained for the 5PPR composite. This output sufficiently powers up 26 nos of light-emitting diodes (LEDs) (Figure 7.14e). Furthermore, to design the material as an additional power source, rectified output was stored in a capacitor of 10 μF capacitance. The NG can proficiently charge the capacitor within a very short gap (115 s) with a saturated value of ~4.2 V (Figure 7.14f, the inserted figure represented the equivalent circuit diagram). The calculated stored energy 88.2 μJ with power 0.76 μW was obtained. To check the reproducibility of the fabricated device, the typical stepwise cyclic charging and discharging behavior of the capacitor was investigated, and the data shows that the capacitor could get charged and discharged repeatedly.

7.5 ZF Nanorod (High Aspect Ratio) and Copolymer PVDF–HFP–Based Nanocomposite

Consistent with earlier results and the associated discussion, incorporation of rod-shaped nanostructure with a high AR can most effectively enhance β-crystallinity with reduced agglomeration leading to expanding the complete performance of the system. Compared to other PVDF copolymers, PVDF–HFP exhibits superiority in terms of significantly higher piezoelectric coefficient ($d31$) and electromechanical coupling factor ($k31$), i.e. $|d31/d33| > 1$[6–8]. Moreover, excellent mechanical characteristics with film-forming ability, low cost, chemical stability, and weather-resilient features are better compared to other PVDF copolymers [6, 30]. Two-pot hydrothermally synthesized ZF-R (~length of 330 nm) was added in PVDF–HFP matrix with different weight ratios to determine the electrical percolation threshold of the system. A maximum 57% β-phase has been obtained with 3 wt% nanocomposites and enhanced dielectric, ferroelectric, and piezoelectric properties (Figure 7.15a–d) has been achieved for the composite depriving of any additional treatment, such as conventional electric poling or annealing.

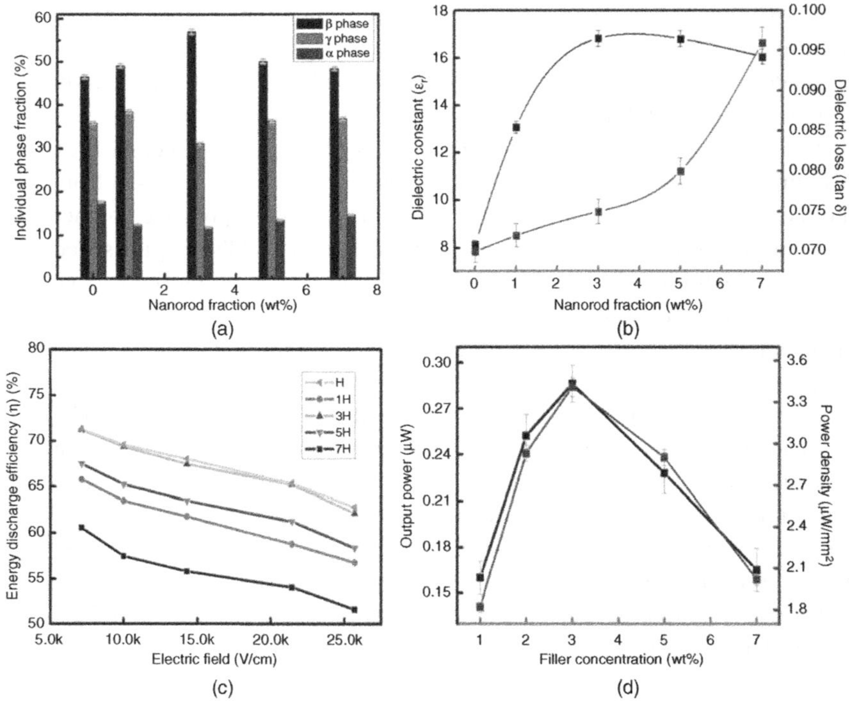

Figure 7.15 Comparative study of (a) individual phase fraction, (b) variation in dielectric permittivity and dielectric loss, (c) energy discharge efficiency, (d) output power and power density of different filler fraction-loaded composites.

7.6 Applications Still Explored and Future Scope

7.6.1 Powering Commercial LEDs in Series Connection and Charging Different Capacitors by Single Finger Tapping

The simplistic hybrid flexible nanocomposite NG, comprising of ZF nanorods and PVDF–HFP, with a high piezo output (Voc), i.e. 43.8 V and 3.41 μW/mm^3 power density, can illuminate red LEDs (Figure 7.16a) under periodic finger impartation and relaxation conditions.

To establish its application as a DC power source, four different capacitors (e.g. 1, 2.2, 4.7, and 10 μF) were connected individually, attached to a full-wave bridge rectifier unit for measurement of charging capability. It was successfully able to charge capacitors under repetitive finger movement within a short period (62 and 100s for 1 and 4.7 μF, respectively), as well as reached their corresponding steady state (Figure 7.16b). A maximum stored energy of 198.45 μJ was reached after charging with a 10 μF capacitor.

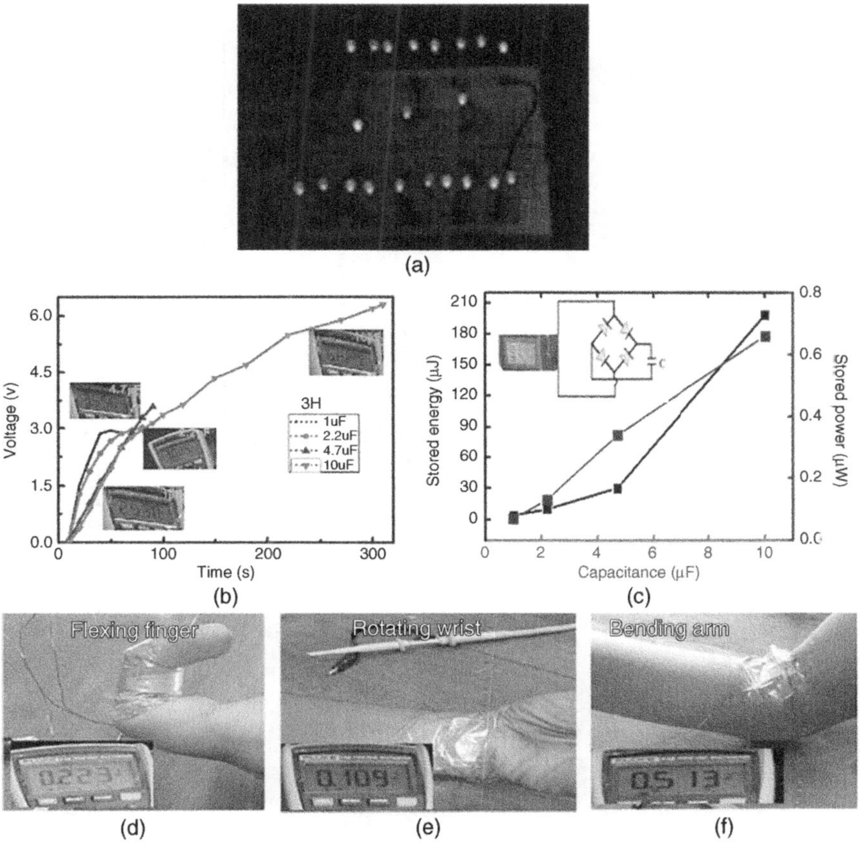

Figure 7.16 (a) Enlighted LEDs by finger impartation, (b,c) capacitor charging response of NG, individual performance of NG by different body movement, (d) finger flexing, (e) wrist rotating, and (f) arm bending and inset figure exhibited corresponding output DC voltage utilizing scavenging mechanical energy.

7.6.2 Energy-Harvesting Performance from Human Body Movement

The wearable NG (ZF/PVDF–HFP) (~36 mm × 30 mm) film was integrated with the human body to scavenge diverse body movements, such as flexing fingers, rotating the wrist, or bending the arm (shown in Figure 7.16c–e). The small perturbations can effectively yield an output response (DC voltage). The inset of each figure represents the response of the output. Comparatively, the arm bending motion have higher DC output due to sustaining higher impact pressure [138].

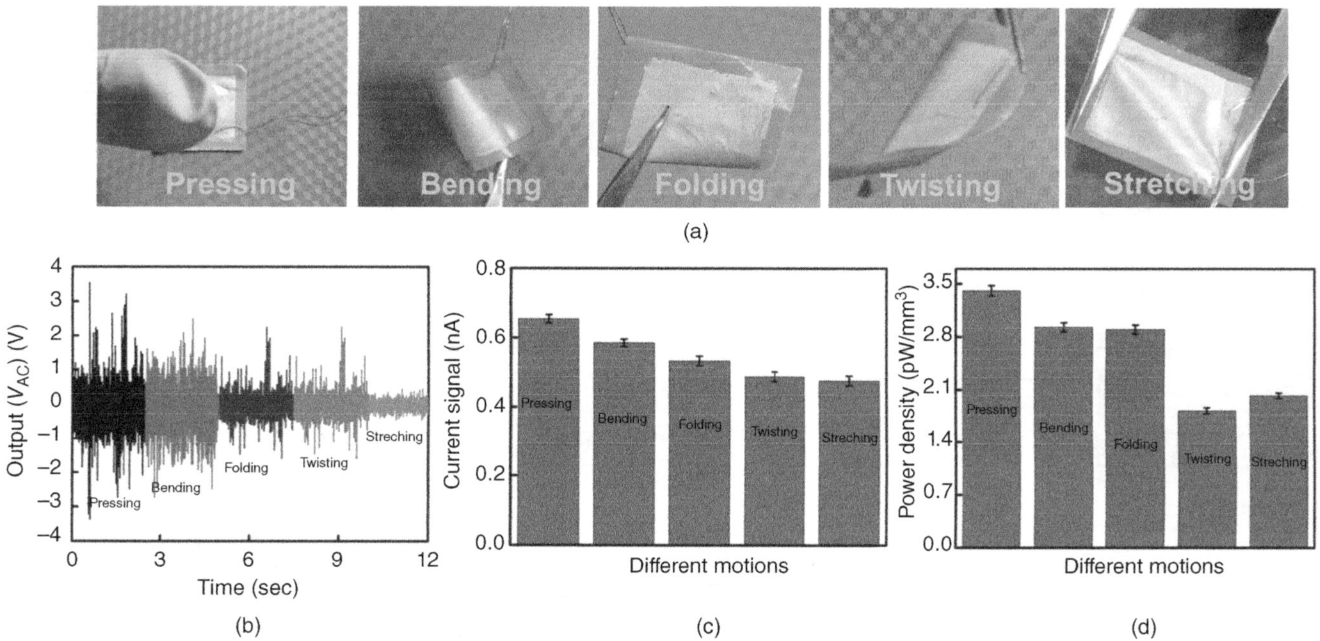

Figure 7.17 (a) Individual performance of HFNG by different motions such as pressing, twisting, folding, stretching, or bending of the film, (b) produced ac open circuit voltage, (c) current signal, (d) power density.

Because of its admirable tractability, NG can work under diverse deformations without any mechanical failure and efficiently harvesting the respective energy from these deformations. The performance of individual NGs by different motions such as pressing, twisting, folding, stretching, or bending the film (schematically represented in Figure 7.17a–d) generated different electrical signals (open-circuit voltage, short-circuit current, and power density).

7.6.3 Energy Harvesting from Ambient Airflow

NGs can also scavenge ambient environmental energy, such as airflow. The demonstration was done under both conditions: in an open atmosphere and in slight turbulence (the air was blown by a blower at a 45° angle from a fixed distance) (schematically represented in Figure 7.18a). The composite was fixed to the edge of a table. With the increasing velocity of the blower, the output voltage increased and maximized for the velocity at ~7.8 m/s (output peak to peak voltage 3.2 V).

7.6.4 Application as Self-Powered Pressure Sensor/Height Monitor

The pressure sensitivity of NG was estimated by dropping a rubber ball (density = 1.67 g/cc) from different elevations on the surface (h = 5, 10, 15, 18, 22, 25 cm) of the NG and measured piezo-responses are shown in Figure 7.17a. The respective impact pressure from different elevations was calculated, and the corresponding maximum AC voltage was plotted. The

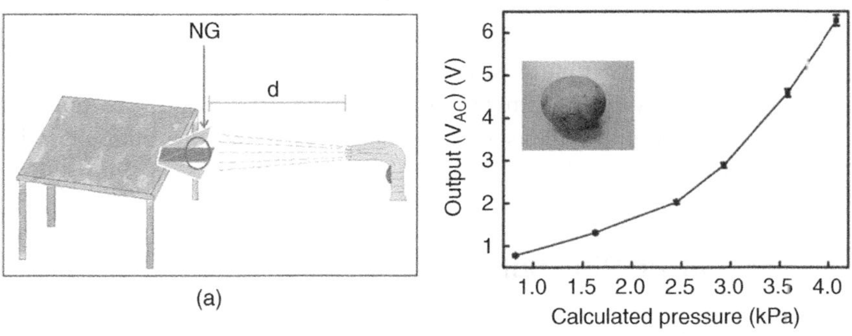

Figure 7.18 (a) Schematic representation of experimental setup of energy harvesting from airflow, (b) the pressure sensitivity of HFNG in terms of the highest output peak voltage (photograph of a rubber ball is in inset).

linearity in the results direct that the NG has the capability of distinguishing even a very minute change in pressure. Therefore, proving the capability of this NG for further exploration for application as a self-powered height monitoring sensor.

7.7 Conclusion

Incorporation of wet-chemically synthesized ZF with tailored shape and size into PVDF or its copolymer matrix is a very good technique to achieve polar phase-enriched nanocomposite with improved dielectric, ferroelectric, and piezoelectric properties. A filler with a high AR value shows more activity than the lower counterpart to polar-phase nucleation. Agglomeration in ZF becomes minimized after surface modification using different modifiers, such as PEG, TEOS, and SDS. Surface modification induced a self-polarization behavior in PVDF and helped to increase electroactive phase fraction and interfacial polarization, which plays a major role in the determination of the electrical properties of the composite. The addition of an external conductive dispersing agent such as PANI also helps to improve dielectric inhomogeneity in the composite and thus acts as a catalyst for polar phase nucleation in PVDF and improves piezo-response throughput. A rod-shaped ZF (optimized shape)/PVDF–HFP-based NG was fabricated with an optimized filler concentration. Piezo-sensitivity of the systems was examined by employing different small-scale physical variations, such as finger impact, body movement, and different motions (such as stretching, bending, flexing, and twisting the film). The NG exhibited the potentiality to illuminate 23 red LEDs, charge up different capacitors (Maximum stored energy 198.45 μJ), scavenge environmental vibrations such as ambient airflow (maximum 3.2 V AC peak-to-peak voltage was obtained), and harvest electricity from muscular motions integrated into the body joints, such as arm, wrist, and finger. Further, the NG was used in self-powered sensor device applications such as height and pressure monitors with desired accuracy levels.

7.8 Future Direction

In future, the major research work can be followed based on this dissertation as proposed here:

- Future research should encompass investigations on large-scale demonstrations for several conceivable areas.

- Demonstration of piezo composite NG in self-powered wireless sensors, sophisticated gadgets, wearable displays, artificial electronic skin, multi-modal sensors, and so on.
- Application of NG in monitoring vehicle speed, machine vibrations, sound vibrations, mimicking the behavior of the human skin, and so on.
- The demonstration of NG as a wearable textile.
- Application of this NG integrating with other energy-harvesting devices, such as photovoltaic cells to construct hybrid NG with high efficiency.
- Other one-dimensional ferrite nanostructure–polymer-based composites can be fabricated using a similar technique.

References

1 Wang, Z.L., Chen, J., and Lin, L. (2015). Progress in triboelectric nanogenerators as a new energy technology and self-powered sensors. *Energy Environ. Sci.* 8: 2250–2282.

2 Lin, Z., Chen, J., and Yang, J. (2016). Recent progress in triboelectric nanogenerators as a renewable and sustainable power source. *J. Nanomater.* 2016: 1–24.

3 Yan, J. and Jeong, Y.G. (2016). High performance flexible piezoelectric nanogenerators based on BaTiO3 nanofibers in different alignment modes. *ACS Appl. Mater. Interfaces* 8: 15700–15709.

4 Panero, S., Romoli, C., Achilli, M. et al. (1995). Impact of household batteries in landfills. *J. Power Sources* 57: 9–12.

5 Adhikary, P., Biswas, A., and Mandal, D. (2016). Improved sensitivity of wearable nanogenerators made of electrospun Eu^{3+} doped P(VDF-HFP)/graphene composite nanofibers for self-powered voice recognition. *Nanotechnology* 27: 495501.

6 Adhikary, P., Garain, S., and Mandal, D. (2015). The co-operative performance of a hydrated salt assisted sponge like P (VDF-HFP) piezoelectric generator: an effective piezoelectric based energy harvester. *Phys. Chem. Chem. Phys.* 17: 7275.

7 Neese, B., Wang, Y., Chu, B. et al. (2007). Piezoelectric responses in poly(vinylidene fluoride/hexafluoropropylene) copolymers. *Appl. Phys. Lett.* 90: 242917.

8 Adhikary, P., Garain, S., Ram, S., and Mandal, D. (2016). Flexible hybrid eu^{3+} doped P(VDF-HFP) nanocomposite film possess hypersensitive electronic transitions and piezoelectric throughput. *J. Polym. Sci. B Polym. Phys.* 54: 2335.

9 Wang, Z.L., Chen, J., and Lin, L. (2015). Progress in triboelectric nanogenerators as a new energy technology and self-powered sensors. *Energy Environ. Sci.* 8: 2250.

10 Martins, P., Lopes, A.C., and Mendez, S.L. (2014). Electroactive phases of poly(vinylidene fluoride): determination, processing and applications. *Prog. Polym. Sci.* 39: 683–706.

11 Zhang, G.Q., Brannum, D., Dong, D.X. et al. (2016). Interfacial polarization-induced loss mechanisms in polypropylene/$BaTiO_3$ nanocomposite dielectrics. *Chem. Mater.* 28: 4646–4660.

12 Satapathy, S., Gupta, P.K., Varma, K.B.R. et al. (2008). *Study on Dielectric Behavior of Lithium Tantalate (LT) Nano Particle filled Poly (vinylidene fluoride) (PVDF) Nano Composite*, vol. 20, 509. International Publishing House Pvt. Ltd.

13 Al-Saygh, A., Ponnamma, D., AlMaadeed, M.A. et al. (2017). Flexible pressure sensor based on PVDF nanocomposites containing reduced graphene oxide-titania hybrid nanolayers. *Polymers* 9: 33–52.

14 (a) Razzaq, H., Nawaz, H., Siddiqa, A. et al. (2016). A brief review on nanocomposites based on PVDF with nanostructured TiO_2 as filler. *Madridge J. Nanotechnol. Nanosci.* 1: 37–43. (b) Devi, P.I. and Ramachandran, K. (2011). Dielectric studies on hybridised PVDF–ZnO nanocomposites. *J. Exp. Nanosci.* 6: 281–293.

15 Huang, X., Liu, H., Zhang, X., and Jiang, H. (2015). High performance all-solid-state flexible micro-pseudocapacitor based on hierarchically nanostructured Tungsten Trioxide composite. *ACS Appl. Mater. Interfaces* 7: 27845–27852.

16 Ramadoss, A.K., Saravanakumar, B., and Kim, S.J. (2015). Vanadium pentoxide/reduced graphene oxide composite as an efficient electrode material for high-performance supercapacitors and self-powered systems. *Energy Technol.* 3: 913–924.

17 Gaur, M.S., Indolia, A.P., Rogachev, A.A., and Rahachou, A.V. (2015). Influence of SiO_2 nanoparticles on morphological, thermal, and dielectric properties of PVDF. *J. Therm. Anal. Calorim.* 122: 1403–1416.

18 Gonçalves, R., Martins, P.M., Caparrós, C. et al. (2013). Nucleation of the electroactive β-phase, dielectric and magnetic response of Poly(vinylidene fluoride) composites with Fe_2O_3 nanoparticles. *J. Non-Cryst. Solids* 361: 93–99.

19 Guo, Y., Liu, Y., Wang, J. et al. (2010). Giant magnetodielectric effect in 0–3 Ni0.5Zn0.5Fe2O4-Poly(vinylidene-fluoride) nanocomposite films. *J. Phys. Chem. C* 114: 13861–13866.

20 Roy, S., Thakur, P., Hoque, N.A. et al. (2016). Enhanced electroactive β-phase nucleation and dielectric properties of PVDF-HFP thin films

influenced by Montmorillonite and Ni(OH)2 nanoparticle modified montmorillonite. *RSC Adv.* 6: 21881–21894.

21 Chinya, I. and Sen, S. (2017). Improved dielectric and touch sensing performance of surface modified zinc ferrite (ZF)/polyvinylidene fluoride (PVDF) composite. *Sensors Actuators A* 267: 301–309.

22 Chinya, I., Pal, A., and Sen, S. (2017). Polyglycolated zinc ferrite incorporated poly(vinylidene fluoride) (PVDF) composites with enhanced piezoelectric response. *J. Alloys Compd.* 722: 829–838.

23 Skinner, D.P., Newnham, R.E., and Cross, L.E. (1978). Flexible composite transducers. *Mater. Res. Bull.* 13: 599–607.

24 Seema, A., Dayas, K.P., and Varghese, J.M. (2007). PVDF-PZT-5H composites prepared by hot press and tape casting techniques. *J. Appl. Polym. Sci.* 106: 146–151.

25 Choi, W., Choi, K., Yang, G. et al. (2016). Improving piezoelectric performance of lead-free polymer composites with high aspect ratio $BaTiO_3$ nanowires. *Polym. Test.* 53: 143–148.

26 Feng, Y., Li, W.L., Hou, Y.F. et al. (2015). Enhanced dielectric properties of PVDF-HFP/$BaTiO_3$-nanowire composites induced by interfacial polarization and wire-shape. *J. Mater. Chem. C* 3: 1250.

27 Arbatti, M., Shan, X.B., and Cheng, Z.Y. (2007). Ceramic–polymer composites with high dielectric constant. *Adv. Mater.* 19: 1369–1372.

28 Dang, Z.M., Zhou, T., Yao, S.H. et al. (2009). Advanced calcium copper titanate/polyimide functional hybrid films with high dielectric permittivity. *Adv. Mater.* 21: 2077–2082.

29 Zepu, W., Nelson, J.K., Jianjun, M. et al. (2012). Effect of high aspect ratio filler on dielectric properties of polymer composites: a study on barium titanate fibers and graphene platelets. *IEEE Trans. Dielectr. Electr. Insul.* 19: 960–967.

30 Chinya, I., Pal, A., and Sen, S. (2019). Flexible, hybrid nanogenerator based on zinc ferrite nanorods incorporated Poly (vinylidene fluoride-co-hexafluoropropylene) nanocomposite for versatile mechanical energy harvesting. *Mater. Res. Bull.* 118: 110515–110525.

31 Yu, K., Niu, Y.J., Bai, Y. et al. (2013). Poly(vinylidene fluoride) polymer based nanocomposites with significantly reduced energy loss by filling with core-shell structured $BaTiO_3/SiO_2$ nanoparticles. *Appl. Phys. Lett.* 102: 102903–102907.

32 Chinya, I., Sasmal, A., Pal, A., and Sen, S. (2019). Flexible piezoelectric energy harvesters using different architectures of ferrite-based nanocomposites. *CrystEngComm* 21: 3478–3488.

33 Yuan, Z.H. and Zhang, L.D. (2001). Synthesis, characterization and photocatalyticactivity of $ZnFe_2O_4/TiO_2$ nanocomposite. *J. Mater. Chem.* 11: 1265–1268.

34 Shanmugavani, A. and Selvan, R.K. (2014). Synthesis of $ZnFe_2O_4$ nanoparticles and their asymmetric configuration with $Ni(OH)_2$ for a pseudocapacitor. *RSC Adv.* 4: 27022–27029.

35 Yao, L., Hou, X., Hu, S. et al. (2013). A facile bubble-assisted synthesis of porous Zn Ferrite hollow microsphere and their excellent performance as an anode in lithium ion battery. *J. Solid State Electrochem.* 17: 2055–2060.

36 Fan, G., Gu, Z., Yang, L., and Li, F. (2009). Nanocrystalline zinc ferrite photocatalysts formed using the colloid mill and hydrothermal technique. *Chem. Eng. J.* 155: 534–541.

37 Yu, S.H. and Yoshimura, M. (2000). Direct fabrication of ferrite MFe_2O_4 (M = Zn, Mg)/Fe composite thin films by soft solution processing. *Chem. Mater.* 12: 3805–3810.

38 Sharifi, I., Shokrollahi, H., and Amiri, S. (2012). Ferrite-based magnetic nanofluids used in hyperthermia applications. *J. Magn. Magn. Mater.* 324: 903–915.

39 Chaudhary, R., Kanwar, R.K., and Kanwar, J.R. (2015). Zinc Ferrite nanoparticles as highly effective magnetic resonance imaging contrast agents with emphasis on atherosclerosis. *Nanomater. Mol. Nanotechnol.* 4: 1000164–1000167.

40 Mekap, A., Das, P.R., and Choudhary, R.N.P. (2013). Dielectric, magnetic and electrical properties of $ZnFe_2O_4$ ceramics. *J. Mater. Sci. Mater. Electron.* 24: 4757–4763.

41 Bhalla, A.S., Nair, K.M., Lloyd, I.K. et al. (1994). Ferroic materials: design, preparation and characteristics. *Ceram. Trans.* 43: 112.

42 Bowen, L.J., Gentilman, R.L., Pham, H.T. et al. (1993). Injection molded fine-scale piezoelectric composite transducers. In: *Proceedings of IEEE Ultrasonics Symposium*, 499–503. IEEE.

43 Gentilman, R.L., Fiore, D., Pham, H. et al. (1995). Manufacturing of 1-3 piezocomposite sonopanel transducers. *Proc. SPIE* 2447: 274–281.

44 Niu, Y., Yu, K., Bai, Y., and Wang, H. (2015). Enhanced dielectric performance of $BaTiO_3$/PVDF composites prepared by modified process for energy storage applications. *IEEE Trans. Ultrason. Ferroelectr. Freq. Control* 62: 108–115.

45 Venkatragavaraj, E., Satish, B., Vinod, P.R., and Vijaya, M.S. (2001). Piezoelectric properties of ferroelectric PZT-polymer composites. *J. Phys. D. Appl. Phys.* 34: 487–492.

46 Ervin, J.D., Brei, D., Hoy, C.A.V. et al. (1996). New fabrication process for active microsized metal/ceramic devices. In: *Proceedings of the ASME Aerospace Division*, vol. 52, 695–697. American Society of Mechanical Engineers.

47 Hoy, C.A.V., Barda, A., Griffth, M., and Halloran, J.W. (1998). Microfabrication of ceramics by coextrusion. *J. Am. Ceram. Soc.* 81: 152–158.

48 Seema, A., Dayas, K.P., and Varghese, J.M. (2007). PVDF-PZT-5H composites prepared by hot press and tape casting techniques. *J. Appl. Polym. Sci.* 106: 146–151.

49 Zhang, H., Lamnawar, K., and Maazouz, A. (2015). Fundamental studies of interfacial rheology at multilayered model polymers for coextrusion process. *AIP Conf. Proc.* 1664: 100008–100012.

50 Han, P., Pang, S., Fan, J. et al. (2013). Highly enhanced piezoelectric properties of PLZT/PVDF composite by tailoring the ceramic Curie temperature, particle size and volume fraction. *Sensors Actuators A* 204: 74–78.

51 Jain, A., Kumar, J.S., Mahapatra, D.R., and Kumar, H.H. (2010). *Proceeding of SPIE Smart Structures/NDE*. San Diego: SPIE Digital Library.

52 Jain, A., Kumar, J.S., and Gayathri, A. (2009). *Proceeding of SPIE Smart Structures/NDE*. San Diego: SPIE Digital Library.

53 Ameduri, B. (2009). From vinylidene fluoride (VDF) to the applications of VDF-containing polymers and copolymers: recent developments and future trends. *Chem. Rev.* 109: 6632–6686.

54 Falcao, A.F.D.O. (2010). Wave energy utilization: a review of the technologies. *Renew. Sustain. Energy Rev.* 14: 899–918.

55 DurgaPrasad, P. and Hemalatha, J. (2021). Multifunctional films of poly(vinylidene fluoride)/$ZnFe_2O_4$ nanofibers for nanogenerator applications. *J. Alloys Compd.* 854: 157189–157198.

56 Prabhakaran, T. and Hemalatha, J. (2014). Flexible Films of β-phase poly(vinylidene fluoride)/$ZnFe_2O_4$ polymer nanocomposite for magnetoelectric device applications. *Sci. Adv. Mater.* 6: 1313–1321.

57 Mackay, M.E., Tuteja, A., Duxbury, P.M. et al. (2006). General strategies for nanoparticle dispersion. *Science* 311: 1740–1743.

58 Dang, Z.M., Yuan, J.K., Zha, J.W. et al. (2012). Fundamentals, processes and applications of high-permittivity polymer–matrix composites. *Prog. Mater. Sci.* 57: 660–723.

59 Skinner, D.P., Newnham, R.E., and Cross, L.E. (1978). Flexible composite transducers. *Mater. Res. Bull.* 13: 599–607.

60 Chinya, I. and Sen, S. (2019). Influence of nanoparticle size on nucleation of electroactive phase and energy storage behavior of zinc ferrite/poly(vinylidene fluoride) nanocomposite. *J. Mater. Sci. Mater. Electron.* 30: 5137–5148.

61 Maharajan, M., Mursalin, M., Narjinary, M. et al. (2014). Synthesis, characterization and vapour sensing properties of nanosized $ZnFe_2O_4$. *Trans. Indian Ceram. Soc.* 73: 102–104.

62 Ghosh, P. (2011). *Polymer Science and Technology: Plastics, Rubbers, Blends and Composites*, 3e. McGraw-Hill Education.

63 Martins, P., Caparros, C., Goncalves, R. et al. (2012). Role of nanoparticle surface charge on the nucleation of the electroactive β-poly(vinylidene fluoride) nanocomposites for sensor and actuator applications. *J. Phys. Chem. C* 116: 15790–15794.

64 Andrew, J.S. and Clarke, D.R. (2008). Enhanced ferroelectric phase content of polyvinylidene difluoride fibers with the addition of magnetic nanoparticles. *Langmuir* 24: 8435–8438.

65 Seema, A., Dayas, K.P., and Varghese, J.M. (2007). PVDF-PZT-5H composites prepared by hot press and tape casting techniques. *J. Appl. Polym. Sci.* 106: 146–151.

66 Choi, Y.J., Yoo, M.J., Kang, H.W. et al. (2013). Dielectric and piezoelectric properties of ceramic-polymer composites with 0–3 connectivity type. *J. Electroceram.* 30: 30–35.

67 Liu, S.H., Xue, S.X., Xiu, S.M. et al. (2016). Surface-modified $Ba(Zr_{0.3}Ti_{0.7})O_3$ nanofibers by Polyvinylpyrrolidone filler for poly(vinylidene fluoride) composites with enhanced dielectric constant and energy storage density. *Sci. Rep.* 6: 26198–26208.

68 Alam, M., Ghosh, S.K., Sultana, A., and Mandal, D. (2015). Lead-free $ZnSnO_3$/MWCNTs-based selfpoled flexible hybrid nanogenerator for piezoelectric power generation. *Nanotechnology* 26: 165403–165409.

69 Li, Z., Fredin, L.A., Tewari, P. et al. (2010). *In-situ* catalytic encapsulation of core-shell nanoparticles having variable shell thickness: dielectric and energy storage properties of high-permittivity metal oxide nanocomposites. *Chem. Mater.* 22: 5154–5164.

70 Chinya, I. and Sen, S. (2018). Surface modified zinc ferrite (ZF)/polyvinylidene fluoride (PVDF) nanocomposite: a novel material for application as a flexible energy harvester. *Mater. Today Proc.* 5: 10047–10053.

71 Chen, G., Yang, W.L., Lin, J. et al. (2017). Geometrical shape adjustment of $KTa_{0.5}Nb_{0.5}O_3$ nanofillers for tailored dielectric properties of $KTa_{0.5}Nb_{0.5}O_3$/PVDF composite. *J. Mater. Chem. C* 5: 8135–8143.

72 Liu, S.H., Xiu, S.M., Shen, B. et al. (2016). Dielectric properties and energy storage densities of poly(vinylidene fluoride) nanocomposite with surface hydroxylated cube shaped Ba0.6Sr0.4TiO3 Nanoparticles. *Polymers* 8: 45–55.

73 Pan, Z.B., Yao, L., Zhai, J. et al. (2017). High-energy-density polymer nanocomposites composed of newly structured one-dimensional $BaTiO_3@Al_2O_3$ nanofibers. *ACS Appl. Mater. Interfaces* 9: 424–433.

74 Deng, F., Zheng, Q.S., Wang, L.F., and Nan, C.W. (2007). Effects of anisotropy, aspect ratio, and nonstraightness of carbon nanotubes on thermal conductivity of carbon nanotube composites. *Appl. Phys. Lett.* 90: 021914–021916.

75 Yao, S.H., Dang, Z.M., Jiang, M.J., and Xu, H.P. (2007). Influence of aspect ratio of carbon nanotube on percolation threshold in ferroelectric polymer nanocomposite. *Appl. Phys. Lett.* 91: 212901–212903.

76 Shingne, N., Geuss, M., Hartmann-Azanza, B. et al. (2013). Formation, morphology and internal structure of one-dimensional nanostructures of the ferroelectric polymer P(VDF-TrFE). *Polymer* 54: 2737–2744.

77 Abdalla, S., Obaid, A., and Marzouki, F.M.A. (2016). Preparation and characterization of poly(vinylidene fluoride): a high dielectric performance nano-composite for electrical storage. *Respir. Physiol.* 6: 617–626.

78 Sharma, T., Naik, S., Langevine, J. et al. (2015). Aligned PVDF-TrFE nanofibers with high-density PVDF nanofibers and PVDF core–shell structures for endovascular pressure sensing. *IEEE Trans. Biomed. Eng.* 62: 188–195.

79 Butt, H.J., Cappella, B., and Kapp, M. (2005). Force measurements with the atomic force microscope: technique, interpretation and applications. *Surf. Sci. Rep.* 59: 1–152.

80 Maji, S., Sarkar, P.K., Mandal, D. et al. (2015). Selforiented b-crystalline phase in the polyvinylidene fluoride ferroelectric and piezo-sensitive ultrathin Langmuir-Schaefer film. *Phys. Chem. Chem. Phys.* 17: 8159–8165.

81 Li, Y., Ge, X., Wang, L. et al. (2013). Dielectric relaxation behavior of PVDF composites with nanofillers of different conductive nature. *Curr. Nanosci.* 9: 679–685.

82 Dang, Z.M., Yuan, J.K., Yao, S.H., and Liao, R.J. (2013). Flexible nanodielectric materials with high permittivity for power energy storage. *Adv. Mater.* 25: 6334–6365.

83 Ortiz, R.P., Facchetti, A., and Marks, T.J. (2010). High-k organic, inorganic and hybrid dielectrics for low-voltage organic field- effect transistors. *Chem. Rev.* 11: 205–239.

84 Balberg, I., Azulay, D., Toker, D., and Millo, O. (2004). Percolation and tunneling in composite materials. *Int. J. Mod. Phys. B* 18: 2091–2121.

85 Hummel, R.E. (2001). Electrical properties of polymers, ceramics, dielectrics, and amorphous materials. In: *Electronic Properties of Materials*, 3e, 166–193. NY: Springer Verlag.

86 Park, K.I., Lee, M., Liu, Y. et al. (2012). Flexible nanocomposite generator made of $BaTiO_3$ nanoparticles and graphitic carbons. *Adv. Mater.* 24: 2999–3004.

87 Li, C., Yu, S., Luo, S. et al. (2016). Enhancement of dielectric performance upto GHz of the composites with polymer encapsulated hybrid $BaTiO_3$–Cu as fillers: multiple interfacial polarizations playing a key role. *RSC Adv.* 6: 36450–36458.

88 Liu, S.H., Xue, S.X., Xiu, S.M. et al. (2016). Surface-modified $Ba(Zr_{0.3}Ti_{0.7})O_3$ nanofibers by polyvinylpyrrolidone filler for poly(vinylidene fluoride) composites with enhanced dielectric constant and energy storage density. *Sci. Rep.* 6: 26198–26208.

89 Liu, S., Xue, S., Zhang, W., and Zhai, J. (2014). Enhanced dielectric and energy storage density induced by surface-modified BaTiO3 nanofibers in poly(vinylidene fluoride) nanocomposites. *Ceram. Int.* 40: 15633–15640.

90 Chinya, I. and Sen, S. (2018). Surface modified zinc ferrite (ZF)/polyvinylidene fluoride (PVDF) nanocomposite: a novel material for application as a flexible energy harvester. *Mater. Today Proc.* 5: 10047–10053.

91 Gao, L., He, J., Hu, J., and Li, Y. (2014). Large enhancement in polarization response and energy storage properties of poly(vinylidene fluoride) by improving the interface effect in nanocomposites. *J. Phys. Chem. C* 118: 831–838.

92 Karan, S.K., Bera, R., Paria, S. et al. (2016). An approach to design highly durable piezoelectric nanogenerator based on self-poled PVDF/AlO-rGO flexible nanocomposite with high power density and energy conversion efficiency. *Adv. Energy Mater.* 6: 1601016–1601027.

93 Liu, S., Xue, S., Xiu, S. et al. (2016). Surface-modified Ba(Zr0.3Ti0.7)O3 nanofibers by polyvinylpyrrolidone filler for poly(vinylidenefluoride) composites with enhanced dielectric constant and energy storage density. *Sci. Rep.* 6: 26198–26209.

94 Martins, P., Caparros, C., Gonçalves, R. et al. (2012). Role of nanoparticle surface charge on the nucleation of the electroactive β-poly(vinylidenefluoride) nanocomposites for sensor and actuator applications. *J. Phys. Chem. C* 116: 15790–15794.

95 Jayakumar, O.D., Mandal, B.P., Majeed, J. et al. (2013). Inorganic–organic multiferroic hybrid films of Fe_3O_4 and PVDF with significant magneto-dielectric coupling. *J. Mater. Chem. C* 1: 3710–3715.

96 Wang, T., Li, W., Luo, L., and Zhu, Y. (2013). Ultrahigh dielectric constant composites based on the oleic acid modified ferroferric oxide nanoparticles and polyvinylidene fluoride. *Appl. Phys. Lett.* 102: 092904–092907.

97 Martins, P., Costa, C.M., and Lanceros-Mendez, S. (2011). Nucleation of electroactive β-phase poly(vinilidene fluoride) with $CoFe_2O_4$ and $NiFe_2O_4$ nanofillers: a new method for the preparation of multiferroic nanocomposites. *Appl. Phys. A Mater. Sci. Process.* 103: 233–237.

98 Deka, S., Joy, P.A., and Kumar, A.P. (2009). Single step synthesis and properties of M/MFe_2O_4 and $PVDF/M/MFe_2O_4$ (M =Co,Ni) magnetic nanocomposites. *Sci. Adv. Mater.* 1: 262–268.

99 Chen, T., Tang, Q., Wang, B. et al. (2015). Dielectric and magnetic properties of poly(vinylidene fluoride) composites doped with pomegranate-like PPY@$NiFe_2O_4$ nanospheres. *Mater. Lett.* 159: 413–416.

100 Adak, B., Chinya, I., and Sen, S. (2016). Enhanced dielectric and energy storage performance of surface treated gallium ferrite/polyvinylidene fluoride nanocomposites. *RSC Adv.* 6: 105137–105145.

101 Mishra, M.K., Moharana, S., and Mahaling, R.N. (2017). Enhanced dielectric properties of poly(vinylidene fluoride)–surface functionalized $BiFeO_3$ composites using sodium dodecyl sulphate as a modulating agent for device applications. *J. Appl. Polym. Sci.* 134: 45040–45049.

102 Moharana, S., Mishra, M.K., Chopkar, M., and Mahaling, R.N. (2016). Enhanced dielectric properties of surface hydroxylated bismuth ferrite poly(vinylidenefluoride-co-hexa fluoropropylene) composites for energy storage devices. *J. Sci. Adv. Mater. Devices* 1: 461–467.

103 Skinner, D.P., Newnham, R.E., and Cross, L.E. (1978). Flexible composite transducers. *Mater. Res. Bull.* 13: 599–607.

104 Abdalla, S., Obaid, A., and Marzouki, F.M.A. (2016). Preparation and characterization of poly(vinylidene fluoride): a high dielectric performance nano-composite for electrical storage. *Respir. Physiol.* 6: 617–626.

105 Tazaki, M., Wada, R., Okabe, M., and Homma, T. (1997). Crystallization and gelation of poly(vinylidene fluoride) in organic solvents. *J. Appl. Polym. Sci.* 65: 1517–1524.

106 Liang, C.L., Mai, Z.H., Xie, Q. et al. (2014). Induced formation of dominating polar phases of poly(vinylidene fluoride): positive ion–CF2 dipole or negative ion–CH2 dipole interaction. *J. Phys. Chem. B* 118: 9104–9111.

107 Sihvola, A. (1999). *Electromagnetic Mixing Formulas and Applications*, 68. London: The Institute of Electrical Engineering.

108 Dang, Z.M., Yuan, J.K., Zha, J.W. et al. (2012). Fundamentals, processes and applications of high-permittivity polymer-matrix composites. *Prog. Mater. Sci.* 57: 660–723.

109 Brosseau, C. (2006). Modelling and simulation of dielectric heterostructures: a physical survey from an historical perspective. *J. Phys. D. Appl. Phys.* 39: 1277–1294.

110 Myroshnychenko, V. and Brosseau, C. (2005). Finite-element modeling method for the prediction of the complex effective permittivity of two-phase random statistically isotropic heterostructures. *J. Appl. Phys.* 97: 044101.

111 Rao, Y., Qu, J., Marinis, T., and Wong, C.P. (2000). A precise numerical prediction of the effective dielectric constant for polymer-ceramic composite based on effective-medium theory. *IEEE Trans. Compon. Packag. Technol.* 23: 680–683.

112 Ying, K.L. and Hsieh, T.E. (2007). Sintering behaviors and dielectric properties of nanocrystalline Barium Titanate. *Mater. Sci. Eng. B-Sol. St. Mater. Adv. Technol.* 138: 241–245.

113 Kerner, E.H. (1956). The electrical conductivity of composite media. *Proc. Phys. Soc. London, Sect. B* 69: 802–807.

114 Mahato, P.K. and Sen, S. (2015). Effect of surface modification of ceramic particles by SDS on the electrical properties of PZT-PVDF and BT-PVDF composites: interface effect. *J. Mater. Sci. Mater. Electron.* 26: 2969–2976.

115 Xia, W. and Zhang, Z.C. (2018). PVDF-based dielectric polymers and their applications in electronic materials. *IET Nanodielectr.* 1: 17–31.

116 Sun, Y., Zhang, Z., and Wong, C.P. (2005). Influence of interphase and moisture on the dielectric spectroscopy of epoxy/silica composites. *Polymer* 46: 2297–2305.

117 Gowdhaman, P., Annamalai, V., and Thakur, O.P. (2016). Piezo, ferro and dielectric properties of ceramic-polymer composites of 0-3 connectivity. *Ferroelectrics* 493: 120–129.

118 Furukawa, T., Ishida, K., and Fukada, E. (1979). Piezoelectric properties in the composite systems of polymers and PZT ceramics. *J. Appl. Phys.* 50: 4904–4912.

119 Rao, Y. and Wong, C.P. (2004). Material characterization of a high-dielectric constant polymer ceramic composite for embedded capacitor for RF applications. *J. Appl. Polym. Sci.* 92: 2228–2231.

120 Liu, S., Xue, S., Xiu, S. et al. (2016). Surface-modified $Ba(Zr_{0.3}Ti_{0.7})O_3$ nanofibers by polyvinylpyrrolidone filler for poly(vinylidene fluoride) composites with enhanced dielectric constant and energy storage density. *Sci. Rep.* 6: 26198–26208.

121 Chinya, I. and Sen, S. (2018). Surface modified zinc ferrite (ZF)/polyvinylidene fluoride (PVDF) nanocomposite: a novel material for application as a flexible energy harvester. *Mater. Today Proc.* 5: 10047–10053.

122 Sarkar, S., Garain, S., Mandal, D., and Chattopadhyay, K.K. (2014). Electro-active phase formation in PVDF–BiVO4 flexible nanocomposite films for high energy density storage application. *RSC Adv.* 4: 48220–48227.

123 Prateek, V., Thakur, K., and Gupta, R.K. (2016). Recent progress on ferroelectric polymer-based nanocomposites for high energy density capacitors: synthesis, dielectric properties, and future aspects. *Chem. Rev.* 116: 4260–4317.

124 Rahimabady, M., Mirshekarloo, M.S., Yao, K., and Lub, L. (2013). Dielectric behaviors and high energy storage density of nanocomposites with core–shell $BaTiO_3$@TiO_2 in poly(vinylidene fluoride-hexafluoropropylene). *Phys. Chem. Chem. Phys.* 15: 16242–16248.

125 Luo, S., Yu, S., Sun, R., and Wong, C.P. (2014). Nano Ag-deposited BaTiO3 hybrid particles as fillers for polymeric dielectric composites: toward high dielectric constant and suppressed loss. *ACS Appl. Mater. Interfaces* 6: 176–182.

126 Li, C., Yu, S., Luo, S. et al. (2016). Enhancement of dielectric performance upto GHz of the composites with polymer-encapsulated hybrid BaTiO3–Cu as fillers: multiple interfacial polarizations playing a key role. *RSC Adv.* 6: 36450–36458.

127 Li, Y.C., Tjong, S.C., and Li, R.K.Y. (2011). Dielectric properties of binary Polyvinylidene fluoride/Barium Titanate nanocomposites and their nanographite doped hybrids. *Express Polym Lett* 5: 526–534.

128 Sultana, A., Alam, M.M., Garain, S. et al. (2015). An effective electrical throughput from PANI supplement ZnS nanorods and PDMS-based flexible piezoelectric nanogenerator for power up portable electronic devices: an alternative of MWCNT filler. *ACS Appl. Mater. Interfaces* 7: 19091–19097.

129 Zhang, X.J., Wang, G.S., Wei, Y.Z. et al. (2013). Polymer-composite with high dielectric constant and enhanced absorption properties based on graphene–CuS nanocomposites and polyvinylidene fluoride. *J. Mater. Chem. A* 1: 12115–12122.

130 Ghosh, S.K., Sinha, T.K., Mahanty, B. et al. (2016). Porous polymer composite membrane based nanogenerator: a realization of self-powered wireless green energy source for smart electronics applications. *J. Appl. Phys.* 12: 174501–174511.

131 Cordero, F. (2015). Elastic properties and enhanced piezoelectric response at morphotropic phase boundaries. *Materials* 8: 8195–8245.

132 Seoul, C., Kim, Y.T., and Baek, C.K. (2003). Electrospinning of poly(vinylidene fluoride)/dimethylformamide solutions with carbon nanotubes. *J. Polym. Sci. B* 41: 1572–1577.

133 Wang, K.F., Liu, J.M., and Ren, Z.F. (2009). Multiferroicity: the coupling between magnetic and polarization orders. *Adv. Phys.* 58: 321–448.

134 Liang, C.L., Mai, Z.H., Xie, Q. et al. (2014). Induced formation of dominating polar phases of poly(vinylidene fluoride): positive ion–CF_2 dipole or negative ion–CH_2 dipole interaction. *J. Phys. Chem. B* 118: 9104–9111.

135 Dakin, T.W. (2006). Conduction and polarization mechanisms and trends in dielectric. *IEEE Electr. Insul. Mag.* 22: 11–28.

136 Hoque, N.A., Thakur, P., Bala, N. et al. (2016). Tunable photoluminescence emissions and large dielectric constant of the electroactive poly(vinylidene fluoride–hexafluoropropylene) thin films modified with SnO2 nanoparticles. *RSC Adv.* 6: 29931–29943.

137 Chinya, I., Sasmal, A., and Sen, S. (2020). Conducting polyaniline decorated in-situ poled ferrite nanorod-PVDF based nanocomposite as piezoelectric energy harvester. *J. Alloys Compd.* 815: 152312–152319.

138 Park, K.I., Lee, M., Liu, Y. et al. (2012). Flexible nanocomposite generator made of BaTiO3 nanoparticles and graphitic carbons. *Adv. Mater.* 24: 2999–3004.

8

Conductive Filler-Based Hybrid Piezoelectric Materials

Ankur Shukla[1]* *and Priyanka Gupta*[2]*

[1] *Indian Institute of Technology Delhi, Huaz Khas, New Delhi, India*
[2] *Department of Textile and Fibre Engineering, Indian Institute of Technology Delhi, Huaz Khas, New Delhi, India*

8.1 Introduction

Inexhaustible energy sources are available in nature and may be harvested on a massive scale. Pursuing renewable, ecologically sustainable, and economically viable energy sources has become progressively more crucial to addressing future energy requirements. A diverse array of renewable energy sources exist within our surroundings, encompassing kinetic energy, solar energy, and bioenergy. Extensive investigation has explored energy-harvesting technologies that can harness energy from easily accessible sources such as vibration, human motion, water, air, heat, light, and chemical reactions.

A major issue, however, in this age of the Internet of Things (IoT) is the harvesting of energy on a significant microgrid scale [1]. With the unprecedented growth of science and technology in the 21st century, the need for sustainable and wearable energy harvesters continuously evolves, resulting in increasingly stringent material requirements.

In recent studies, there has been a significant emphasis on developing kinetic energy harvesting through three primary transduction mechanisms: electromagnetic, piezoelectric, and triboelectric. Piezoelectric transduction has garnered the most attention among these mechanisms due to its advantageous characteristics, such as superior power densities, high-energy conversion efficiency, simpler architectural designs, and scalability [2, 3]. Consequently, there has been a significant surge in the

*Equal contributors.

Hybrid Materials for Piezoelectric Energy Harvesting and Conversion, First Edition.
Edited by S. Wazed Ali, Satyaranjan Bairagi, and Shahid Ul Islam.
© 2024 John Wiley & Sons, Inc. Published 2024 by John Wiley & Sons, Inc.

utilization of piezoelectric materials in various domains, such as sensors [4–6], actuators [7], nanogenerators [8–14], MEMS devices [15, 16], portable electronics [17, 18], composites [19], and biomedicines [1, 20–23].

The utilization of PZT based devices in various consumer applications, including smart systems, medical imaging devices, automobiles, and sound generators, has prompted apprehension owing to the potential toxicity associated with the presence of lead (Pb) [24]. PZT-based ceramics are also not suitable for high-temperature applications. In contrast, lead-free ceramics such as $KNbO_3$ and $NaNbO_3$ are biocompatible, allowing for their versatile use in sensors and actuators that can be directly implanted into living organisms. Furthermore, the properties of these lead-free ceramics can be easily customized, making them a preferable alternative to Pb-based PZT ceramics. Polymers, on the other hand, offer high flexibility, excellent stability, and biocompatibility, making them well-suited for integration into flexible devices.

Piezoelectric materials help harvest electrical energy when subjected/exposed to specific mechanical stimuli, such as stress, elongation, tension, vibration, and pressure [25]. They convert the mechanical energy to electrical energy, as shown in Figure 8.1. The phenomenon of energy conversion is observed in piezoelectric materials due to their molecular structure, which is arranged in a manner that leads to the manifestation of an electric dipole, resulting in a local charge separation. Applying mechanical pressure on such materials causes a deformation of the dipole, leading to the generation of a charge that can be harnessed to power diverse devices. Piezoelectric materials are commonly used in sensor applications and are present in various devices, such as accelerometers, microphones, and load cells. A significant proportion of contemporary research in power harvesting utilizing piezoelectric materials has been directed toward enhancing the efficacy of piezoelectric power harvesting systems.

Lead zirconate titanate, commonly called PZT, is the prevailing piezoelectric material utilized in power harvesting applications. PZT is a piezoceramic type known for its ability to convert mechanical energy into electrical energy. Despite its widespread use, the inherent brittleness of PZT constrains the strain it can endure without incurring damage [27]. Piezoceramics are prone to fatigue crack propagation under high-frequency cyclic loading. Poly(vinylidene fluoride) (PVDF) is another type of piezoelectric material characterized by its significant flexibility compared to PZT. The materials exhibiting piezoelectric behavior are often embedded with fillers such as ceramics, Ag or silver nanowires (AgNWs), graphene, reduced graphene oxide (rGO), and carbon nanotubes (CNTs), which improve their performance in terms of dielectric constant and piezoelectric

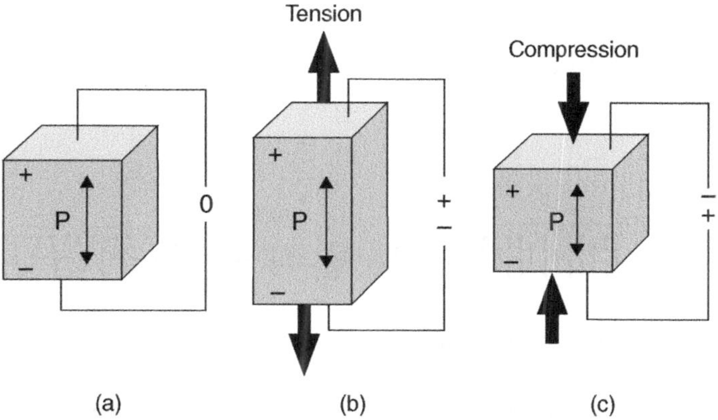

Figure 8.1 Schematic representation of piezoelectric effect. Source: Reused with permission from [26].

properties. [27, 28] also demonstrated the use of semiconducting polymer PEDOT/PSS by coating it over a PVDF film. Though ceramics offer higher dielectric constant, they result in certain limitations, such as high weight, low flexibility, and poor mechanical performance [29, 30]. Conductive fillers can tune the polymer microstructure, enhancing the dielectric and piezoelectric properties. The application of piezoelectric materials using conductive fillers is a fast-growing area of research. Thus, there is a requirement to review the recent research studies that could pave the path for future research studies. The present chapter discusses in detail the various conductive fillers used for the fabrication of piezoelectric materials and their effect on piezoelectric properties.

8.2 Piezoelectricity: A Brief Overview

Piezoelectricity is generated when a force is applied to deform the piezoelectric material. Initially, in the absence of an applied electric field, the dipole moments inherent in the piezoelectric material exhibit a random alignment between two electrodes (Figure 8.2a) [31]. Without an externally applied electric field, the dipole moments inherent in the piezoelectric material display a random orientation between two electrodes (Figure 8.2b). Without an external force, the device remains in equilibrium and does not acquire any electric signal. However, the material becomes polarized due to the compressive strain when a compressive force is applied in the vertical direction. This polarization induces a piezoelectric potential between the electrodes. During the progression of this procedure, a signal is obtained as an output

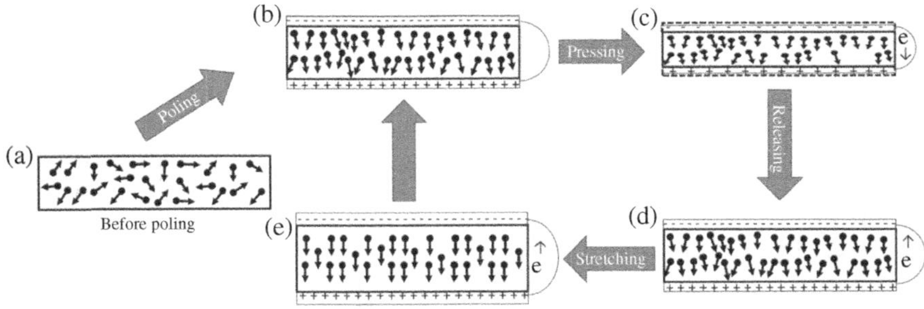

Figure 8.2 Working mechanism of a piezoelectric energy harvester.

(Figure 8.2c). When the applied force is released, a minor tensile force arises, generating a reverse piezoelectric potential (Figure 8.2d,e) [21]. In the context of energy harvesting, the generation of charge (Q) across the opposing surfaces of a piezoelectric material with a given area (A) in response to the application of mechanical stress ($\Delta\sigma$) can be mathematically described by the following equation [32]:

$$Q = d_{33} \times A \times \Delta\sigma$$

where the material's piezoelectric coefficient (pC/N) in 33-mode is denoted by d_{33}.

8.3 Role of Conductive Fillers in Piezoelectric Materials

Conductive fillers play a significant role in enhancing the properties and performance of piezoelectric materials. These materials exhibit the phenomenon of generating electric charges when subjected to mechanical stress, and they also undergo deformation when exposed to an electric field. Conductive fillers are added to piezoelectric polymers or ceramics to modify their electrical and mechanical properties. Some of the roles and benefits of conductive fillers in piezoelectric materials include:

(i) *Improved electrical conductivity:* Conductive fillers, such as CNTs, graphene, or metals, can significantly increase the electrical conductivity of the piezoelectric material. This enhanced conductivity can lead to better charge generation and distribution, improving the piezoelectric response.

(ii) *Electromechanical coupling:* Conductive fillers can improve the electromechanical coupling coefficient of piezoelectric materials.

This coefficient represents the efficiency of the conversion between mechanical and electrical energy. Higher coupling coefficients result in more effective energy conversion.

(iii) *Enhanced piezoelectric properties:* Conductive fillers can modify the piezoelectric properties of the base material. They can enhance the piezoelectric constants, such as d_{33} and d_{31}, which are measures of the material's sensitivity to mechanical stress and electrical field, respectively. This enhancement leads to improved sensitivity and responsiveness in sensing and actuation applications.

(iv) *Reduced electrical impedance:* Conductive fillers can help lower the electrical impedance of piezoelectric materials. This is particularly important for applications requiring low impedance, such as high-frequency transducers or medical ultrasound devices. Lower impedance ensures efficient energy transfer between the electrical circuit and the piezoelectric material.

(v) *Improved thermal conductivity:* Many conductive fillers have higher thermal conductivity than the base piezoelectric materials. Adding these fillers can help dissipate heat more effectively, preventing overheating and maintaining stable performance under demanding operating conditions.

(vi) *Mechanical reinforcement:* Conductive fillers can enhance the mechanical properties of the piezoelectric material. This reinforcement can improve durability, toughness, and resistance to mechanical stresses. As a result, the material can withstand repeated deformations without experiencing significant fatigue.

(vii) *Uniform electric field distribution:* Conductive fillers can help distribute the applied electric field more uniformly across the piezoelectric material. This uniform distribution minimizes the occurrence of electrical field concentrations that could lead to premature material failure or reduced performance.

(viii) *Tailored properties:* By adjusting the type, concentration, and distribution of conductive fillers, researchers and engineers can tailor the properties of piezoelectric materials to suit specific application requirements. This flexibility allows for the creation of materials optimized for different purposes.

Conductive fillers have a crucial role in enhancing materials' electrical, mechanical, and piezoelectric properties, making them more effective for various applications that rely on piezoelectric effects. Numerous scholars have conducted investigations into the impact of piezoelectric fillers on various performance parameters. Several researchers have shown the benefits of including conductive fillers into the polymer matrix of a piezoelectric composite [33–36], where the electrically conductive filler increases the conductivity of the polymer matrix.

The presence of conductive fillers directly affects the poling efficiencies of the piezoelectric material they are embedded into. The amount of fillers added to a matrix affects its volume conductivity, thus affecting the poling efficiency of the matrix under an electric field. Without conductive fillers, most of the electric field is lost in the matrix only [19]. In addition, the conductive fillers allow more charge to easily reach the surface of the electrodes. Carbon fiber is used as a conductive filler, as it increases the piezoelectric coefficient and improves the polymer damping properties of the reinforced polymers [37]. As observed by Bairagi and Ali, adding 0.1% CNT to the PVDF/KNN-based electrospun fiber web increased the output voltage by 32.8%. The presence of CNTs in an electrospinning solution also increases the mechanical stretching of the fiber produced [38].

The addition of 0.2% to the volume of a polyamide/PZT composite system results in a significant increase in dielectric permittivity along with an increase in conductivity, where a synergistic behavior of CNT particles with PZT particles was observed by [39]. Adding CNTs further improved the piezoelectricity and absorption of sound waves in a low-frequency region [40]. Banerjee and Cook-Chennault examined the impact of varying volume fractions and sizes (micron and nano) of conductive filler (specifically aluminum) on the loss factor (tan δ) and strain coefficient ($d33$) of composite piezoelectric materials. They concluded that the differences in loss factor and strain coefficients between composites containing micron- and nano-sized aluminum particles might be attributed to many variables, including agglomeration, contact resistance, and excessive air gaps [41].

8.4 Conductive Filler-Based Piezoelectric Materials

Conductive filler-based piezoelectric materials typically consist of a piezoelectric matrix material embedded with conductive fillers. The piezoelectric matrix generates mechanical strain when subjected to mechanical stress, resulting in a piezoelectric potential. The conductive fillers facilitate electrical conductivity within the material, allowing the generated electrical charges to be extracted. Various conductive fillers can be used in these materials, such as carbon-based materials (CNTs, graphene, and carbon black [CB]), metallic nanoparticles (silver, gold, and copper), and conducting polymers (polypyrrole and polyaniline). The choice of filler depends on factors such as desired electrical conductivity, mechanical properties, and compatibility with the matrix material. Piezoelectric materials can be categorized into four distinct groups, including crystals, films, ceramics, and polymers [40].

There have been reports of several different types of piezoelectric materials containing semiconductors (ZnO, ZnS, and GaN), ceramics (PZT, $BaTiO_3$, $NaNbO_3$, $KNbO_3$, KNN, and $ZnSnO_3$), polymers (PVDF and its copolymers, polyamides, and parylene-C), and natural materials such as bones, hair, collagen fibrils, and peptides. The improved contact between the electrodes and the material, as a result of the conductive filler, increases the usability of piezoelectric materials. Additionally, higher poling efficiency is produced by the presence of conductive fillers.

Ceramic piezoelectric materials exhibit high electroactive potential and high piezoelectric coefficient but low durability and high toxicity. On the other hand, electroactive polymers, such as PVDF have benefits, including being lightweight, stretchable, conformable, having high biocompatibility, and having low acoustic impedance, making them appropriate for usage in flexible sensors, sonars, and actuators [42].

8.4.1 Carbon-Based Piezoelectric Materials

Due to its piezoelectric and pyroelectric properties, high thermal stability, elasticity, and chemical resistance, PVDF has been the most suitable material for piezoelectric applications. It exists in five different crystalline forms known as α, β, δ, γ, and ε phase, of which the β-phase has the highest dielectric constant [43]. CNTs have a wide range of potential applications due to their electrical, thermal, and mechanical properties, as well as their low density, high aspect ratio, and high stiffness. With the addition of a small amount of CNTs, the electrical conductivity of polymer doped with CNTs exhibits a dramatic increase [44]. CNTs have extremely small dimensions and a coiled and twisted structure, frequently resulting in the particles' agglomeration when embedded in a polymer matrix. Therefore, for using CNTs as filler, different methods such as ultrasonication, ball milling, calendaring, and functionalization of CNTs are used to prevent their agglomeration [45]. For better adhesion between CNT and the polymer while embedding, various surface modification methods, such as grafting, must be done [46–48]. The addition of CNT fillers in the PVDF matrix remarkably enhances piezoelectric properties, and thermal, mechanical, and pyroelectric properties (Figure 8.3) [49, 50]. The application of poling and stress to the films helps in the alignment of polymer chains normal to the direction of applied

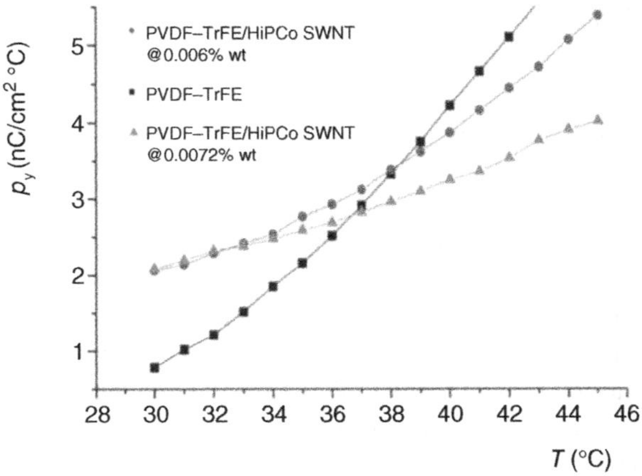

Figure 8.3 Pyroelectric coefficient for PVDF piezoelectric nanogenerator doped with different fillers. Source: Reprinted with permission from Levi et al. [49].

stress, thus, resulting in higher β-phase content and higher piezoelectric properties when MWCNT is added to PVDF composite [51].

CNT is frequently combined with other fillers to form a three-component system in which PVDF contains two distinct filler types that confer multiple properties on the composite. They are frequently referred to as hybrid piezoelectric materials. Bairagi and Ali [38, 52] fabricated a nanogenerator using electrospinning technique, wherein CNT and KNN nanorods were combined to form piezoelectric nanogenerators (PENG). It has been observed that the response of piezoelectric materials containing both KNN and CNTs is superior to that of materials containing only KNN nanorods. This is because nanorods have a greater specific surface area than particles, resulting in the accelerated formation of β-crystals, forming a greater number of directed dipoles that enhance the piezoelectric response. The output voltage of the PENG based on KNN nanorod with PVDF polymer-based nanocomposite material produced by solution casting was ~3.4 V, and that obtained using melt spinning methods was ~3.7 V [52, 53]. Bairagi and Ali also demonstrated the effectiveness of electrospinning for developing KNN/PVDF-based PENGs, which could produce an output of ~17.5 V. The enhanced voltage output of the electrospun nanocomposite fibers is attributed to the concurrent in situ polarization of the PVDF polymer chains and the KNN nanorods, which take place in a more intense electric field. In the electrospinning procedures, the heightened electric field enables effective mechanical elongation of the PVDF polymer chains, thereby impacting the piezoelectric capabilities of the nanogenerator. In the case of an electrospun polymer, the filler loading may be less than with other techniques that guarantee the necessary flexibility of the generated nanogenerator [54, 55].

For further improvement in harvesting efficiency, [38] added CNT to KNN/PVDF-based material and used electrospinning to fabricate fibrous webs. Conductive filler allows a separate charge to easily flow toward the nanogenerator's surfaces or electrodes. The hybrid material significantly improved the output voltage to 23.24 V and a current of 9 A for PVDF/3% KNN/0.1% CNTs. The inclusion of CNTs in the PVDF/KNN solution resulted in a nucleation effect in the electrospun fibers. This effect was verified by determining the beta fraction value of the hybrid materials (84% for PVDF/3% KNN/0.1% CNTs). Moreover, the incorporation of CNTs into the solution of PVDF/KNN/CNTs during the process of electrospinning resulted in the formation of a conductive pathway within the final piezoelectric device. This inclusion of CNTs also improved the ability of the nanofibrous web to withstand mechanical stretching.

For rigid damping materials, Ma and Wang [35] examined the microstructure and characteristics of PMN–PZT piezoceramics with epoxy-based composites containing CNTs. The researchers determined that a consistent network of electrical conductivity was established within the range of the percolation threshold (1.0–1.5 g CNTs per 100 g epoxy). These materials showed superior dynamic mechanical loss factors compared to those that did not contain semiconductive inclusions and those below the percolation threshold [35]. Tian and Wang [56] also created and tested epoxy/multi-walled CNT/piezoelectric ceramic composites. They also concluded that the integration of PZT and CNT has improved loss factors with higher CNT content ([56]).

Graphene is a two-dimensional, thinnest carbon material with sp^2 hybridized carbon atoms packed closely. The addition of graphene particles to polymer matrices is known to enhance different properties such as tensile strength, thermal conductivity, electrical conductivity, electromagnetic interference (EMI) shielding, and low coefficient of thermal expansion [57, 58]. In a recent study, the graphene layer was used to deposit a vertical ZnO nanotube attached to a polyimide (PI) film to fabricate high-resolution and highly sensitive pressure sensors [59]. When graphene is oxidized using oxygen-containing functional groups, the surface of graphene oxide (GO) shows wrinkles or ripples that appear as two-dimensional structures over an otherwise plane surface of the GO layer. By reducing the functional groups present in GO, the properties of graphene and GO can be found in the resulting rGO [60]. The chemically reduced graphene oxide (CrGO) in the presence of a dielectric matrix increases the energy output when exposed to a cyclic mechanical strain. Kumar et al. [25] embedded the CrGO in a room-temperature vulcanized rubber. They observed the energy output increased with the number of cycles of the applied mechanical strain after the initial stabilization of the material (Figure 8.4) [25].

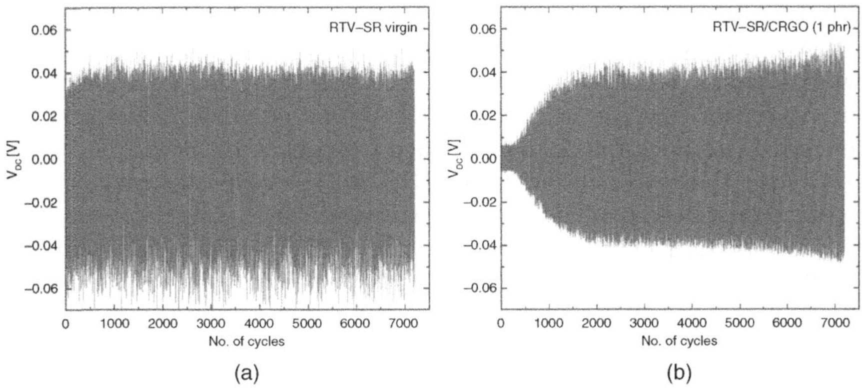

Figure 8.4 (a) Output voltages for RTV–SR silicone rubber composite, and (b) output voltages for RTV–SR/CrGO (1phr). Source: Reproduced from Kumar et al. [25]/Express Polymer Letters.

8.4.2 Metal-Based Piezoelectric Materials

Silver is often used as conductive filler in the form of AgNW to fabricate piezoelectric devices. AgNWs can be functionalized using different polymers where AgNw acts as an electrode, and the coated polymer would help in sensing [61]. AgNW offers high conductivity and flexibility, making it suitable for fabricating flexible electrodes for better contacts for PU-based piezoelectric devices [62]. Few other applications of AgNWs as electrodes involve their applications by spraying the AgNW solution or AgNWs dispersion in clay on the flexible piezoelectric devices [63].

Chiu and Kao [64] researched conductive composites containing silver particles in liquid epoxy resin. They explored the impact of introducing an additional element, CB or carbon fibrils (CFs). In the epoxy–silver–CB composites, they noticed large clusters of Ag particles and no visible CB networks. This was attributed to interference between the two conductive networks, where CB obstructed Ag particle dispersion. They also noted that Ag particles disrupted CB network connectivity. However, in the epoxy–silver–CFs composites, CFs acted as conductive bridges, linking dispersed Ag clusters and enhancing composite conductivity [64]. Karttunen et al. [65] observed that when nano-sized silver particles were incorporated into a thermoplastic elastomer and formed slight clusters, the point at which the conductive network formed (percolation threshold) was found to be between 13% and 16%, and the electrical resistance measured $2.0 \times 10^{-1}\ \Omega\,\text{cm}$. Conversely, when these silver particles heavily clustered together, the percolation threshold increased to 20%, resulting in higher electrical resistance, which was measured at $2.9 \times 10^{3}\ \Omega\,\text{cm}$.

ZnO has emerged as a prevalent material for developing flexible piezoelectric devices. However, there have been controversies regarding the linear response and the deformation of piezoelectric devices. The same was resolved by fabricating a thin layer of epitaxial ZnO over the CeO_2/Hastelloy substrate. Hastelloy is an alloy of nickel, molybdenum, chromium, iron, and silicon. The ultrathin (~32 nm) layer of ZnO could effectively produce a voltage up to 1 V for a change of 8.14 pm vertically [66]. The high orientation properties of ZnO have enabled it to be used in many piezoelectric applications. Recently, a hybrid piezo/triboelectric (H–P/TENGs) generator, wherein the orientation of ZnO plays a deciding role in increasing the output power density of the nanogenerator by up to 17 times [67]. When used with TiO_2, the coatings prepared via hydrothermal analysis exhibit both piezoelectric and antimicrobial properties. A higher piezoelectric output (22.2 pm/V) is obtained when TiO_2 particles are mixed with hydroxyapatite, a bioactive ceramic, and ferroelectric properties [68]. Multi-element doping is often done to prepare lead-free piezoelectric ceramics that produce excellent piezoelectric properties of 523 pC/N (Yan et al., 2022). As the concentration of barium titanate increases, the response also increases. The electrospun fiber mat prepared using polyvinylidene fluoride–trifluoroethylene doped with barium titanate exhibited a linear response with external force and electrical output [69]. The inclusion of conducting metal oxides imparts multifunctional properties to piezoelectric materials and increases piezoelectric output.

8.4.3 Polymer-Based Piezoelectric Materials

Polymers, which comprise carbon-based materials, exhibit a piezoelectric phenomenon due to their molecular composition and alignment. Compared to ceramics, polymers are considerably softer and demonstrate moderate strain coefficients (d_{33}) and voltage coefficients (g_{33}). Nevertheless, their distinctive characteristics make them suitable for various energy-harvesting applications. Piezoelectricity can be observed in amorphous or non-crystalline polymers with dipoles within their molecular structure. The alignment of these dipoles occurs through a process known as poling, which is carried out at temperatures exceeding the glass transition temperature (T_g) of the polymer. Table 8.1 shows the piezoelectric coefficients of some polymers used for piezoelectric applications.

8.4.3.1 PVDF-Based Polymeric Materials

Different types of PVDF-based polymers are used for piezoelectric device fabrication, as shown in Figure 8.5. High d_{33} values could be achieved using

Table 8.1 Piezoelectric coefficients of different polymers.

Polymer name	d_{33}	d_{31}	d_{14}	References
PVDF	−24 to −34	8–22	—	[70, 71]
P(VDF–TrFE)	24–40	12–25	—	([72]; [73])
P(VDF–CTFE)	140	—	—	[74]
P(VDF–HFP)	24	30–43	—	([75]; [76]; [77])
Polyamide 11	4	14 at 100–200 °C	—	([78]; [79])
Polyimide	2.5–16.5	—	—	[16]
Polylactic acid (PLA)	—	1.58	9.82	([80]; [81])
Cellulose	5.7 ± 1.2	1.88–30.6	−35–60	([82]; [83]; [84])
Polyurethane	—	27.2	—	[85]
Polyurea	19 at 60 °C	10	—	([81]; [86])
Polyacrylonitrile (PAN)	—	2	—	[87]
Parylene-C	2	—	—	[16]
Liquid crystal polymers	−70	—	—	[32]

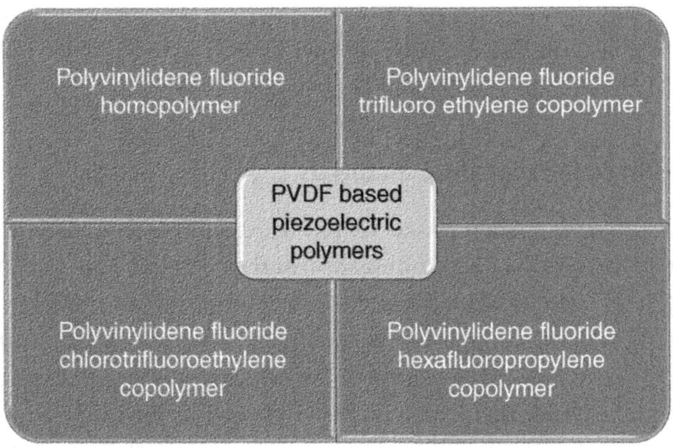

Figure 8.5 PVDF-based piezoelectric polymers.

PVDF nanofibers spun using electrospinning, resulted in almost four times the d_{33} value (−63.25 pC/N in a single fiber) compared to that of PVDF film (≅15 pC/N). The output voltages from 5 to 30 mV and current from 0.5 to 3 nA, with a peak voltage of 0.2 mV and a current of 35 nA could be achieved using parallelly connected 500 nanofibers, subjected to repeated mechanical straining [21]. Hollow electrospun nanofibers could result in a maximum voltage up to 76 mV and a current of 39 nA due to the formation of a high β phase when subjected to continuous stretching and releasing at a strain of 0.05% and a frequency of 7 Hz ([88]). The dipoles could be aligned perpendicular to the substrate using the Langmuir–Blodgett technique, which leads to the direct formation of β phase over ultrathin PVDF films. This is due to the formation of hydrogen bonds between the PVDF film and the water molecules, helping to achieve d_{33} values up to 49 pC/N ([89]).

Another semicrystalline polymer is formed by copolymerization of polyvinylidene fluoride and trifluoroethylene, also known as P(VDF–TrFE). The inclusion of TrFE in PVDF leads to more β-phase formation, and the piezoelectric properties occur at temperatures below the Curie temperature, where the crystalline phase undergoes reorientation in response to a high-intensity poling field. To further increase the F(β) up to 66.33%, consecutive annealing–pressing–poling could be done over P(VDF–TrFE) films [90]. Annealing itself could increase the crystallinity up to 19% when performed at 130 °C for 48 h due to the increased transition from α to β phase. Lee et al. [91] conducted a study in which they utilized micropatterning techniques to create pyramid-shaped and trigonal line-shaped structures on P(VDF–TrFE) material (J.-H. [91]). The micropatterned P(VDF–TrFE)-based PENG exhibited a significantly higher output (4.4 V and 3.4 μA) compared to the flat film PENG (1 V and 0.1 μA) when subjected to vertical compressive force. The increased performance can be attributed to the greater level of strain that is generated in the micropatterned films when subjected to the same applied force. In a separate study, nanowires of PVDF and P(VDF–TrFE) (70/30) were fabricated by employing a template-wetting technique within the channels of APA membranes [92]. The inclusion of TrFE with PVDF strongly impacts the piezoelectric properties of the resulting PENG.

In order to obtain a higher strain response, chlorofluorotrifluoroethylene (CFTE) is often copolymerized with VDF to obtain P(VDF–CTFE) [93]. However, a higher electric field is required to generate a higher strain response. A small amount of up to 12% CFTE used during copolymerization demonstrates a significant electrostrictive strain response exceeding 5% and a d_{33} value of 140 pC/N when subjected to polarization at an electric field strength of 70 MV/m ([74]). It is often copolymerized with hexafluoropropylene (HFP) to impart elastomeric properties to the PVDF film. At low concentrations, HFP imparts thermoplasticity to the polymer film, but at slightly higher concentrations, the film exhibits elastomeric behavior [94–97]. When drawn uniaxially, such films also show higher d_{31} values of about 30 pC/N, whereas the films produced via extrusion followed by stretching and poling showed d_{33} value of up to 24 pC/N [75, 76].

8.4.3.2 Non-PVDF-Based Piezoelectric Polymers

Several polymers exhibit piezoelectric properties, including PLA, polyureas, PAN, and polyurethanes. PLA is gaining traction as a favorable energy-harvesting choice due to its exceptional piezoelectric characteristics. Carbonyl groups in PLA induce polarity, resulting in a d_{14} value of 10 pC/N even without poling [80, 98]. A poly-L-lactic acid-based device resulted in a maximum power of 14.7 μW and a voltage of 9.05 V that could charge a 3.2 V battery and a d_{14} value of 9.57 pC/N after poling [99]. Electrospinning PLLA fibers could also fabricate strain sensing and energy harvesting using human body movement [100]. Polyurea is another elastomeric class of polymers that could impart piezoelectric and pyroelectric properties after poling. When attached to a cantilever, they can even be used for power generation with an output efficiency of 0.233% at a load resistance of 1 MΩ [101]. Also, aliphatic polyureas have been reported to have lower piezoelectric coefficients compared to aromatic ureas, along with high dependence on poling temperature when an odd number of $-CH_3$ groups are present [87]. Polyurethanes are often used for producing foams that could have high d_{33} coefficients of up to 244 pC/N upon the addition of polar doping molecules, whereas the polyurethane isotropic films could exhibit d_{31} values of 38 pC/N [102, 103]. When subjected to force, the polymer foams can undergo significant alterations in volume. In contrast, dipole-doped polymers exhibit a significantly higher degree of piezo responsiveness than conventional polymers and ceramics. Another class of polymers that exhibit low piezoelectricity are polyamides or nylon. However, as low as 0.5 pC/N d_{33} value could be observed in a poled and stretched nylon film [104]. Odd-numbered nylons have a net dipole moment leading to piezoelectric and pyroelectric properties [98]. PAN, on the other hand, shows a high piezoelectric effect when PAN nanofibers are produced using electrospinning method due to the high amount of sawtooth planar configuration. The PAN nanofibers could produce higher voltage than PVDF under the same conditions, which might be improved further by controlling the poling temperature (W. [103]). In the case of high-temperature applications, PI has proved to work well for piezoelectric applications due to its high T_g value [87].

8.5 Applications

Conductive filler-based hybrid piezoelectric materials have a wide range of piezoelectric properties, good mechanical flexibility, and easier and cheaper ways to make sensors and energy harvesters that work on the millimeter and micrometer scale. Due to their numerous properties, these materials are used in various applications, such as vibrational energy harvesters to extract power from vibrations and impedance matching, acoustic transducers, piezoelectric tactile sensors, and inertial sensors. Piezoelectric materials, particularly composites, are advantageous for low-frequency vibration energy collection due to their flexibility and dependability.

The incorporation of conductive fillers can enhance the overall performance of piezoelectric materials. Electrical conductivity enables efficient charge transfer, improving piezoelectric responses, and better signal detection in sensing applications. Conductive fillers can also improve the mechanical properties of the composite material. They can enhance stiffness, strength, and toughness, crucial for maintaining structural integrity in real-world applications. These materials can be integrated into wireless sensor networks for structural health monitoring (SHM), environmental sensing, and other applications. The harvested energy can power the sensors, reducing the need for manual maintenance and battery replacements. Conductive filler-based piezoelectric materials offer a versatile platform for developing advanced functional materials with synergistic properties [105].

8.5.1 Sensing and Actuation

The ability of conductive filler-based hybrid piezoelectric materials to provide mechanical responsiveness and electrical functionality drives their integration in sensing and actuation applications. Both sensors and detectors can be made using these materials. High-performance vibration, strain, and pressure sensors can be made using these materials. This is useful for applications ranging from seismic detectors to pressure sensors ([106]). Figure 8.6 shows the various applications of these materials in sensing and actuation areas.

Pressure sensors utilize hybrid piezoelectric materials to generate electrical signals through the piezoelectric effect when mechanical pressure or force is exerted on the material. The presence of conductive fillers improves the sensitivity and effectiveness of signal detection. Utilizing these materials, strain and vibration sensors are created to track changes in mechanical deformation. Accurate and real-time measurements are produced due to the transfer of electrical charges by the conductive fillers. Touch and tactile sensors incorporate hybrid piezoelectric materials to detect touch, pressure, or deformation. This technology is very helpful in robotics, prosthetics, and touch screens. Mice and other acoustic sensors can also use hybrid piezoelectric materials based on conductive fillers. The ability of materials

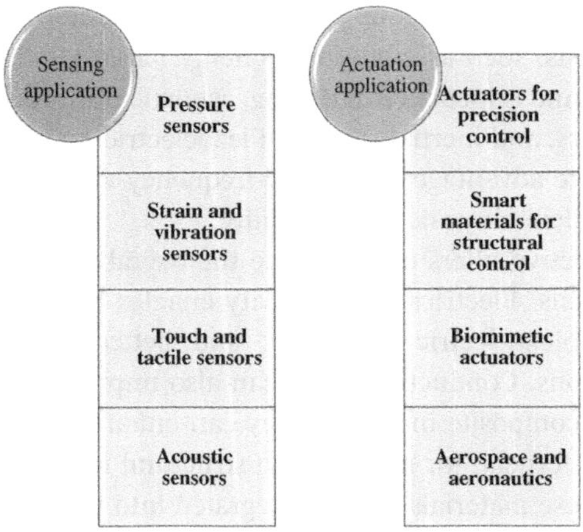

Figure 8.6 Application of conductive filler-based hybrid piezoelectric materials in sensing and actuation.

to transform sound waves into electrical impulses makes sound analysis and detection easier [107–109].

These substances can function as piezoelectric actuators, converting electrical inputs into predetermined mechanical motions. Hybrid piezoelectric materials are used as actuators to produce precise and accurately calibrated mechanical displacements. Devices requiring high precision, also known as precision positioning systems, such as microscopes, nanomanipulators, and microfluidic systems, can utilize these actuators. Utilizing these materials in construction allows for active vibration control, mechanical oscillation damping, and modification of structural properties. This is particularly useful for civil engineering, automotive, and aerospace applications. Biomimetic actuators mimic natural movements and are created using hybrid piezoelectric materials. These components generate realistic movement in soft robotics, prosthetics, and medical equipment. By changing their shape in response to electrical inputs, these materials are used in various applications, such as adaptive aerodynamics and morphing wing designs, which can enhance aerodynamic performance [108, 110–112].

8.5.2 Energy Harvesting

Conductive filler-based hybrid piezoelectric materials have garnered significant attention and found numerous applications in energy harvesting. Energy harvesting converts environmental mechanical energy, such as vibrations, motions, or deformations, into usable electrical energy. Due to their electrical conductivity and piezoelectric properties, these materials are perfectly suited for efficient energy conversion. These materials can capture mechanical energy and convert it into electrical energy to power small devices, attributable to the piezoelectric effect and electrical conductivity. Energy-harvesting devices utilize hybrid piezoelectric materials that incorporate conductive fillers. Electrical energy, which can be stored or used to power low-energy electronic devices, is generated from mechanical vibrations or deformations. The materials produce electrical charges due to vibrations, deformations, or movements. These electrical charges can be collected and stored to power electronics or recharge batteries [113–116]. Figure 8.7 demonstrates the application and utilization of these materials for energy harvesting.

Hybrid piezoelectric materials, which incorporate conductive fillers, can be utilized in devices that capture vibrations from various sources, such as industry, transportation, or human activity. The materials produce electrical charges when they are vibrated due to the piezoelectric effect. The conductive fillers facilitate charge transfer, thereby enhancing the effectiveness of energy harvesting [117]. These components can create flooring systems that harness energy from foot traffic and convert it into electricity. The mechanical forces generated by foot movement in densely populated areas, such as

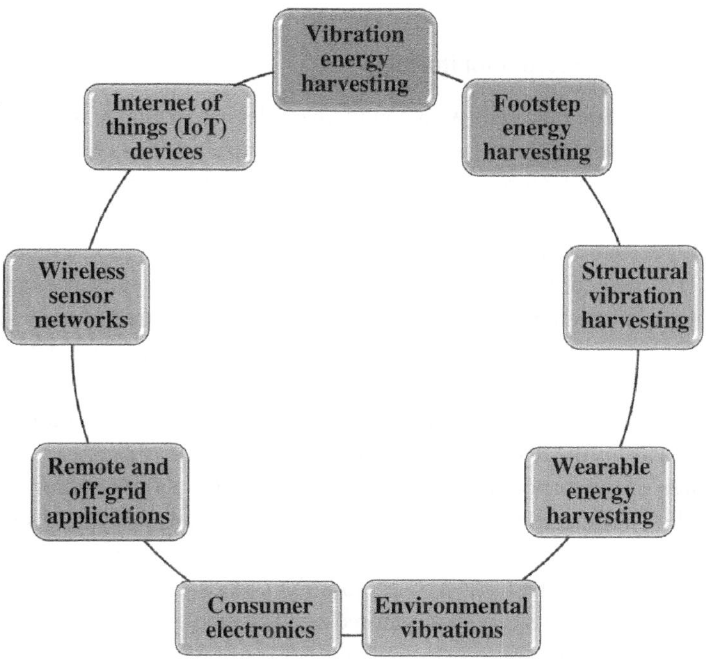

Figure 8.7 Use of conductive filler-based hybrid piezoelectric materials in energy harvesting.

malls or subway stations, can be captured and stored to provide power for low-energy appliances or lighting [118, 119].

Ambient vibrations caused by wind, traffic, or other factors can be converted into electrical energy by incorporating these materials into building structures or infrastructure, such as bridges and highways. With this strategy, it may be possible to generate sustainable energy from existing buildings. These substances can absorb energy from atmospheric vibrations, ocean waves, and wind in urban environments. When subjected to these vibrations, they produce electrical charges, which can be collected and stored for various purposes. These materials can also provide a sustainable energy solution by converting mechanical energy from the environment into electrical power in remote or off-grid areas where conventional power sources are scarce or unavailable [14, 120, 121]. Hybrid piezoelectric materials, based on conductive fillers, are suitable for wearable technology that harnesses power from human movement. They can be incorporated into clothing or accessories to power medical implants, wearable electronics, and health monitoring equipment [122, 123]. Consumer electronics, such as smartwatches and fitness trackers, can utilize energy-harvesting materials to obtain some or all of their power from the user's movements.

8.5.3 Structural Health Monitoring

Hybrid piezoelectric materials with conductive fillers are valuable in SHM, which assesses the durability and security of structures, such as buildings, bridges, aircraft, and more. These materials can monitor the structural health and identify potential damage or stress by being integrated into constructions such as bridges or buildings. These materials have unique qualities for detecting and assessing structural damage, deformation, and changes in mechanical characteristics [124–126]. Wearable medical devices utilize these materials as part of their construction. They can detect physiological factors, such as breathing patterns, muscle contractions, and heartbeats [127]. Figure 8.8 demonstrates the application and utilization of these materials for the SHM area.

Structures can incorporate hybrid piezoelectric materials to function as sensors. The piezoelectric capabilities of the materials cause electrical signals to be produced whenever the structure is subjected to mechanical stress or deformation due to factors, such as vibrations, loads, or environmental conditions. These signals can be monitored and assessed to detect the presence of cracks, delamination, corrosion, or other forms of damage. By monitoring changes in the electrical response, piezoelectric materials can detect the progression of cracks. The stress distribution in the material changes

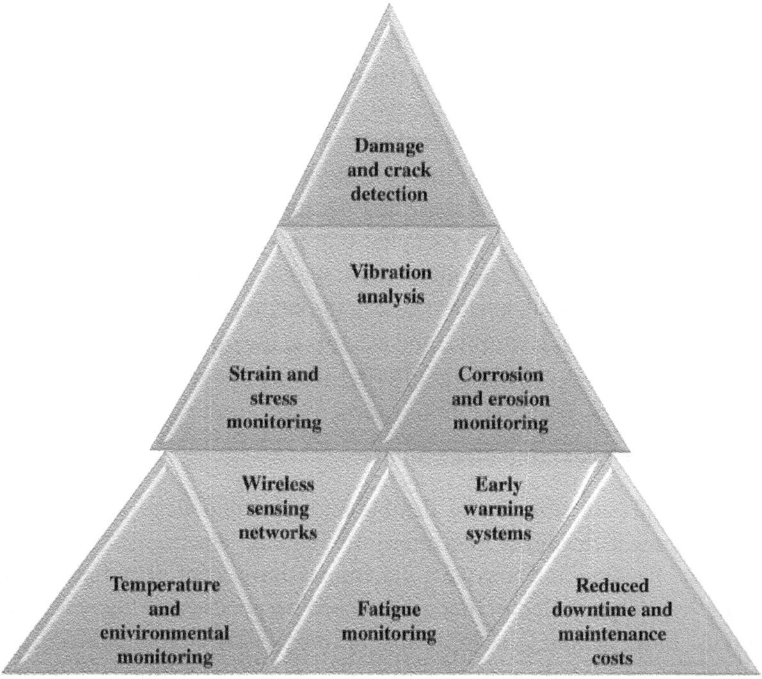

Figure 8.8 Different applications of conductive filler-based hybrid piezoelectric material in the structural health monitoring sector.

when a fracture spreads, which is reflected in the electrical signals produced by the piezoelectric sensors [128–130].

Engineers can track strains and stress in real-time by incorporating conductive filler-based hybrid piezoelectric materials at specific locations within a structure. Changes in strain distribution can provide insights into the structural stability and load-bearing capacity of the system [131]. These substances can act as vibration sensors to monitor the dynamic behavior of a structure. Analyzing the vibrational frequencies and modes can determine the structural condition and potential abnormalities.

Conductive filler-based hybrid piezoelectric materials can be used to monitor the corrosion or erosion of structures by observing changes in the mechanical characteristics of materials. In response to material loss, changes in stiffness or modulus are reflected in the electrical response. Over time, cyclic loading on a structure can cause fatigue damage. Hybrid piezoelectric materials can assist in monitoring the accumulation of fatigue damage by detecting changes in mechanical characteristics and structural behavior.

These materials can also monitor the temperature and other environmental factors that affect structural health. The piezoelectric sensors can detect mechanical strains caused by temperature changes. Wireless sensor networks can incorporate conductive filler-based hybrid piezoelectric materials for distributed and remote SHM. This makes it possible to collect and analyze data across large or inaccessible structures in real-time.

Early warning systems can be created because structural health can be continuously monitored. A catastrophic failure can be avoided by setting off alarms for maintenance or intervention whenever there is a significant deviation from expected behavior. These materials aid in optimizing maintenance plans, reducing downtime, and extending the operational lifespan of buildings by providing constant monitoring and precise damage assessment.

8.5.4 Flexible/Wearable Electronics

These materials are suitable for flexible and wearable electronics due to their mechanical flexibility and ability to generate electricity from body movement. Flexible and wearable electronics have a lot of potential for utilizing hybrid piezoelectric materials based on conductive fillers. These materials have advantages in terms of piezoelectricity and electrical conductivity, making them ideal for constructing devices that can absorb energy from mechanical movements while still being flexible and adaptable [123, 132, 133]. Figure 8.9 shows some applications of these materials.

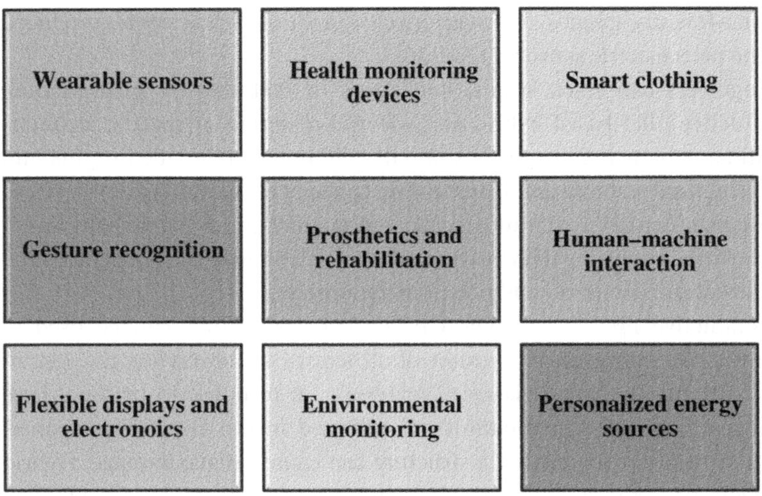

Figure 8.9 Conductive filler-based hybrid piezoelectric materials in flexible/wearable electronics.

Wearable technology can incorporate conductive filler-based hybrid piezoelectric materials to harvest mechanical energy from walking, bending, and stretching. The various components of the wearable device, such as sensors, displays, and communication modules, can be powered using the captured energy once it has been converted into electrical energy. These substances can act as sensors to track user interactions, environmental variables, and physiological characteristics. They can be used to create motion sensors that monitor gestures and movements, as well as pressure sensors that detect touch, force, or changes in pressure. Piezoelectric sensors integrated into flexible electronics enable the detection and interpretation of wearers' motions and movements. This applies to user interface designs, virtual reality, and gaming.

Flexible and wearable electronics can track vital indicators, such as heart rate, breathing rate, and body temperature using conductive filler-based hybrid piezoelectric materials. Longer device operation without the need for regular battery replacements is made possible by the ability to generate electricity from body movements. These components can be integrated into textiles to create smart clothing that responds to user actions or environmental changes. For instance, smart jackets with integrated sensors and energy-harvesting capabilities can generate power for heating components or offer real-time feedback.

These materials can enhance the comfort and functionality of wearers of prosthetics and assistive technology. They can be built into braces or limbs to track movements and provide feedback, which aids in rehabilitation. Conductive filler-based hybrid piezoelectric materials can be used to develop interactive surfaces that respond to touch or pressure. This opens up new possibilities for human–machine interaction in various contexts. Using these materials makes it possible to create flexible displays, touchscreens, and electronic components that can conform to curved or uneven surfaces. This creates opportunities for cutting-edge device designs.

Hybrid piezoelectric wearables utilize conductive fillers and can also monitor environmental factors, such as radiation exposure or pollution levels. This enables consumers to have real-time access to information about their surroundings. People can generate their power while moving by integrating these materials into their clothing or accessories, reducing their reliance on external power sources for electronic devices.

8.5.5 Electromechanical Devices

Micro-electromechanical systems (MEMS) devices employ these materials for mechanical actuation and control. They are useful components in

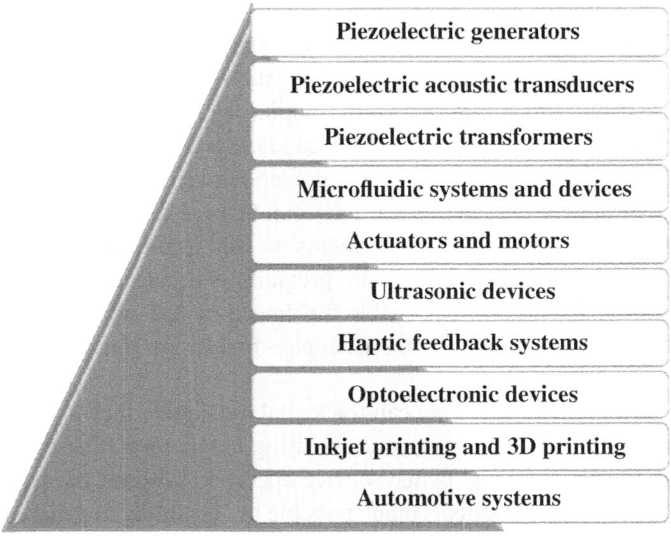

Figure 8.10 Applications of conductive filler-based hybrid piezoelectric materials in electromechanical devices.

MEMS technology due to their small size and ability to generate controlled movements. Due to their unique synergistic piezoelectric and conductive properties, conductive filler-based hybrid piezoelectric materials are utilized in various electromechanical systems. These materials are essential for developing machinery that can efficiently convert mechanical energy into electrical energy and vice versa [134]. Figure 8.10 shows a few applications of these materials in electromechanical devices.

Piezoelectric generators, also known as energy harvesters, can be constructed using hybrid piezoelectric materials that incorporate conductive fillers. These gadgets convert mechanical vibrations from machinery, footfalls, or the environment into useful electrical energy. These generators can supply power to wireless sensors, low-energy electronic devices, and batteries or even assist with energy harvesting for larger systems. These materials can be used to create acoustic transducers that utilize the piezoelectric effect to convert electrical signals into sound waves or vice versa. They can be employed in audio devices, such as speakers and microphones, ultrasound devices, underwater communication systems, and resonators in various applications. Using hybrid piezoelectric materials, high-frequency transformers that effectively transfer electrical voltage levels can be made. They are used in inverters, high-frequency electronics, and power supply circuits.

These materials are used in microfluidics to precisely control fluids on a microscopic scale. Electrical signal applications can induce mechanical vibrations that regulate the flow of liquids through microchannels, allowing for various applications in lab-on-a-chip systems and drug delivery devices. These materials can be used as actuators in devices that require controlled mechanical movement, converting electrical signals into mechanical motion and vice versa. The conductive fillers enhance the responsiveness of the actuator, enabling precise and controlled movements in various applications, including robotics, precision positioning systems, MEMS, and automation.

In electronic devices such as smartphones and game controllers, haptic feedback can be generated using conductive filler-based hybrid piezoelectric materials. The ability of the materials to produce vibrations or mechanical sensations enhances user engagement and interaction. Modulators and tunable lenses are two examples of optoelectronic devices that can incorporate conductive filler-based hybrid piezoelectric materials. The material's refractive index can be modified by applying mechanical stress, thereby altering its optical properties. These materials can be incorporated into printing techniques to create smart structures, electronic devices, or sensors on surfaces. They enable the fabrication of complex structures with integrated functionalities. These materials can be used in automotive applications, such as active suspension systems, tire pressure sensors, and energy recovery systems that capture the energy from moving and bouncing vehicles.

8.6 Summary

This chapter discusses the importance of harnessing inexhaustible energy sources in nature to meet future energy demands sustainably. It highlights the diversity of renewable energy sources, such as kinetic, solar, and bioenergy, and ongoing research into energy-harvesting technologies. The text emphasizes the need for microgrid-scale energy harvesting, especially in the IoT, and discusses the development of kinetic energy-harvesting mechanisms, focusing on piezoelectric transduction due to its efficiency and versatility. It also addresses concerns about the toxicity of lead-based piezoelectric materials, leading to the exploration of lead-free ceramics and polymers as alternatives. The principle of piezoelectric energy conversion, involving the deformation of materials to generate electrical charges, is explained.

It also highlights the crucial role of conductive fillers in enhancing the properties and performance of piezoelectric materials. It explains that these materials generate electric charges when subjected to mechanical stress and deform when exposed to an electric field. This chapter also mentions studies on the influence of conductive filler properties, such as size and volume fraction, on various performance parameters. It provides examples of materials and their performance improvements due to the addition of conductive fillers. Conductive filler-based piezoelectric materials involve a matrix embedded with conductive fillers to enhance their properties. These materials generate electric charges under mechanical stress and employ conductive fillers for charge extraction. Various fillers such as carbon-based materials, metallic nanoparticles, and conducting polymers are used based on specific requirements. These materials can be crystals, films, ceramics, or polymers. These materials have diverse sensors, energy-harvesting, and flexible electronics applications. Conductive fillers enhance their performance and make them suitable for various technological applications.

8.7 Challenges

Conductive filler-based hybrid piezoelectric materials exhibit significant potential for various applications. However, several research problems must be addressed before realizing their potential. Despite their promising characteristics, piezoelectric materials based on conductive fillers face challenges such as long-term stability, uniform dispersion of fillers, and maintaining a balance between conductivity and piezoelectric performance. Ongoing research aims to redefine these materials' properties, fabrication techniques, and applications for a wide range of innovative technologies. Researchers are attempting to optimize fillers' distribution, content, and compatibility with various matrix materials to overcome the aforementioned obstacles. Figure 8.11 illustrates some of the difficulties. To address these challenges, interdisciplinary research involving materials science, engineering, physics, and chemistry is necessary. Advances in nanotechnology, improved understanding of material interfaces, and innovative manufacturing techniques will be necessary to overcome these challenges and fully harness the potential of conductive filler-based hybrid piezoelectric materials.

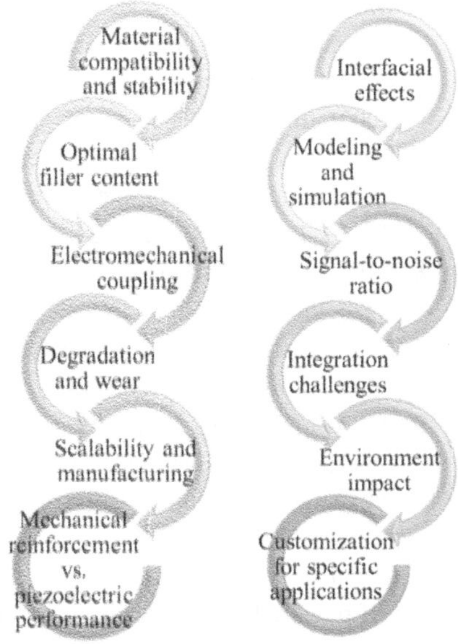

Figure 8.11 Challenges in the research area of conductive filler-based hybrid piezoelectric materials.

References

1 Surmenev, R.A., Orlova, T., Chernozem, R.V. et al. (2019). Hybrid lead-free polymer-based nanocomposites with improved piezoelectric response for biomedical energy-harvesting applications: a review. *Nano Energy* 62: 475–506. https://doi.org/10.1016/j.nanoen.2019.04.090.

2 Deng, W., Jin, L., Chen, Y. et al. (2018). An enhanced low-frequency vibration ZnO nanorod-based tuning fork piezoelectric nanogenerator. *Nanoscale* 10 (2): 843–847. https://doi.org/10.1039/C7NR07325A.

3 Bowen, C.R., Kim, H.A., Weaver, P.M., and Dunn, S. (2014). Piezoelectric and ferroelectric materials and structures for energy harvesting applications. *Energy Environ. Sci.* 7 (1): 25–44. https://doi.org/10.1039/C3EE42454E.

4 Tian, M., Xu, L., Dan, H., and Yang, Y. (2022). Extended linear detection range of a $Bi_{0.5}Na_{0.5}TiO_3$ thin film-based self-powered UV photodetector via current and voltage dual indicators. *Nanoscale Horiz.* 7 (10): 1240–1249. https://doi.org/10.1039/D2NH00204C.

5 Tressler, J.F., Alkoy, S., and Newnham, R.E. (1998). Piezoelectric sensors and sensor materials. *J. Electroceram.* 2 (4): 257–272. https://doi.org/10.1023/A:1009926623551.

6 Wang, Z., Guo, S., Li, H. et al. (2019). The semiconductor/conductor interface piezoresistive effect in an organic transistor for highly sensitive pressure sensors. *Adv. Mater.* 31 (6): 1805630. https://doi.org/10.1002/adma.201805630.

7 Wang, Z.L., Yang, R., Zhou, J. et al. (2010). Lateral nanowire/nanobelt based nanogenerators, piezotronics and piezo-phototronics. *Mater. Sci. Eng. R. Rep.* 70 (3): 320–329. https://doi.org/10.1016/j.mser.2010.06.015.

8 Burk, M.H., Schröder, S., Moormann, W. et al. (2020). Fabrication of diazocine-based photochromic organic thin films via initiated chemical vapor deposition. *Macromolecules* 53 (4): 1164–1170. https://doi.org/10.1021/acs.macromol.9b02443.

9 Hu, Y. and Wang, Z.L. (2015). Recent progress in piezoelectric nanogenerators as a sustainable power source in self-powered systems and active sensors. *Nano Energy* 14: 3–14. https://doi.org/10.1016/j.nanoen.2014.11.038.

10 Karan, S.K., Maiti, S., Lee, J.H. et al. (2020). Recent advances in self-powered tribo-/piezoelectric energy harvesters: all-in-one package for future smart technologies. *Adv. Funct. Mater.* 30 (48): 2004446. https://doi.org/10.1002/adfm.202004446.

11 Liu, C., Hua, B., You, S. et al. (2015). Self-amplified piezoelectric nanogenerator with enhanced output performance: the synergistic effect of micropatterned polymer film and interweaved silver nanowires. *Appl. Phys. Lett.* 106 (16): 163901. https://doi.org/10.1063/1.4918986.

12 Sarker, M.R., Saad, M.H.M., Riaz, A. et al. (2022). A bibliometric analysis of low-cost piezoelectric micro-energy harvesting systems from ambient energy sources: current trends, issues and suggestions. *Micromachines* 13 (6): Article 6. https://doi.org/10.3390/mi13060975.

13 Tamang, A., Ghosh, S.K., Garain, S. et al. (2015). DNA-assisted β-phase nucleation and alignment of molecular dipoles in PVDF film: a realization of self-poled bioinspired flexible polymer nanogenerator for portable electronic devices. *ACS Appl. Mater. Interfaces* 7 (30): 16143–16147. https://doi.org/10.1021/acsami.5b04161.

14 Vatansever, D., Hadimani, R.L., Shah, T., and Siores, E. (2011). An investigation of energy harvesting from renewable sources with PVDF and PZT. *Smart Mater. Struct.* 20 (5): 055019. https://doi.org/10.1088/0964-1726/20/5/055019.

15 Cook-Chennault, K.A., Thambi, N., and Sastry, A.M. (2008). Powering MEMS portable devices—a review of non-regenerative and regenerative power supply systems with special emphasis on piezoelectric energy harvesting systems. *Smart Mater. Struct.* 17 (4): 043001. https://doi.org/10.1088/0964-1726/17/4/043001.

16 Ramadan, K.S., Sameoto, D., and Evoy, S. (2014). A review of piezoelectric polymers as functional materials for electromechanical transducers. *Smart Mater. Struct.* 23 (3): 033001. https://doi.org/10.1088/0964-1726/23/3/033001.

17 Mitcheson, P.D., Yeatman, E.M., Rao, G.K. et al. (2008). Energy harvesting from human and machine motion for wireless electronic devices. *Proc. IEEE* 96 (9): 1457–1486. https://doi.org/10.1109/JPROC.2008.927494.

18 Wang, Z.L., Zhu, G., Yang, Y. et al. (2012). Progress in nanogenerators for portable electronics. *Mater. Today* 15 (12): 532–543. https://doi.org/10.1016/S1369-7021(13)70011-7.

19 Stuber, V.L., Mahon, T.R., van der Zwaag, S., and Groen, P. (2019). The effect of the intrinsic electrical matrix conductivity on the piezoelectric charge constant of piezoelectric composites. *Mater. Res. Express* 7 (1): 015703. https://doi.org/10.1088/2053-1591/ab5bb3.

20 Hwang, G.-T., Kim, Y., Lee, J.-H. et al. (2015). Self-powered deep brain stimulation via a flexible PIMNT energy harvester. *Energy Environ. Sci.* 8 (9): 2677–2684. https://doi.org/10.1039/C5EE01593F.

21 Mahapatra, S.D., Mohapatra, P.C., Aria, A.I. et al. (2021). Piezoelectric materials for energy harvesting and sensing applications: roadmap for future smart materials. *Adv. Sci.* 8 (17): 2100864. https://doi.org/10.1002/advs.202100864.

22 Motamedi, A.S., Mirzadeh, H., Hajiesmaeilbaigi, F. et al. (2017). Piezoelectric electrospun nanocomposite comprising Au NPs/PVDF for nerve tissue engineering. *J. Biomed. Mater. Res. A* 105 (7): 1984–1993. https://doi.org/10.1002/jbm.a.36050.

23 Shung, K.K., Cannata, J.M., and Zhou, Q.F. (2007). Piezoelectric materials for high frequency medical imaging applications: a review. *J. Electroceram.* 1 (19): 141–147. https://doi.org/10.1007/s10832-007-9044-3.

24 Maeder, M. D., Damjanovic, D., & Setter, N. (2023). *Lead Free Piezoelectric Materials.*

25 Kumar, V., Mandal, T.K., Kumar, A. et al. (2022). Multifunctional properties of chemically reduced graphene oxide and silicone rubber composite: facile synthesis and application as soft composites for piezoelectric energy harvesting. *Express Polym Lett* 16 (9): 978–995. https://doi.org/10.3144/expresspolymlett.2022.71.

26 Soin, N., Anand, S.C., and Shah, T.H. (2016). Energy harvesting and storage textiles. In: *Handbook of Technical Textiles*, vol. 2016, 357–396. Woodhead Publishing https://doi.org/10.1016/B978-1-78242-465-9.00012-4.

27 Lee, C.-C., Ho, J.-C., Hu, T.-S., and Wang, Y.-W. (2005). Printing organic thin-film transistor technology. *IEEE VLSI-TSA International Symposium on VLSI Technology, 2005. (VLSI-TSA-Tech)*, 49–52. https://doi.org/10.1109/VTSA.2005.1497077

28 Lee, C.S., Joo, J., Han, S., and Koh, S.K. (2004). Multifunctional transducer using poly (vinylidene fluoride) active layer and highly conducting poly (3,4-ethylenedioxythiophene) electrode: actuator and generator. *Appl. Phys. Lett.* 85 (10): 1841–1843. https://doi.org/10.1063/1.1784890.

29 Dang, Z.-M., Wang, L., Yin, Y. et al. (2007). Giant dielectric permittivities in functionalized carbon-nanotube/electroactive-polymer nanocomposites. *Adv. Mater.* 19 (6): 852–857. https://doi.org/10.1002/adma.200600703.

30 Shen, Y., Lin, Y., Li, M., and Nan, C.-W. (2007). High dielectric performance of polymer composite films induced by a percolating interparticle barrier layer. *Adv. Mater.* 19 (10): 1418–1422. https://doi.org/10.1002/adma.200602097.

31 Ren, X., Fan, H., Zhao, Y., and Liu, Z. (2016). Flexible lead-free BiFeO3/PDMS-based nanogenerator as piezoelectric energy harvester. *ACS Appl. Mater. Interfaces* 8 (39): 26190–26197. https://doi.org/10.1021/acsami.6b04497.

32 Wan, C. and Rhys Bowen, C. (2017). Multiscale-structuring of polyvinylidene fluoride for energy harvesting: the impact of molecular-, micro- and macro-structure. *J. Mater. Chem. A* 5 (7): 3091–3128. https://doi.org/10.1039/C6TA09590A.

33 Chen, G.-H., Wu, D.-J., Weng, W.-G., and Yan, W.-L. (2001). Preparation of polymer/graphite conducting nanocomposite by intercalation polymerization. *J. Appl. Polym. Sci.* 82 (10): 2506–2513. https://doi.org/10.1002/app.2101.

34 Hori, M., Aoki, T., Ohira, Y., and Yano, S. (2001). New type of mechanical damping composites composed of piezoelectric ceramics, carbon black and epoxy resin. *Compos. A: Appl. Sci. Manuf.* 32 (2): 287–290. https://doi.org/10.1016/S1359-835X(00)00141-X.

35 Ma, M. and Wang, X. (2009). Preparation, microstructure and properties of epoxy-based composites containing carbon nanotubes and PMN-PZT piezoceramics as rigid piezo-damping materials. *Mater. Chem. Phys.* 116 (1): 191–197. https://doi.org/10.1016/j.matchemphys.2009.03.009.

36 Tsantzalis, S., Karapappas, P., Vavouliotis, A. et al. (2007). Enhancement of the mechanical performance of an epoxy resin and fiber reinforced epoxy resin composites by the introduction of CNF and

PZT particles at the microscale. *Compos. A: Appl. Sci. Manuf.* 38 (4): 1076–1081. https://doi.org/10.1016/j.compositesa.2006.04.015.

37 Liu, X., Xiong, C., Sun, H. et al. (2006). Piezoelectric and dielectric properties of PZT/PVC and graphite doped with PZT/PVC composites. *Mater. Sci. Eng. B* 127 (2): 261–266. https://doi.org/10.1016/j.mseb.2005.10.022.

38 Bairagi, S. and Ali, S.W. (2020b). Investigating the role of carbon nanotubes (CNTs) in the piezoelectric performance of a PVDF/KNN-based electrospun nanogenerator. *Soft Matter* 16 (20): 4876–4886. https://doi.org/10.1039/D0SM00438C.

39 Carponcin, D., Dantras, E., Dandurand, J. et al. (2014). Electrical and piezoelectric behavior of polyamide/PZT/CNT multifunctional nanocomposites: polyamide/PZT/CNT multifunctional nanocomposites. *Adv. Eng. Mater.* 16 (8): 1018–1025. https://doi.org/10.1002/adem.201300519.

40 Wu, C.M. and Chou, M.H. (2016). Polymorphism, piezoelectricity and sound absorption of electrospun PVDF membranes with and without carbon nanotubes. *Compos. Sci. Technol.* 127: 127–133. https://doi.org/10.1016/j.compscitech.2016.03.001.

41 Banerjee, S. and Cook-Chennault, K.A. (2012). An investigation into the influence of electrically conductive particle size on electromechanical coupling and effective dielectric strain coefficients in three phase composite piezoelectric polymers. *Compos. A: Appl. Sci. Manuf.* 43 (9): 1612–1619. https://doi.org/10.1016/j.compositesa.2012.03.014.

42 Kang, J.H., Sauti, G., Park, C. et al. (2015). Multifunctional electroactive nanocomposites based on piezoelectric boron nitride nanotubes. *ACS Nano* 9 (12): 11942–11950. https://doi.org/10.1021/acsnano.5b04526.

43 Begum, S., Kausar, A., Ullah, H., and Siddiq, M. (2016). Exploitation of carbon nanotubes in high performance polyvinylidene fluoride matrix composite: a review. *Polym.-Plast. Technol. Eng.* 55 (2): 199–222. https://doi.org/10.1080/03602559.2015.1055505.

44 Taheri, S., Nakhlband, E., and Nazockdast, H. (2013). Microstructure and multiwall carbon nanotube partitioning in polycarbonate/acrylonitrile-butadiene-styrene/multiwall carbon nanotube nanocomposites. *Polym.-Plast. Technol. Eng.* 52 (3): 300–309. https://doi.org/10.1080/03602559.2012.748807.

45 Ma, P.-C., Siddiqui, N.A., Marom, G., and Kim, J.-K. (2010). Dispersion and functionalization of carbon nanotubes for polymer-based nanocomposites: a review. *Compos. A: Appl. Sci. Manuf.* 41 (10): 1345–1367. https://doi.org/10.1016/j.compositesa.2010.07.003.

46 Ghorbani, M. and Eisazadeh, H. (2012). Polypyrrole/carbon nanotube and polyaniline/carbon nanotube nanocomposites: synthesis and characterization. *Polym.-Plast. Technol. Eng.* 51 (13): 1367–1371. https://doi.org/10.1080/03602559.2012.702362.

47 Homenick, C.M., Lawson, G., and Adronov, A. (2007). Polymer grafting of carbon nanotubes using living free-radical polymerization. *Polym. Rev.* 47 (2): 265–290. https://doi.org/10.1080/15583720701271237.

48 Liu, P. (2005). Modifications of carbon nanotubes with polymers. *Eur. Polym. J.* 41 (11): 2693–2703. https://doi.org/10.1016/j.eurpolymj.2005.05.017.

49 Levi, N., Czerw, R., Xing, S. et al. (2004). Properties of polyvinylidene difluoride–carbon nanotube blends. *Nano Lett.* 4 (7): 1267–1271. https://doi.org/10.1021/nl0494203.

50 Manna, S., Batabyal, S.K., and Nandi, A.K. (2006). Preparation and characterization of silver–poly(vinylidene fluoride) nanocomposites: formation of piezoelectric polymorph of poly(vinylidene fluoride). *J. Phys. Chem. B* 110 (25): 12318–12326. https://doi.org/10.1021/jp061445y.

51 Mohammadi, B., Yousefi, A.A., and Bellah, S.M. (2007). Effect of tensile strain rate and elongation on crystalline structure and piezoelectric properties of PVDF thin films. *Polym. Test.* 26 (1): 42–50. https://doi.org/10.1016/j.polymertesting.2006.08.003.

52 Bairagi, S. and Ali, S.W. (2020a). Poly (vinylidine fluoride) (PVDF)/potassium sodium niobate (KNN) nanorods based flexible nanocomposite film: influence of KNN concentration in the performance of nanogenerator. *Org. Electron.* 78: 105547. https://doi.org/10.1016/j.orgel.2019.105547.

53 Bairagi, S. and Ali, S.W. (2019b). A unique piezoelectric nanogenerator composed of melt-spun PVDF/KNN nanorod-based nanocomposite fibre. *Eur. Polym. J.* 116: 554–561. https://doi.org/10.1016/j.eurpolymj.2019.04.043.

54 Bairagi, S. and Ali, S.W. (2019a). Influence of high aspect ratio lead-free piezoelectric fillers in designing flexible fibrous nanogenerators: demonstration of significant high output voltage. *Energy Technol.* 7 (10): 1900538. https://doi.org/10.1002/ente.201900538.

55 Teka, A., Bairagi, S., Shahadat, M. et al. (2018). Poly(vinylidene fluoride) (PVDF)/potassium sodium niobate (KNN)–based nanofibrous web: a unique nanogenerator for renewable energy harvesting and investigating the role of KNN nanostructures. *Polym. Adv. Technol.* 29 (9): 2537–2544. https://doi.org/10.1002/pat.4365.

56 Tian, S. and Wang, X. (2008). Fabrication and performances of epoxy/multi-walled carbon nanotubes/piezoelectric ceramic composites as rigid piezo-damping materials. *J. Mater. Sci.* 43 (14): 4979–4987. https://doi.org/10.1007/s10853-008-2734-7.

57 Eda, G. and Chhowalla, M. (2009). Graphene-based composite thin films for electronics. *Nano Lett.* 9 (2): 814–818. https://doi.org/10.1021/nl8035367.

58 Dreyer, D.R., Park, S., Bielawski, C.W., and Ruoff, R.S. (2010). The chemistry of graphene oxide. *Chem. Soc. Rev.* 39 (1): 228–240. https://doi.org/10.1039/B917103G.

59 Park, J., Ghosh, R., Song, M.S. et al. (2022). Individually addressable and flexible pressure sensor matrixes with ZnO nanotube arrays on graphene. *NPG Asia Mater.* 14 (1): 1–13. https://doi.org/10.1038/s41427-022-00386-4.

60 Matsumoto, Y., Koinuma, M., Kim, S.Y. et al. (2010). Simple photoreduction of graphene oxide nanosheet under mild conditions. *ACS Appl. Mater. Interfaces* 2 (12): 3461–3466. https://doi.org/10.1021/am100900q.

61 Shin, S.-H., Park, D.H., Jung, J.-Y. et al. (2017). Ferroelectric zinc oxide nanowire embedded flexible sensor for motion and temperature sensing. *ACS Appl. Mater. Interfaces* 9 (11): 9233–9238. https://doi.org/10.1021/acsami.7b00380.

62 Xue, B., Zhang, F., Zheng, J., and Xu, C. (2021). Flexible piezoelectric device directly assembled through the continuous electrospinning method. *Smart Mater. Struct.* 30 (4): 045006. https://doi.org/10.1088/1361-665X/abe3a8.

63 Kim, T., Cui, Z., Chang, W.-Y. et al. (2020). Flexible 1–3 composite ultrasound transducers with silver-nanowire-based stretchable electrodes. *IEEE Trans. Ind. Electron.* 67 (8): 6955–6962. https://doi.org/10.1109/TIE.2019.2937063.

64 Chiu, F.-C. and Kao, G.-F. (2012). Polyamide 46/multi-walled carbon nanotube nanocomposites with enhanced thermal, electrical, and mechanical properties. *Compos. A: Appl. Sci. Manuf.* 43 (1): 208–218. https://doi.org/10.1016/j.compositesa.2011.10.010.

65 Karttunen, M., Ruuskanen, P., Pitkänen, V., and Albers, W.M. (2008). Electrically conductive metal polymer nanocomposites for electronics applications. *J. Electron. Mater.* 37 (7): 951–954. https://doi.org/10.1007/s11664-008-0451-2.

66 Li, Y., Feng, J., Zhao, Y. et al. (2022). Ultrathin flexible linear-piezoelectric ZnO thin film actuators: tuning the piezoelectric responses by in-plane epitaxial strain. *Appl. Surf. Sci.* 599: 153969. https://doi.org/10.1016/j.apsusc.2022.153969.

67 Lim, Y.P., Koay, J.S.C., Zhao, J. et al. (2022). Modulating ZnO growth structures for maximum power output of hybrid piezo/triboelectric nanogenerator. *Adv. Funct. Mater.* 32 (46): 2206750. https://doi.org/10.1002/adfm.202206750.

68 Sánchez-Hernández, A.K., Lozano-Rosas, R., Gervacio-Arciniega, J.J. et al. (2022). Piezoelectric and mechanical properties of hydroxyapatite/titanium oxide composites. *Ceram. Int.* 48 (16): 23096–23103. https://doi.org/10.1016/j.ceramint.2022.04.289.

69 Luo, Y., Zhao, L., Luo, G. et al. (2022). All electrospun fabrics based piezoelectric tactile sensor. *Nanotechnology* 33 (41): 415502. https://doi.org/10.1088/1361-6528/ac7ed5.

70 Martins, M., Ribeiro, S., Ribeiro, C. et al. (2013). Effect of poling state and morphology of piezoelectric poly(vinylidene fluoride) membranes for skeletal muscle tissue engineering. *RSC Adv.* 3 (39): 17938–17944. https://doi.org/10.1039/C3RA43499K.

71 Nalwa, H.S. (1995). *Ferroelectric polymers: Chemistry: Physics, and Applications*. CRC Press.

72 Lovinger, A.J. (1983). *Science* 220: 1115–1121. Google scholar there is no corresponding record for this reference.

73 Martins, P., Lopes, A.C., and Lanceros-Mendez, S. (2014). Electroactive phases of poly(vinylidene fluoride): determination, processing and applications. *Prog. Polym. Sci.* 39 (4): 683–706. https://doi.org/10.1016/j.progpolymsci.2013.07.006.

74 Li, Z., Wang, Y., and Cheng, Z.-Y. (2006). Electromechanical properties of poly(vinylidene-fluoride-chlorotrifluoroethylene) copolymer. *Appl. Phys. Lett.* 88 (6): 062904. https://doi.org/10.1063/1.2170425.

75 Huan, Y., Liu, Y., Yang, Y., and Wu, Y. (2007). Influence of extrusion, stretching and poling on the structural and piezoelectric properties of poly (vinylidene fluoride-hexafluoropropylene) copolymer films. *J. Appl. Polym. Sci.* 104 (2): 858–862. https://doi.org/10.1002/app.25603.

76 Künstler, W., Wegener, M., Seiß, M., and Gerhard-Multhaupt, R. (2001). Preparation and assessment of piezo- and pyroelectric poly (vinylidene fluoride-hexafluoropropylene) copolymer films. *Appl. Phys. A Mater. Sci. Process.* 73: 641–645. https://doi.org/10.1007/s003390100991.

77 Neese, B., Wang, Y., Chu, B. et al. (2007). Piezoelectric responses in poly(vinylidene fluoride/hexafluoropropylene) copolymers. *Appl. Phys. Lett.* 90 (24): 242917. https://doi.org/10.1063/1.2748076.

78 Frübing, P., Kremmer, A., Gerhard-Multhaupt, R. et al. (2006). Relaxation processes at the glass transition in polyamide 11: from rigidity to viscoelasticity. *J. Chem. Phys.* 125 (21): 214701. https://doi.org/10.1063/1.2360266.

79 Takase, Y., Lee, J.W., Scheinbeim, J.I., and Newman, B.A. (1991). High-temperature characteristics of nylon-11 and nylon-7 piezoelectrics. *Macromolecules* 24 (25): 6644–6652. https://doi.org/10.1021/ma00025a014.

80 Bernard, F., Gimeno, L., Viala, B. et al. (2017). Direct piezoelectric coefficient measurements of PVDF and PLLA under controlled strain and stress. *Proceedings* 1 (4): 4. https://doi.org/10.3390/proceedings1040335.

81 Fukada, E. (1998). New Piezoelectric Polymers. *Jpn. J. Appl. Phys.* 37 (5S): 2775. https://doi.org/10.1143/JJAP.37.2775.

82 Csoka, L., Hoeger, I.C., Rojas, O.J. et al. (2012). Piezoelectric effect of cellulose nanocrystals thin films. *ACS Macro Lett.* 1 (7): 867–870. https://doi.org/10.1021/mz300234a.

83 Hassan, S.H., Voon, L., Velayutham, T.S. et al. (2018). Review of cellulose smart material: biomass conversion process and progress on cellulose-based electroactive paper. *J. Renewable Mater.* 6 (1): 1–25. https://doi.org/10.7569/JRM.2017.634173.

84 Kim, J., Yun, S., Mahadeva, S.K. et al. (2010). Paper actuators made with cellulose and hybrid materials. *Sensors* 10 (3): 3. https://doi.org/10.3390/s100301473.

85 Lebrun, L., Guyomar, D., Guiffard, B. et al. (2009). The characterisation of the harvesting capabilities of an electrostrictive polymer composite. *Sensors Actuators A Phys.* 153 (2): 251–257. https://doi.org/10.1016/j.sna.2009.05.009.

86 Morimoto, M., Koshiba, Y., Misaki, M., and Ishida, K. (2015). Thermal stability of piezoelectric properties and infrared sensor performance of spin-coated polyurea thin films. *Appl. Phys. Express* 8 (10): 101501. https://doi.org/10.7567/APEX.8.101501.

87 Harrison, J.S. and Ounaies, Z. (2002). *Encyclopedia of Polymer Science and Technology*. Wiley.

88 Liu, Z.H., Pan, C.T., Lin, L.W. et al. (2013). Direct-write PVDF nonwoven fiber fabric energy harvesters via the hollow cylindrical near-field electrospinning process. *Smart Mater. Struct.* 23 (2): 025003. https://doi.org/10.1088/0964-1726/23/2/025003.

89 Li, X., Lim, Y.-F., Yao, K. et al. (2013). Ferroelectric poly(vinylidene fluoride) homopolymer nanotubes derived from solution in anodic alumina membrane template. *Chem. Mater.* 25 (4): 524–529. https://dci.org/10.1021/cm3028466.

90 Bae, J.-H. and Chang, S.-H. (2015). Characterization of an electroactive polymer (PVDF-TrFE) film-type sensor for health monitoring of composite structures. *Compos. Struct.* 131: 1090–1098. https://doi.org/10.1016/j.compstruct.2015.06.075.

91 Lee, J.-H., Yoon, H.-J., Kim, T.Y. et al. (2015). Micropatterned P(VDF-TrFE) film-based piezoelectric nanogenerators for highly sensitive self-powered pressure sensors. *Adv. Funct. Mater.* 25 (21): 3203–3209. https://doi.org/10.1002/adfm.201500856.

92 Cauda, V., Stassi, S., Bejtka, K., and Canavese, G. (2013). Nanoconfinement: an effective way to enhance PVDF piezoelectric properties. *ACS Appl. Mater. Interfaces* 5 (13): 6430–6437. https://doi.org/10.1021/am4016878.

93 Ameduri, B. (2009). From vinylidene fluoride (VDF) to the applications of VDF-containing polymers and copolymers: recent developments and future trends. *Chem. Rev.* 109 (12): 6632–6686. https://doi.org/10.1021/cr800187m.

94 Ahmed, T.S., DeSimone, J.M., and Roberts, G.W. (2008). Continuous copolymerization of vinylidene fluoride with hexafluoropropylene in supercritical carbon dioxide: high-hexafluoropropylene-content amorphous copolymers. *Macromolecules* 41 (9): 3086–3097. https://doi.org/10.1021/ma702526u.

95 Améduri, B., Boutevin, B., and Kostov, G. (2001). Fluoroelastomers: synthesis, properties and applications. *Prog. Polym. Sci.* 26 (1): 105–187. https://doi.org/10.1016/S0079-6700(00)00044-7.

96 Apostolo, M., Arcella, V., Storti, G., and Morbidelli, M. (1999). Kinetics of the emulsion polymerization of vinylidene fluoride and hexafluoropropylene. *Macromolecules* 32 (4): 989–1003. https://doi.org/10.1021/ma980683r.

97 Drobny, J.G. and Moore, A.L. (2005). *Fluoroelastomers Handbook: The Definitive User's Guide*. William Andrew.

98 Mishra, S., Unnikrishnan, L., Nayak, S.K., and Mohanty, S. (2019). Advances in piezoelectric polymer composites for energy harvesting applications: a systematic review. *Macromol. Mater. Eng.* 304 (1): 1800463. https://doi.org/10.1002/mame.201800463.

99 Zhao, C., Zhang, J., Wang, Z.L., and Ren, K. (2017). A poly(l-lactic acid) polymer-based thermally stable cantilever for vibration energy harvesting applications. *Adv. Sustainable Syst.* 1 (9): 1700068. https://doi.org/10.1002/adsu.201700068.

100 Zhu, J., Jia, L., and Huang, R. (2017). Electrospinning poly(l-lactic acid) piezoelectric ordered porous nanofibers for strain sensing and energy harvesting. *J. Mater. Sci. Mater. Electron.* 28 (16): 12080–12085. https://doi.org/10.1007/s10854-017-7020-5.

101 Koyama, D. and Nakamura, K. (2010). Electric power generation using vibration of a polyurea piezoelectric thin film. *Appl. Acoust.* 71 (5): 439–445. https://doi.org/10.1016/j.apacoust.2009.11.009.

102 Fukada, E., Date, M., Ochiai, T., Koyama, S., and Sasaki, Y. (1999). Piezoelectricity in polyurethane films induced by electrostriction and bias electric field. *10th International Symposium on Electrets (ISE 10). Proceedings (Cat. No.99 CH36256)*, 655–658. https://doi.org/10.1109/ISE.1999.832131

103 Wang, W., Zheng, Y., Sun, Y. et al. (2021). High-temperature piezoelectric conversion using thermally stabilized electrospun polyacrylonitrile membranes. *J. Mater. Chem. A* 9 (36): 20395–20404. https://doi.org/10.1039/D1TA04400A.

104 Kawai, H. (1969). The piezoelectricity of poly (vinylidene fluoride). *Jpn. J. Appl. Phys.* 8 (7): 975. https://doi.org/10.1143/JJAP.8.975.

105 Elnabawy, E., Hassanain, A.H., Shehata, N. et al. (2019). Piezoelastic PVDF/TPU nanofibrous composite membrane: fabrication and characterization. *Polymers* 11 (10): 10. https://doi.org/10.3390/polym11101634.

106 Wu, X.-H., Chen, C., Shen, Y.-P., and Tian, X.-G. (2002). A high order theory for functionally graded piezoelectric shells. *Int. J. Solids Struct.* 39 (20): 5325–5344. https://doi.org/10.1016/S0020-7683(02)00418-3.

107 Wang, G., Liu, T., Sun, X.-C. et al. (2018). Flexible pressure sensor based on PVDF nanofiber. *Sensors Actuators A Phys.* 280: 319–325. https://doi.org/10.1016/j.sna.2018.07.057.

108 Wang, S., Rong, W., Wang, L. et al. (2019). A survey of piezoelectric actuators with long working stroke in recent years: classifications, principles, connections and distinctions. *Mech. Syst. Signal Process.* 123: 591–605. https://doi.org/10.1016/j.ymssp.2019.01.033.

109 Zhou, H., Wang, Z., Zhao, W. et al. (2021). Robust and sensitive pressure/strain sensors from solution processable composite hydrogels enhanced by hollow-structured conducting polymers. *Chem. Eng. J.* 403: 126307. https://doi.org/10.1016/j.cej.2020.126307.

110 Li, J., Zhou, X., Zhao, H. et al. (2015). Design and experimental performances of a piezoelectric linear actuator by means of lateral motion. *Smart Mater. Struct.* 24 (6): 065007. https://doi.org/10.1088/0964-1726/24/6/065007.

111 Yoo, S.-Y., Ha, J.-Y., Yoon, S.-J., and Choi, J.-W. (2013). High-power properties of piezoelectric hard materials sintered at low temperature for multilayer ceramic actuators. *J. Eur. Ceram. Soc.* 33 (10): 1769–1773. https://doi.org/10.1016/j.jeurceramsoc.2013.02.014.

112 You, Y.-M., Liao, W.-Q., Zhao, D. et al. (2017). An organic-inorganic perovskite ferroelectric with large piezoelectric response. *Science* 357 (6348): 306–309. https://doi.org/10.1126/science.aai8535.

113 Briscoe, J., Jalali, N., Woolliams, P. et al. (2013). Measurement techniques for piezoelectric nanogenerators. *Energy Environ. Sci.* 6 (10): 3035–3045. https://doi.org/10.1039/C3EE41889H.

114 Lallart, M. and Guyomar, D. (2010). Piezoelectric conversion and energy harvesting enhancement by initial energy injection. *Appl. Phys. Lett.* 97 (1): 014104. https://doi.org/10.1063/1.3462304.

115 Shin, Y.-H., Jung, I., Noh, M.-S. et al. (2018). Piezoelectric polymer-based roadway energy harvesting via displacement amplification module. *Appl. Energy* 216: 741–750. https://doi.org/10.1016/j.apenergy.2018.02.074.

116 Zhang, Y., Liu, X., Wang, G. et al. (2020). Enhanced mechanical energy harvesting capability in sodium bismuth titanate based lead-free piezoelectric. *J. Alloys Compd.* 825: 154020. https://doi.org/10.1016/j.jallcom.2020.154020.

117 Sun, J., Guo, H., Ribera, J. et al. (2020). Sustainable and biodegradable wood sponge piezoelectric nanogenerator for sensing and energy harvesting applications. *ACS Nano* 14 (11): 14665–14674. https://doi.org/10.1021/acsnano.0c05493.

118 Kim, S.H., Kim, J.W., Park, J.-J. et al. (2019). Predicting energy expenditure during gradient walking with a foot monitoring device: model-based approach. *JMIR MHealth UHealth* 7 (10): e12335. https://doi.org/10.2196/12335.

119 Marshiana, D., Sherine, M.E., Sunitha, N., and Vinothkumar, C. (2016). Footstep power production using piezoelectric sensors. *Res. J. Pharm. Technol.* 9 (7): 831. https://doi.org/10.5958/0974-360X.2016.00157.8.

120 Gholikhani, M., Roshani, H., Dessouky, S., and Papagiannakis, A.T. (2020). A critical review of roadway energy harvesting technologies. *Appl. Energy* 261: 114388. https://doi.org/10.1016/j.apenergy.2019.114388.

121 Tao, K., Yi, H., Yang, Y. et al. (2020). Origami-inspired electret-based triboelectric generator for biomechanical and ocean wave energy harvesting. *Nano Energy* 67: 104197. https://doi.org/10.1016/j.nanoen.2019.104197.

122 Mo, X., Zhou, H., Li, W. et al. (2019). Piezoelectrets for wearable energy harvesters and sensors. *Nano Energy* 65: 104033. https://doi.org/10.1016/j.nanoen.2019.104033.

123 Mokhtari, F., Spinks, G.M., Fay, C. et al. (2020). Wearable electronic textiles from nanostructured piezoelectric fibers. *Adv. Mater. Technol.* 5 (4): 1900900. https://doi.org/10.1002/admt.201900900.

124 Meyers, F.N., Loh, K.J., Dodds, J.S., and Baltazar, A. (2013). Active sensing and damage detection using piezoelectric zinc oxide-based nanocomposites. *Nanotechnology* 24 (18): 185501. https://doi.org/10.1088/0957-4484/24/18/185501.

125 Wait, J.R., Park, G., and Farrar, C.R. (2005). Integrated structural health assessment using piezoelectric active sensors. *Shock. Vib.* 12 (6): 389–405.

126 Yan, S., Wu, J., Sun, W. et al. (2013). Development and application of structural health monitoring system based on piezoelectric sensors. *Int. J. Distrib. Sens. Netw.* 9 (11): 270927. https://doi.org/10.1155/2013/270927.

127 Guo, W., Tan, C., Shi, K. et al. (2018). Wireless piezoelectric devices based on electrospun PVDF/$BaTiO_3$ NW nanocomposite fibers for human motion monitoring. *Nanoscale* 10 (37): 17751–17760. https://doi.org/10.1039/C8NR05292A.

128 Gama, A.L. and Morikawa, S.R.K. (2008). Monitoring fatigue crack growth in compact tension specimens using piezoelectric sensors. *Exp. Mech.* 48 (2): 247–252. https://doi.org/10.1007/s11340-007-9086-0.

129 Ray, M.C. and Pradhan, A.K. (2006). The performance of vertically reinforced 1–3 piezoelectric composites in active damping of smart structures. *Smart Mater. Struct.* 15 (2): 631. https://doi.org/10.1088/0964-1726/15/2/047.

130 Zou, F., Benedetti, I., and Aliabadi, M.H. (2013). A boundary element model for structural health monitoring using piezoelectric transducers. *Smart Mater. Struct.* 23 (1): 015022. https://doi.org/10.1088/0964-1726/23/1/015022.

131 Huo, Z., Wang, X., Zhang, Y. et al. (2020). High-performance Sb-doped p-ZnO NW films for self-powered piezoelectric strain sensors. *Nano Energy* 73: 104744. https://doi.org/10.1016/j.nanoen.2020.104744.

132 Huang, S., Tang, G., Huang, H. et al. (2018). Enhanced piezo-response in copper halide perovskites based PVDF composite films. *Sci. Bull.* 63 (19): 1254–1259. https://doi.org/10.1016/j.scib.2018.07.014.

133 Lund, A., Rundqvist, K., Nilsson, E. et al. (2018). Energy harvesting textiles for a rainy day: woven piezoelectrics based on melt-spun PVDF microfibres with a conducting core. *npj Flexible Electron.* 2 (1): 1. https://doi.org/10.1038/s41528-018-0022-4.

134 Fernandes, E., Martin, B., Rua, I. et al. (2018). Design, fabrication, and testing of a low frequency MEMS piezoelectromagnetic energy harvester. *Smart Mater. Struct.* 27 (3): 035017. https://doi.org/10.1088/1361-665X/aaaba5.

9

Semiconductive Filler-Based Hybrid Piezoelectric Materials

Swagata Banerjee[1], Aiswarya Baburaj[2], Akshaya Aliyana[3], Satyaranjan Bairagi[4], and Naveen Kumar Sindhughatta Kalaiah[2]

[1]Department of Textile and Fibre Engineering, Indian Institute of Technology Delhi, Huaz Khas, New Delhi, India
[2]Department of Electronics, Mangalore University, Mangaluru, India
[3]Research Institute for Flexible Materials, School of Textiles and Design, Heriot-Watt University, Edinburgh, UK
[4]Materials and Manufacturing Research Group, James Watt School of Engineering, University of Glasgow, Scotland, UK

9.1 Introduction

In recent decades, remarkable technological advancements have transformed the world, introducing sophisticated sensor technologies that have significantly improved our lives. Virtually every aspect of human existence has become automated, driven by intelligent devices that rely on a continuous power source for seamless operation. This trend has led to a substantial increase in energy demand across various spheres of our daily lives. Traditional power sources, such as batteries, have been relied upon to provide reliable power until their depletion. However, their disposal or recharging processes can be laborious and inconvenient. Consequently, there is a pressing need for intelligent alternatives that can offer continuous and efficient power without harming the environment [1]. One promising solution lies in harnessing the principles of piezoelectricity [2] and triboelectricity [3]. These emerging energy-harvesting techniques capitalize on the conversion of wasted mechanical energy in our surroundings into electrical power. They are cost-effective, environmentally friendly, and straightforward, making them ideal for meeting the growing energy demands. By utilizing these nonconventional energy-harvesting methods, we can effectively tap into unused energy sources, helping to bridge the gap

Hybrid Materials for Piezoelectric Energy Harvesting and Conversion, First Edition.
Edited by S. Wazed Ali, Satyaranjan Bairagi, and Shahid Ul Islam.
© 2024 John Wiley & Sons, Inc. Published 2024 by John Wiley & Sons, Inc.

between energy supply and demand while minimizing the impact on the environment.

This chapter provides a concise survey of techniques and strategies employed to enhance the properties of piezoelectric materials, with a specific emphasis on semiconductor-based substances. Furthermore, it explores the wide-ranging applications of these materials across diverse scientific domains, showcasing their potential in interdisciplinary research and innovation. By shedding light on these advancements, the chapter underscores the expansive scope and promising prospects of semiconductor-based piezoelectric materials in various fields of study.

9.2 Piezoelectric Materials

Piezoelectricity is displayed by a special class of materials, called piezoelectric materials. These materials possess a non-centrosymmetric crystal structure, which forms the basis behind showing such interesting behavior. Under neutral conditions, the charge centers of the positive and negative ions in the crystal lattice coincide with each other. Hence, it shows no charge separation. On application of external force, the charge centers of the opposite charges separate, giving rise to a difference in potential between the material surfaces. This is called the direct piezoelectric effect [4]. On the other hand, the conversion of electrical input to mechanical output is termed the converse piezoelectric effect. Piezoelectric materials can be broadly classified into single crystals, ceramics, polymers, and composites. Single crystals have very high electromechanical coupling coefficients; however, they are expensive enough to use. Based on composition, ceramics can be divided into lead-based and lead-free ceramics. Recently, due to the deleterious effect of lead on the environment and humans, there has been an increased acceptance of lead-free materials in comparison to their lead-based counterparts. Ceramics have very good piezoelectric coefficients; however, their brittle nature makes it difficult to extend their applications to flexible requirements. Polymers have good voltage coefficients and are flexible materials. However, they have inferior piezoelectric coefficients compared to ceramics. The idea of composite seeks to amalgamate the properties of its constituent materials to develop an entity with unique properties. The use of polymers and ceramics together in the form of composites helps to utilize the advantage of both materials while compensating for their individual shortcomings.

Most piezoelectric materials are rarely used in their pure form. They either lack certain properties or encounter manufacturing or synthesis problems. Modifications of these materials through physical and chemical means serve to improve upon the properties of such materials and solve the manufacturing-related issues. Sintering and modified sintering techniques might be used as physical means to enhance the properties of piezoelectric materials. Chemical means employing doping the materials with some foreign elements to develop a novel material with unique piezoelectric properties.

9.3 Semiconductor-Modified Hybrid Piezoelectric Materials

9.3.1 Semiconductor-Modified Lead-Based Hybrid Piezoelectric Materials

Lead-based piezoelectric materials were among the earliest discoveries in the field of piezoelectric ceramics, showing great promise for future advancements. Pure lead titanate ($PbTiO_3$) is rarely used as a stand-alone piezoelectric material, as it is typically modified or incorporated into solid solutions to exhibit its piezoelectric properties. This ferroelectric perovskite material has a tetragonal structure at room temperature and undergoes a Curie transition around 493 °C. Phase transitions in lead titanate can be induced by changes in temperature and hydrostatic pressure. The remarkable piezoelectric constants observed in lead-based ceramics, compared to their lead-free counterparts, can be attributed to the hybridization of Ti–O interactions influenced by the presence of lead. The hybridization of the Ti—O bond is crucial in the ferroelectric nature of $PbTiO_3$. In this material, the 6s state of lead and the 2p state of oxygen strongly hybridize, influencing the Ti–O interactions and resulting in significant strain within the perovskite structure. Lead zirconate ($PbZrO_3$) is another ceramic material with antiferroelectric properties. Solid solutions of $PbTiO_3$ and $PbZrO_3$ exhibit solubility across all compositions, giving rise to materials with intriguing properties. Pb $(Zr_{1-x}Ti_x)O_3$ (PZT) is one such solid solution that displays fascinating phase transitions from tetragonal to rhombohedral structures at room temperature, depending on the composition. This composition-dependent structural transition, known as the morphotropic phase boundary (MPB), is associated with a high piezo response due to the ease of dipole orientations at these boundaries. Lead zirconate titanate (PZT) is widely available in the form of polycrystalline ceramics and thin films, while reports on PZT bulk crystals are rare. Doping is commonly employed to synthesize new materials with enhanced piezoelectric properties. Doping by external elements can create lead or cationic vacancies, resulting in the formation of soft PZT ceramics with characteristics such as improved piezoelectric coefficient, lower mechanical quality factor, lower coercive field, and higher losses. On the other hand, substitutions that lead to oxygen vacancies yield hard PZT materials with lower losses, permittivity, resistivity, and higher mechanical quality factors. Soft PZT materials are easier to polarize and exhibit dipole behavior compared to hard PZT ceramics [5].

Though PZT ceramics have appreciable piezoelectric properties, their mechanical properties are not as commendable. Often, improvement in mechanical property is achieved at the expense of its electrical attributes. In a study, yttria-stabilized zirconia (YSZ) was used to modify PZT through conventional solid-state sintering, in order to improve upon its mechanical properties while maintaining its electrical characteristics [6]. With the increase in YSZ content, the tetragonality ratio decreased for the modified ceramics, owing to the diffusion of the Zr^{4+} ions into the B-site of the PZT lattice, gradually transitioning toward the rhombohedral Zr-rich phase of PZT. The improved mechanical properties of the YSZ-modified PZT ceramics are due to the toughening effect caused by martensitic transformation. The piezoelectric coefficients of the modified ceramics (d_{33} = 283 pC/N) were found comparable to that of pristine PZT (d_{33} = 285 pC/N) [7]. The properties of $0.90Pb(Zr,Ti)O_3$–$0.03Pb(Fe_{2/3}W_{1/3})O_3$–$0.07Pb(Mn_{1/3}Nb_{2/3})O_3$ were modified by doping with excess of Pb_3O_4 and WO_3. The modified ceramics existed in the tetragonal phase. The improved electrical and piezoelectric properties of the modified ceramics were a result of two factors. The incorporation of W^{6+} in the lattice led to the creation of lead vacancies, and hence characteristics of soft piezoelectric material were obtained. Further, the addition of Pb_3O_4 and WO_3 led to the formation of a liquid phase PbO–WO_3. This helps in the effective sintering of the ceramics leading to an improved densified structure. The improvement in the densification of the ceramics enhances their electrical properties [8]. In another study, 0.69 $Pb(Zr_{0.47}Ti_{0.53})O_3$–$0.31[Pb(Zn_{0.4}Ni_{0.6})_{1/3}Nb_{2/3}]O_3$ ceramic was doped with CuO to investigate its effect on electrical and piezoelectric properties. The addition of CuO as a sintering aid brought down the sintering temperature from 1175 to 1025 °C. Further, the structure of the ceramic changed from rhombohedral to tetragonal with the addition of CuO. Cu^{2+} ions replace the Nb^{5+} ions in the crystal lattice creating oxygen vacancies to balance out the charge. This imparted hard characteristics to the doped ceramics. It had a pinning effect on the domain walls that increased the coercive electric field of the doped ceramics. The CuO-doped ceramics had a charge coefficient of 502 pC/N, which was remarkably high compared to the pristine samples [9]. Sb_2O_3 has often been used as a dopant in lead-based ceramics to improve its piezoelectric properties, often at the cost of increased loss factors. However, a work based on Sb_2O_3 doped $(Pb_{0.92}Sr_{0.08})(Zr_{0.533}Ti_{0.443}Nb_{0.024})O_3$ showed an improvement in the piezoelectric properties without any increment in its loss factors. Sb^{5+} ion acted as a soft as well as a hard dopant. The addition of Sb_2O_3 raised the rhombohedral phase content of the doped ceramic that imparted higher piezoelectric strains under high electric field. This resulted in enhanced piezoelectric properties of the modified ceramic composition. The oxygen vacancies were still the dominant carriers despite the addition of dopant. The properties of the modified ceramic were maintained even at high temperatures. This suggests the implementation of such novel materials for extreme temperature conditions [10].

9.3.2 Semiconductor-Modified Lead-Free Hybrid Piezoelectric Materials

The adverse impact of lead-based materials on environmental health has raised significant concerns. It is imperative to seek alternative solutions that can effectively replace lead-based materials without compromising their properties and applications. Consequently, there is a growing focus on the development of lead-free piezoelectric ceramics as substitutes for their lead-based counterparts. In recent times, extensive research and development efforts have been directed toward exploring, enhancing, and optimizing lead-free piezoelectric materials to match the piezoelectric performance of lead-based materials. The aim is to identify and refine lead-free alternatives that can effectively replicate the piezoelectric activity exhibited by lead-based materials. Potassium sodium niobate (KNN) belongs to the class of alkali niobates, which has now become the forerunner of lead-free ceramics. Potassium niobate undergoes phase transition from rhombohedral to orthorhombic to tetragonal as temperature increases from sub-zero. It turns to a cubic phase at a temperature of 435 °C. Sodium niobate on the other hand is orthorhombic and ferroelectric under room temperature. The KNN perovskite has monoclinic phase at room temperature with an orthorhombic symmetry of its unit cell. It has a Curie temperature of around 420 °C. Enhanced piezoelectric properties of KNN are obtained near its MPB. However, even these properties are inferior when compared to the piezoelectric properties of PZT. Thus, appropriate chemical and physical methods have been employed to tailor the properties of KNN to be at par with that of PZT [4].

As a means of chemically modifying KNN, it was doped to produce 0.965 $(K_{0.48}Na_{0.52})$ $(Nb_{0.96}Sb_{0.04})O_3$–$0.035Bi_{0.5}Na_{0.5}Zr_{0.15}Hf_{0.75}O_3$ (KNNS-BNZH) ceramic through conventional solid-state synthesis method. This was then mixed with polyvinyl alcohol (PVA) to form disks, which were then sintered to remove PVA from the ceramic disks, followed by poling of the ceramic disks using a DC electric field. The phase transition study with temperature shows that a mixed-phase composition turned into tetragonal phase structure with increase in temperature. The Curie temperature of the doped ceramic was found to be 235 °C. It was further observed that the dielectric permittivity and dielectric fluctuations of the materials increased and were small with rise in temperature, respectively. The dielectric loss on the other hand was about 23% at Curie temperature, which was insignificant at temperatures lower than that. Properties such as Curie temperature and coercive electric field of the modified ceramics were like the range of soft PZT ceramic materials. Thus, these doped materials could serve as an alternative ceramic in applications where soft PZT is used [11].

Comparing the structural phase boundary of the promising lead-based PZT and lead-free KNN, it has been found that the hybridization of the A-site and B-site ions of the crystal lattice affect the rhombohedral (R)–tetragonal (T) phase boundary [12]. Often, the synthesis of KNN through hot-pressing method results in contamination from the carbon impurities of the graphite mold. To prevent this, a hexagonal boron nitride layer was introduced between the mold surface and the piezoceramic. Further, addition of sodium and Er_2O_3 to KNN reflected some significant results. The addition of Na helped to compensate for the sodium lost during the synthesis and sintering process of KNN. The doped ceramic had a tetragonal structure. The function of Er_2O_3 acted as a sintering aid that obstructed the abnormal grain growth due to the hexagonal boron nitride layer [13]. It has been studied that the coupling ability of the dopants used has profound influence on phase evolution of KNN, thus affecting their piezoelectric properties. The coupling ability of lithium and antimony is weak, and hence they decrease the transition temperature from orthorhombic to tetragonal phase. Further, due to the intense vibration of NbO_6 octahedron, they tend to stabilize the tetragonal phase at low temperatures. However, dopants such as silver and tantalum with high coupling ability, cause intense planar twists of the NbO_6 octahedron, which leads to the higher displacement of the centrally placed Nb [14].

Barium titanate ($BaTiO_3$) is another lead-free piezoelectric ceramic and among the first discovered ferroelectric materials. It is known for its high piezoelectric coefficient, dielectric constant, and low dielectric loss. However, its low Curie temperature of 120 °C, brings concerns while extending the usage temperature range of the material. On increasing the temperature from sub-zero to the Curie temperature, the ceramic undergoes phase transition from rhombohedral to orthorhombic to tetragonal and finally cubic beyond 120 °C [15]. To improve the inherent piezoelectric properties of $BaTiO_3$, it has been chemically modified through various kinds of dopants [16]. Yttrium-doped $BaTiO_3$ ceramics were prepared by mixing yttrium oxide (Yr_2O_3) with $BaTiO_3$ in various amounts through planetary milling followed by drying. It was observed that as the concentration of dopant increases, the microstructure of the ceramic changes from densely sintered to a porous structure with interconnected pores. At some intermediate concentrations, microstructure with defined grains and pores have been observed owing to breakage of the grains. Further, as the concentration of Y^{3+} increases, it replaces the Ba^{2+} ions in the lattice that leads to an increase in negative carriers in the ceramic. It also increases the precipitation of cation-rich species near the grain boundaries. This results in the improvement of n-type semiconductor behavior of the doped ceramic. However, at lower concentrations of Y^{3+}, the n-type behavior is reduced due to the porous microstructure that provides diffusion paths for the carriers [17]. The n-type semiconductor behavior of doped $BaTiO_3$ ceramics depends on the donor dopant and also on certain intrinsic defects introduced in the structure [18]. In a study, Fe^{3+}/Nb^{5+}-doped $BaTiO_3$ ceramics were studied for their dielectric and ferroelectric properties. It was observed that there was a transition from soft to hard dielectric behavior in the doped ceramic with the increase in Nb^{5+} content. Interestingly, Fe^{3+}/Nb^{5+} did not form a pair on being doped. They rather separated out occupying B-sites in the lattice giving rise to different defect dipoles. This results in distributed regions of soft and hard characteristics in the sub-lattices [19]. However, identifying the chemical state of the ions in the doped ceramic has been a challenge in this work. Later on, the same group went on to explore this possibility using X-ray photoelectron spectroscopy (XPS). The results indicated that the Ti ion in the doped ceramic existed in more than one component, while only one component was observed in the case of the pristine undoped ceramic. Further, multiple types of barium and oxygen species were found to exist in both the pristine and doped samples [20]. The method of incorporating niobium in $BaTiO_3$ also has an influence on the electrical properties of the material. It was found that niobium applied as a coating on $BaTiO_3$ ceramics formed a core–shell structure, while chloride mixing through hydrothermal reaction resulted in proper grains with niobium included in the lattice. Moreover, the shell consisted of aggregates of niobium instead of atomic layer deposition, in the case of the coated samples. These samples further possessed a distributed time constant, while the hydrothermal samples had narrower time constant [21].

Bismuth-based layered structure materials have also been explored as lead-free piezoelectric material. The two layers in the structure were identified as $(Bi_2O_2)^{2+}$ and $(A_{m-1}B_mO_{3m+1})^{2-}$ [22]. The high Curie temperature and low dielectric losses are some of the advantages of this class of ceramics. However, disadvantages such as difficulty in poling and inferior spontaneous polarization call for modifications of this material to improve upon its shortcomings [4]. Doping with materials such as lanthanide ions [23], copper–antimony [24], copper–niobium [25], and cerium [26–28] has been carried out to enhance the piezoelectric activity of these piezoelectric ceramic materials.

9.4 Semiconductive Filler-Based Hybrid Piezoelectric Energy Harvesters

The combination of piezoelectric materials and a semiconductor has also been studied to improve the piezoelectric properties of the composite. In a study, molybdenum disulfide (MoS_2), reduced graphene oxide (rGO), and polyvinylidene fluoride (PVDF) were used in combination with a similar objective. MoS_2 nanosheets were synthesized and mixed with rGO to form MoS_2–rGO. This was then mixed with PVDF polymer to prepare solution-cast films of PVDF–rGO–MoS_2. The structural characterization of the nanocomposite films revealed an enhanced electroactive phase content of the polymer [30]. This enhanced β-phase content was attributed to the interaction between the polymer chains and MoS_2–rGO. The free π-electrons of rGO interact with the CH_2 groups of PVDF, while MoS_2 attracts the CF_2 dipoles of the PVDF polymer chains. These interactions together lead to the stabilization of the β phase of PVDF. The enhanced β-phase content of PVDF and the piezoelectric nature of MoS_2 led to an improved output voltage and short-circuit current of 2.4 V and 0.68 μA, respectively (Figure 9.1a, b) [31, 32].

Zinc oxide (ZnO) is a very promising piezoelectric material that has been explored in the form of strain sensors, actuators, and other advanced electronic applications [33]. The issue with ZnO and its nanostructures is their brittle nature and high rigidity. They have often been formed into composites with flexible materials to utilize their piezoelectric property, while overcoming their fragile nature. ZnO nanostructures were grown on paper using a low-temperature solvothermal method. A strain sensor was fabricated from the ZnO–paper nanocomposite by plasma sputtering a thin layer of gold on both sides of the sheet. The strain sensor gave a stable output under static as well as dynamic load testing. This sensor also was sensitive to different frequencies and responded quickly to external stimuli. The output power of the fabricated device is low; however, it possesses good

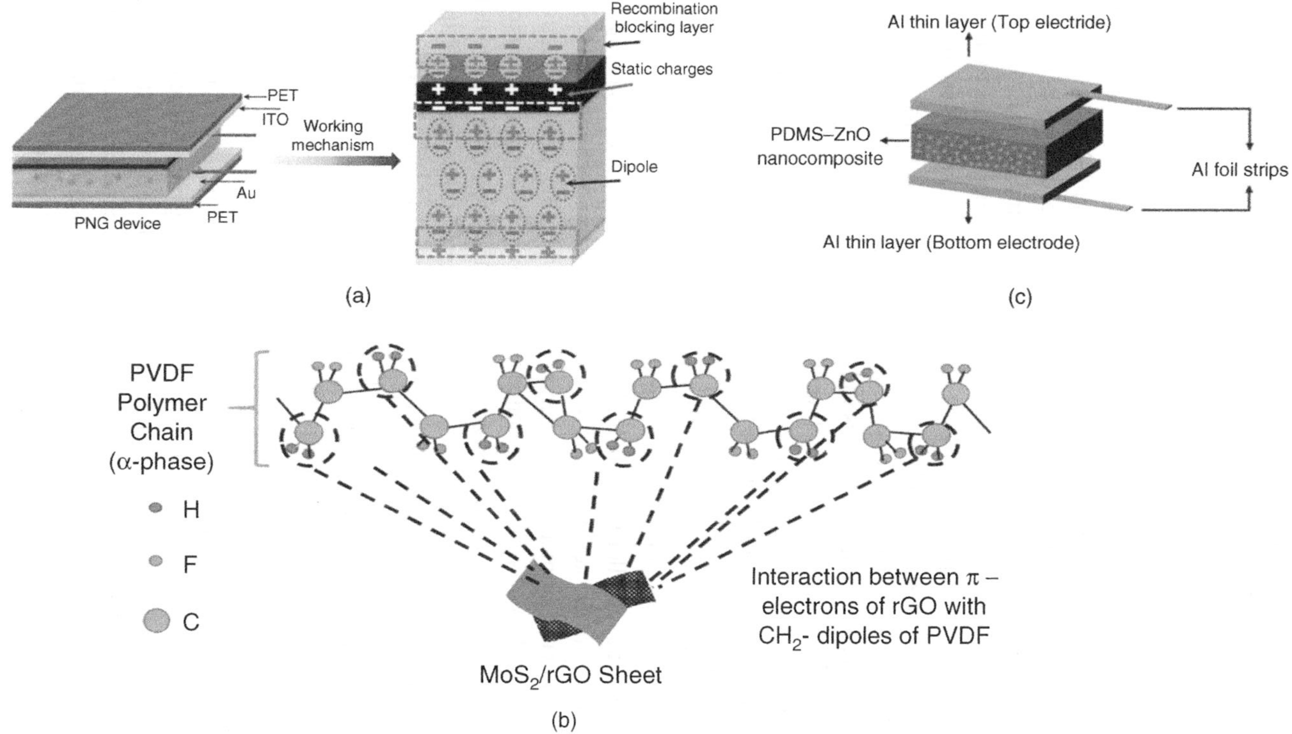

Figure 9.1 (a) Working principle of tri-layered PENG. Source: Yaqoob et al. [29]/Elsevier, (b) interaction between PVDF and MoS_2 and rGO. Source: Faraz et al. [31]/Elsevier, (c) PDMS and ZnO nanocomposite device. Source: Jeronimo et al. [35]/Elsevier. All essential copyrights and permissions received.

sensitivity to strain [34]. The geometry of the nanostructures of ZnO and their respective concentrations in nanocomposites have a profound effect on the overall piezoelectric activity of the material composition. Referring to the geometry of nanoflowers and nanoparticles, it can be inferred that nanoflowers have a spatial distribution of geometry, while nanoparticles do not show such distribution. Moreover, the modulus of nanoflowers was found to be lower than that of nanoparticles, making them less stiff and, hence, more responsive to the applied stress. Thus, PDMS–ZnO-based nanocomposites (Figure 9.1c) showed a higher piezoelectric output voltage of 4.2 V compared to 1.86 V for nanoflowers and nanoparticles of ZnO, respectively. Further, nanoflowers of ZnO provided good connectivity in comparison to nanoparticles of ZnO that helped to achieve better piezoelectric activity in ZnO nanoflower-based nanocomposites [35]. In another study, KNN and ZnO nanorods have been incorporated into a PVDF matrix to produce electrospun nanofibrous web. The synergistic effect of KNN and ZnO helped to achieve a high output voltage and current of 25 V and 1.81 μA, respectively. Through the structural analysis of these nanocomposite membranes, it was ascertained that the incorporated filler, KNN and ZnO, acted as β-nucleating agents of the PVDF polymer. The increased electroactive phase content of the polymer also helped in achieving the appreciable performance of the nanogenerator fabricated from the nanocomposite [36].

Semiconductive filler-based hybrid piezoelectric energy harvesters have significance in the advancement of energy-harvesting technology. These devices provide numerous key benefits by combining semiconductive materials with piezoelectric elements. They improve charge distribution and conductivity inside the piezoelectric material, resulting in greater total energy conversion efficiency. This advancement also broadens the operational frequency range, allowing for more efficient energy harvesting from a wider variety of mechanical vibrations. As a result, these hybrid harvesters show great promise for powering a variety of applications, including wearable electronics, distant sensors, and IoT devices, in situations where traditional power sources are impracticable. Their ability to convert ambient mechanical energy into usable electrical power adds to self-sufficient and sustainable solutions, eliminating dependency on traditional energy sources and batteries while boosting environmental and technological advancements. Piezoelectric materials are currently being extensively studied for their ability to convert mechanical energy into electrical energy, owing to their exceptional energy density and flexibility in integration within systems. The piezoelectric effect, which involves the generation of an electric charge in response to mechanical stress or strain in certain materials, offers notable advantages over alternative methods, such as electromagnetic and electrostatic induction. Compared to these alternative methods, piezoelectric materials demonstrate higher energy density, enabling them to efficiently convert a larger quantity of mechanical energy into electrical energy, resulting in increased power output. Moreover, the unique characteristics of piezoelectric materials facilitate their seamless integration into diverse systems and structures. They can be embedded or attached to objects and surfaces, allowing for energy harvesting from various sources of mechanical vibrations, including machinery, vehicles, infrastructure, and even human movements. The capacity of piezoelectric materials to capture energy from ambient vibrations and convert it into electrical energy makes them particularly attractive in scenarios where conventional power sources are impractical or inaccessible. Their utilization holds significant potential for powering remote or self-sustaining systems, where the availability of alternative energy sources is limited. Their superior energy density and ease of integration position piezoelectric materials as a promising technology for advancing energy harvesting in various applications and sectors [37–40].

Piezoceramics and polymers are both utilized in energy harvesting; however, they offer different characteristics and power outputs. Piezoceramics provide higher output power in the milliwatt range, while polymers typically yield lower output power in the micro- or nanowatt range. However, polymers possess the advantage of quick responsiveness, making them suitable for energy harvesters operating at very low input frequencies below 10 Hz or in scenarios with large amplitude excitations [4]. Their flexibility and mechanical properties enable the efficient conversion of mechanical energy into electrical energy under such conditions. Although the power output may be lower compared to piezoceramics, polymers expand the possibilities for energy harvesting in environments where low-frequency vibrations or significant amplitude excitations are present. The choice between piezoceramics and polymers depends on the specific requirements of the energy-harvesting system, with piezoceramics favored for higher power output needs and polymers preferred for applications involving low-frequency vibrations or large amplitude excitations.

ZnO is a very promising piezoelectric material that has been explored in the form of strain sensors, actuators, and other advanced electronic applications [33]. The issue with ZnO and its nanostructures is their brittle nature and high rigidity. They have often been formed into composites with flexible materials to utilize their piezoelectric property, while overcoming their fragile nature. ZnO nanostructures were grown on paper by a low-temperature solvothermal method. A strain sensor was fabricated from the ZnO-paper nanocomposite by plasma sputtering a thin layer of gold on both sides of the sheet. The strain sensor gave a stable output under static as well as dynamic load testing. This sensor also was sensitive to different frequencies and responded quickly to external stimuli. The output power of the fabricated device is low; however, it possesses good sensitivity to strain [34]. The geometry of ZnO nanostructures and their respective concentrations in nanocomposites have a profound effect on the overall piezoelectric activity of the material composition. Referring to the geometry of nanoflowers and nanoparticles, it can be inferred that nanoflowers have a spatial distribution of geometry, while nanoparticles do not show such distribution. Moreover, the modulus of nanoflowers was found to be lower than nanoparticles, making them less stiff and hence more responsive to the applied stress. Thus, PDMS–ZnO-based nanocomposites (Figure 9.1c) showed a higher piezoelectric output voltage of 4.2 V compared to 1.86 V for nanoflowers and nanoparticles of ZnO, respectively. Further, nanoflowers of ZnO provided good connectivity in comparison to nanoparticles of ZnO that helped to achieve better piezoelectric activity in ZnO nanoflower-based nanocomposites [35]. In another study, KNN and ZnO nanorods were incorporated into a PVDF matrix to produce electrospun nanofibrous web. The synergistic effect of KNN and ZnO helped to achieve a high output voltage and current of 25 V and 1.81 μA, respectively. Through the structural analysis of these nanocomposite membranes, it was ascertained that the incorporated filler, KNN, and ZnO, acted as β-nucleating agents for the PVDF polymer. The increased electroactive phase content of the polymer also helped in achieving the appreciable performance of the nanogenerator fabricated from the nanocomposite [36, 41, 42]. Similarly, Parangusan et al. conducted a study on a tri-layer sandwiching structure composed of a composite material called Ni-doped ZnO/PVDF–HFP (poly(vinylidene fluoride-hexafluoropropylene). They added different weight percentages of Ni-doped ZnO nanoparticles to the PVDF–HFP matrix, resulting in increased crystallization of the beta (β) phase. This, in turn, led to improved piezoelectric and dielectric properties with an increased concentration of the filler material. The study demonstrated that the composite material achieved its maximum output voltage of 1.2 V at 0.5 wt% for Ni-doped ZnO [43–45]. Semiconductor filler-based piezoelectric materials have the potential to enhance energy-harvesting properties. When these materials are subjected to mechanical stress or vibrations, they generate electrical energy through the piezoelectric effect. The presence of semiconductor fillers in the piezoelectric matrix can further enhance this effect by improving the material's piezoelectric properties. Semiconductor fillers can contribute to increased charge generation and separation within the material, resulting in higher output voltages and improved energy conversion efficiency.

The unique properties of semiconductor materials, such as their bandgap and carrier mobility, can influence the energy-harvesting capabilities of the composite material.

MoS_2 is a semiconductor that has gained considerable attention for its piezoelectric characteristics. It has a rare S–Mo–S trilayer structure, similar to graphene. Recently, several piezoelectric composite studies based on MoS_2 have been carried out. ZnO has been grown on MoS_2 nanosheets to form a composite nanostructure. It was followed by its incorporation into PVDF polymer to form a composite film. The inclusion of such nanostructures into the polymer incremented its electroactive phase content. It could produce a maximum open-circuit voltage of 6.22 V at 100 kPa, which further increased to 17.56 V at 300 kPa. The ability of the nanogenerator to harvest energy from human motion indicates its suitability as a wearable energy harvester [46]. Abundant natural polymers such as cellulose have weak piezoelectric properties. Often, piezoelectric reinforcements aid in enhancing the electrical properties of such natural biopolymers. Cellulose was oxidized with tetramethylpiperidinyl-1-oxyl, to improve its dispersion in solvent. A good dispersion helps in the preparation of flexible films. Piezoelectric films of oxidized cellulose nanofibrils with exfoliated MoS_2 nanosheets were prepared. It was observed that the cellulose nanofibrils resided between the nanosheets forming a strong brick-mortar structure. A striking high d_{33} value of 31 pC/N was observed for the composite films prepared. The maximum open-circuit voltage and short-circuit current attained were 4.1 V and 0.21 μA, respectively [47]. Further, incorporation of $BaTiO_3$ nanoparticles along with MoS_2 nanosheets in the PVDF polymer, led to outstanding mechanical and piezoelectric properties. A d_{33} value as high as 45 pC/N was achieved for this ternary composite piezoelectric film. A stable electrical output of 8.2 V and 0.48 μA was achieved for this ternary composition [48]. Few layers of MoS_2 prepared by liquid exfoliation technique were introduced into PVDF polymer to form composite films. The compatibility between the two constituents led to enhanced piezoelectric properties of the composite. However, the piezoelectric response decreased at higher levels of MoS_2 loading owing to the aggregation of filler. The mechanical properties improved in terms of elongation and breaking strength with the increase in the filler content. The high surface area of few layers of MoS_2, provided an effective way to unload stress, thus contributing to the enhancement of mechanical properties of the composite film [49]. MoS_2 flakes synthesized hydrothermally have been incorporated into PVDF polymer at different levels of loading to study their effect on the piezoelectric properties of the composite. The reinforcement could effectively enhance the β-phase content of PVDF from 54% to around

84%. A maximum open-circuit voltage of 13.8 V was obtained for 7% reinforcement-loaded PVDF composite film [50].

Among other semiconductors, titanium dioxide (TiO_2) has also been explored in piezoelectric composites. Solution-cast films of TiO_2 nanoparticles incorporated into different percentages of poly(vinylidene fluoride-*co*-hexafluoropropene) (PVDF–HFP) were prepared. It was found that the semiconductor nanoparticles helped to increase the electroactive phase content of the polymer. Further, the interfacial polarization between the filler and polymer matrix helped to improve the capacitance value of the composite films. It produced a peak voltage of 9.69 V under finger-tapping conditions [51]. Further studies on varying TiO_2 content in the PVDF–HFP matrix have shown that a minimum of 1 wt% loading of filler is necessary to stabilize the electroactive phase of the polymer. Reinforcement content below the mentioned % leads to the formation of a mixture of α and β phases of the polymer [52]. A similar study varying TiO_2 nanofillers in PVDF has also been carried out [53]. Nanoarrays of ZnO/TiO_2 synthesized hydrothermally followed by a low-temperature liquid-phase method were assembled on a titanium substrate. These nanostructures had good piezoelectric properties. The properties of the nanoarray were enhanced with the reduction in their dimensions. It further showed additional properties of antibacterial activity and biocompatibility [54].

Semiconductive filler-based hybrid piezoelectric energy harvesters, which exemplify the synergy of superior materials and inventive engineering, are at the forefront of the energy-harvesting revolution. These devices display notable gains in energy conversion efficiency, charge distribution, and operational frequency range by fusing piezoelectric elements with semiconductive fillers. Their ability to transform ambient mechanical energy into usable electrical power holds enormous promise for a sustainable and decentralized energy future. These harvesters enable increased energy autonomy, decreased environmental impact, and technological innovation by powering wearable gadgets, sensors, and IoT technologies. Semiconductive filler-based hybrid piezoelectric energy harvesters are positioned to play a crucial role in building a more energy-efficient and sustainable future as research and development continue to enhance their design and use.

9.5 Applications

Semiconductive filler-based piezoelectric materials possess remarkable electrical and mechanical properties that make them highly versatile in numerous applications. These materials find utility in energy-harvesting devices, effectively converting mechanical vibrations into electrical energy to power small electronic devices. They also serve as invaluable sensors and actuators, enabling precise measurement of physical parameters, such as pressure, force, and acceleration, and facilitating precise control of mechanical motion. In the medical field, piezoelectric materials are extensively employed in applications, such as ultrasound imaging, drug delivery systems, and implantable devices, showcasing their significance in advancing healthcare technologies. Within the realm of microelectronics, these materials are essential for the development of micro–electro–mechanical systems (MEMS), enabling the creation of miniaturized sensors, actuators, and valves. Furthermore, piezoelectric materials play a vital role in structural health monitoring (SHM), robotics, and haptic feedback systems, ensuring safe and reliable operation in various industrial and technological environments. Their unique properties contribute to enhanced performance, increased efficiency, and improved user experience in a wide range of industrial and scientific settings [55].

In the realm of implantable devices, piezoelectric sensors serve as invaluable tools for early warning systems. These sensors can provide doctors with crucial information well in advance, even before routine check-ups, allowing for timely interventions. For example, cardiomechanical electric sensors utilizing $(Na_{0.5},K_{0.5})NbO_3$ (NKN) thin film with a resonance frequency of 10 Hz have been employed as highly sensitive sensors for monitoring cardiovascular activity [56]. The versatility of piezoelectric ceramic sensors extends to vision correction as well. They can be used in applications such as vision correction lenses or intraocular lenses, where they help improve visual acuity by correcting refractive errors [57]. In recent years, the use of polymers has gained significant attention in the development of synthetic skin for prosthetic applications and robotics. By modifying the properties of polymers, researchers aim to mimic human characteristics and enhance sensitivity. These advancements in polymer-based sensors enable synthetic skin to respond more accurately to external stimuli, offering improved haptic feedback and a more lifelike experience in various applications [58].

By tailoring the sensors to specific pressure ranges, such as intraocular pressure, cranial pressure, blood pressure, and other bodily movements, make them well-suited for precise monitoring in different healthcare scenarios. The utilization of semiconductor filler-based piezoelectric sensors holds significant potential for advancing healthcare technologies and improving patient care by providing more reliable and accurate data for diagnosis and treatment. In a study by Genchi et al. [59], the application of poly(vinylidene fluoride-trifluoroethylene) (P(VDF–TrFE)) and $BaTiO_3$ nanoparticles in lead-free piezoelectric nanocomposite films was investigated for its ability to stimulate and promote the differentiation of SH–SY5Y neuroblastoma cells. The researchers found that the composite film, containing P(VDF–TrFE) and $BaTiO_3$ nanoparticles, exhibited enhanced piezoelectric properties compared to the pure P(VDF–TrFE) film. As a result, the composite film induced the emission of significantly longer neurites in the neuroblastoma cells. Importantly, the composite-type piezoelectric materials used in the study demonstrated benign mechanical properties and compatible piezoelectric responses in the presence of biological matter. This suggests that these materials hold potential for applications in biocompatible systems, where the mechanical and electrical properties are crucial in the interaction with biological tissues or cells [59, 60]. The findings of this study contribute to the exploration of lead-free piezoelectric nanocomposite films as a viable option for stimulating and promoting cellular differentiation in neuroblastoma cells and potentially other biomedical applications. The utilization of semiconductor filler-based piezoelectric sensors in healthcare applications shows great promise. They offer the potential for improved sensitivity, biocompatibility, and tailored functionality, enabling accurate and reliable monitoring of bodily signals for enhanced healthcare outcomes.

Researchers have explored the use of inorganic piezoelectric materials, such as lead zirconate titanate (PZT) [61] and ZnO [62], to fabricate piezoelectric composites. By incorporating these inorganic fillers into the polymer matrix, the performance of piezoelectric polymer sensors can be significantly enhanced. In the study conducted by Niu et al. [61], a stretchable ceramic/polymer piezoelectric composite was developed by blending PZT with solid silicone rubber. The composite exhibited improved piezoelectric properties due to the incorporation of PZT, allowing for enhanced sensing capabilities. By optimizing the filler mass ratio and deformation parameters, the researchers were able to achieve significant advancements in the composite's performance. In a separate study by Gu and coworkers [62], a flexible ZnO/PVDF nanocomposite fibrous membrane was created using electrospinning. The researchers demonstrated that the synergistic effect between the rod-like piezoelectric nanofillers (ZnO) and the electroactive β-crystals of PVDF played a crucial role in enhancing the piezoelectric behavior of the nanocomposite. This combination of materials resulted in improved sensitivity and responsiveness to mechanical stimuli, making the ZnO/PVDF nanocomposite an attractive option for flexible and wearable sensor applications. Similarly, Ferreira et al. [63] developed $BaTiO_3$/epoxy composites for use as piezoelectric sensors in a SHM system. The fabrication process involved direct mechanical mixing; however, the presence of air voids resulting from high filler concentrations negatively impacted the density and electromechanical properties of the composites. Nevertheless, the composite material exhibited promising electromechanical and mechanical properties, making it a suitable candidate for SHM applications. By leveraging the unique properties of inorganic fillers and their interactions with the polymer matrix, these piezoelectric composites offer improved physical and piezoelectric characteristics compared to their individual components [37, 64–66].

Hybrid piezoelectric energy harvesters based on semiconductive fillers provide a versatile and sustainable solution for a wide range of applications. These energy harvesters demonstrate their potential to revolutionize how we generate electricity in a variety of fields, from powering wearable electronics and remote sensors to supporting IoT devices and SHM. They contribute to self-sufficiency, decreased environmental impact, and enhanced efficiency across industries by harvesting ambient mechanical energy, offering a future where devices and systems may run consistently without the limits of existing power sources.

9.6 Conclusion

In summary, energy harvesting involves the conversion of ambient energy into electrical energy using diverse sources. By incorporating semiconductive fillers, such as graphene and metal oxide nanoparticles into piezoelectric materials, hybrid materials with superior energy-harvesting capabilities are created. These hybrid materials surpass traditional piezoelectric materials by exhibiting enhanced piezoelectric coefficients, thanks to strong interactions between the fillers and matrix, as well as aligned fillers within the matrix. This enables efficient conversion of mechanical energy into electrical energy, making them valuable for capturing vibrations across various applications. In the field of biomedicine, semiconductor filler-based materials offer advantages, such as small size, biocompatibility, and flexibility. Piezoelectric sensors based on these materials find applications in cardio monitoring, vision correction, and prosthetics. Moreover, in SHM, semiconductive filler-based piezoelectric sensors enhance the capabilities of structural monitoring. By incorporating semiconductive fillers into piezoelectric materials, their performance is improved, leading to tailored functionality and increased sensitivity in diverse applications. Overall, hybrid piezoelectric materials hold tremendous promise in the realms of energy harvesting, biomedical devices, and SHM. They offer significant advancements in terms of performance, customization, and sensitivity, paving the way for transformative developments in these fields.

References

1 Gokhool, S., Bairagi, S., Kumar, C., and Mulvihill, D.M. (2022). Reflections on boosting wearable triboelectric nanogenerator performance via interface optimisation. *Results Eng.* 17: 100808. https://doi.org/10.1016/j.rineng.2022.100808.

2 Curie, P. (2021). Fundamentals of piezoelectricity. In: *Lead-Free Piezoelectric Materials* (ed. J.-F. Li), 1–18. Wiley.

3 Zhu, G., Peng, B., Chen, J. et al. (2015). Triboelectric nanogenerators as a new energy technology: from fundamentals, devices, to applications. *Nano Energy* 14: 126–138. https://doi.org/10.1016/j.nanoen.2014.11.050.

4 Banerjee, S., Bairagi, S., and Wazed Ali, S. (2021). A critical review on lead-free hybrid materials for next generation piezoelectric energy harvesting and conversion. *Ceram. Int.* 47 (12): 16402–16421. https://doi.org/10.1016/j.ceramint.2021.03.054.

5 Damjanovic, D. (2008). Lead-based piezoelectric materials. In: *Piezoelectric and Acoustic Materials for Transducer Applications* (ed. A. Safari and E.K. Akdoğan), 59–79. Springer.

6 Wongdamnern, N., Triamnak, N., Ngamjarurojana, A. et al. (2008). Comparative studies of dynamic hysteresis responses in hard and soft PZT ceramics. *Ceram. Int.* 34 (4): 731–734. https://doi.org/10.1016/j.ceramint.2007.09.048.

7 Huang, J., Yao, M., Lin, J., and Yao, X. (2022). Enhanced mechanical properties and excellent electrical properties of PZT piezoelectric ceramics modified by YSZ. *Mater. Lett.* 307: 131006. https://doi.org/10.1016/j.matlet.2021.131006.

8 Zong, X., Yang, Z., Li, H., and Yuan, M. (2006). Effects of WO_3 addition on the structure and electrical properties of Pb_3O_4 modified PZT-PFW-PMN piezoelectric ceramics. *Mater. Res. Bull.* 41 (8): 1447–1454. https://doi.org/10.1016/j.materresbull.2006.02.002.

9 Kim, B.S., Ji, J.H., and Koh, J.H. (2021). Improved strain and transduction values of low-temperature sintered CuO-doped PZT-PZNN soft piezoelectric materials for energy harvester applications. *Ceram. Int.* 47 (5): 6683–6690. https://doi.org/10.1016/j.ceramint.2020.11.008.

10 Fei, X., Lin, X., Wang, C. et al. (2022). Sb_2O_3-modified lead zirconate titanate piezoelectric ceramics with enhancing piezoelectricity and low loss. *Am. Ceram. Soc.* 106: 501–512.

11 Qiao, L., Li, G., Tao, H. et al. (2020). Full characterization for material constants of a promising KNN-based lead-free piezoelectric ceramic. *Ceram. Int.* 46 (5): 5641–5644. https://doi.org/10.1016/j.ceramint.2019.11.009.

12 Wang, X., Meng, C., and Wang, Y. (2021). Insight for the construction of R-T phase boundary in KNN piezoceramics from the view of energy band structure and electron density. *Ceram. Int.* 47 (20): 28500–28504. https://doi.org/10.1016/j.ceramint.2021.07.006.

13 Rutkowski, P., Huebner, J., Graboś, A. et al. (2020). Dense KNN polycrystals doped by Er_2O_3 obtained by hot pressing with hexagonal boron nitride. *Materials* 13: 1–12.

14 Wang, X., Zhang, G., and Wang, Y. (2022). Effects of coupling ability of dopants on NbO_6 vibration and piezoelectric properties of KNN lead-free ceramics. *J. Phys. Chem. Solids* 164: 110633. https://doi.org/10.1016/j.jpcs.2022.110633.

15 Vijatovi, M.M., Bobi, J.D., and Stojanovi, B.D. (2008). History and challenges of barium titanate: part I. *Sci. Sintering* 40: 155–165. https://doi.org/10.2298/SOS0802155V.

16 Wang, Y., Shi, S., Dong, Q. et al. (2021). Electrospun lanthanum-doped barium titanate ceramic fibers with excellent dielectric performance. *Mater. Charact.* 172: 110859. https://doi.org/10.1016/j.matchar.2020.110859.

17 Huang, C.M. (2011). Relationship between the evolutions of the microstructure and semiconductor properties of yttrium-doped barium titanate ceramics. *J. Phys. D: Appl. Phys.* 44: 1–9. https://doi.org/10.1088/0022-3727/44/34/345403.

18 Wernicke, R. (1978). The influence of kinetic processes on the electrical conductivity of donor-doped. *Phys. Status Solidi A* 139: 139–144.

19 Unruan, M., Sareein, T., Tangsritrakul, J. et al. (2008). Changes in dielectric and ferroelectric properties of Fe^{3+}/Nb^{5+} hybrid-doped barium titanate ceramics under compressive stress. *J. Appl. Phys.* 104: 1–4. https://doi.org/10.1063/1.3042228.

20 Srisombat, L., Ananta, S., Singhana, B. et al. (2013). Chemical investigation of Fe^{3+}/Nb^{5+}-doped barium titanate ceramics. *Ceram. Int.* 39: S591–S594. https://doi.org/10.1016/j.ceramint.2012.10.142.

21 Hanady Yaseen, Y.T. and Baltianski, S. (2006). Effect of incorporating method of niobium on the properties of doped barium titanate ceramics. *J. Am. Ceram. Soc.* 1589 (107): 1584–1589. https://doi.org/10.1111/j.1551-2916.2006.00966.x.

22 Villafuerte-Castrejón, M.E., Morán, E., Reyes-Montero, A. et al. (2016). Towards ead-free piezoceramics: facing a synthesis challenge. *Materials* 9 (1): 21.

23 Hou, D., Fan, H., Chen, Y. et al. (2022). Effects of lanthanide ions on the structure and electrical properties of Aurivillius Bi_3TiNbO_9 high temperature piezoelectric ceramics. *J. Alloys Compd.* 921: 166065. https://doi.org/10.1016/j.jallcom.2022.166065.

24 Li, X., Zhu, L., Huang, P. et al. (2020). Reduction of oxygen vacancy concentration and large enhancement of electrical performances in Cu/Sb co-doped $Bi_4Ti_3O_{12}$ high temperature piezoelectric ceramics. *J. Appl. Phys.* 127: 1–11. https://doi.org/10.1063/1.5125478.

25 Li, X., Chen, Z., Sheng, L. et al. (2019). Remarkable piezoelectric activity and high electrical resistivity in Cu/Nb co-doped $Bi_4Ti_3O_{12}$ high temperature piezoelectric ceramics. *J. Eur. Ceram. Soc.* 39 (6): 2050–2057. https://doi.org/10.1016/j.jeurceramsoc.2019.01.042.

26 Zhang, F. and Shi, W. (2022). Effects of cerium on structures and electrical properties of (Nb, Ta) modified $Bi_4Ti_3O_{12}$ piezoelectric ceramics. *J. Am. Ceram. Soc.* 4161–4170. https://doi.org/10.1111/jace.18393.

27 Sun, L., Chen, Q., Wu, J. et al. (2014). Dielectric and piezoelectric properties of cerium-doped. *Ceram. Int.* 40 (9): 14159–14163. https://doi.org/10.1016/j.ceramint.2014.06.002.

28 El Achaby, M., Arrakhiz, F.Z., Vaudreuil, S. et al. (2012). Piezoelectric β-polymorph formation and properties enhancement in graphene oxide – PVDF nanocomposite films. *Appl. Surf. Sci.* 258 (19): 7668–7677. https://doi.org/10.1016/j.apsusc.2012.04.118.

29 Yaqoob, U., Uddin, A.S.M.I., and Chung, G.S. (2017). A novel tri-layer flexible piezoelectric nanogenerator based on surface-modified graphene and PVDF-$BaTiO_3$ nanocomposites. *Appl. Surf. Sci.* 405: 420–426. https://doi.org/10.1016/j.apsusc.2017.01.314.

30 Duan, F., Liao, Y., Zeng, Z. et al. (2019). Graphene-based nanocomposite strain sensor response to ultrasonic guided waves. *Compos. Sci. Technol.* 174: 42–49. https://doi.org/10.1016/j.compscitech.2019.02.011.

31 Faraz, M., Singh, H.H., and Khare, N. (2022). A progressive strategy for harvesting mechanical energy using flexible PVDF-rGO-MoS_2 nanocomposites film-based piezoelectric nanogenerator. *J. Alloys Compd.* 890: 161840. https://doi.org/10.1016/j.jallcom.2021.161840.

32 Song, S., Li, Y., Wang, Q., and Zhang, C. (2021). Boosting piezoelectric performance with a new selective laser sintering 3D printable PVDF/graphene nanocomposite. *Compos. Part A: Appl. Sci. Manuf.* 147: 106452. https://doi.org/10.1016/j.compositesa.2021.106452.

33 Il Park, K., Jeong, C.K., Kim, N.K., and Lee, K.J. (2016). Stretchable piezoelectric nanocomposite generator. *Nano Convergence* 3 (1): 1–12. https://doi.org/10.1186/s40580-016-0072-z.

34 Gullapalli, H., Vemuru, V.S.M., Kumar, A. et al. (2010). Flexible piezoelectric ZnO – paper nanocomposite strain sensor. *Small* 6 (5): 1641–1646. https://doi.org/10.1002/smll.201000254.

35 Jeronimo, K., Koutsos, V., Cheung, R., and Mastropaolo, E. (2021). PDMS-ZnO piezoelectric nanocomposites for pressure sensors. *Sensors* 21 (17): https://doi.org/10.3390/s21175873.

36 Bairagi, S. and Ali, S.W. (2020). A hybrid piezoelectric nanogenerator comprising of KNN/ZnO nanorods incorporated PVDF electrospun nanocomposite webs. *Int. J. Energy Res.* 44 (7): 5545–5563. https://doi.org/10.1002/er.5306.

37 Zhang, C., Fan, W., Wang, S. et al. (2021). Recent progress of wearable piezoelectric nanogenerators. *ACS Appl. Electron. Mater.* 3 (6): 2449–2467. https://doi.org/10.1021/acsaelm.1c00165.

38 Mahapatra, B., Patel, K.K., and Patel, P.K. (2020). A review on recent advancement in materials for piezoelectric/triboelectric nanogenerators. *Mater. Today Proc.* 46: 5523–5529. https://doi.org/10.1016/j.matpr.2020.09.261.

39 Zhang, J., He, Y., Boyer, C. et al. (2021). Recent developments of hybrid piezo-triboelectric nanogenerators for flexible sensors and energy harvesters. *Nanoscale Adv.* 3 (19): 5465–5486. https://doi.org/10.1039/d1na00501d.

40 Wang, Z.L., Wang, X., Song, J. et al. (2008). Piezoelectric nanogenerators for self-powered nanodevices. *IEEE Pervasive Comput.* 7 (1): 49–55. https://doi.org/10.1109/MPRV.2008.14.

41 Chou, X., Zhu, J., Qian, S. et al. (2018). All-in-one filler-elastomer-based high-performance stretchable piezoelectric nanogenerator for kinetic energy harvesting and self-powered motion monitoring. *Nano Energy* 53: 550–558. https://doi.org/10.1016/j.nanoen.2018.09.006.

42 Jeong, C.K., Lee, J., Han, S. et al. (2015). A hyper-stretchable elastic-composite energy harvester. *Adv. Mater.* 27 (18): 2866–2875. https://doi.org/10.1002/adma.201500367.

43 Zhou, H., Zhang, Y., Qiu, Y. et al. (2020). Stretchable piezoelectric energy harvesters and self-powered sensors for wearable and implantable devices. *Biosens. Bioelectron.* 168: 112569. https://doi.org/10.1016/j.bios.2020.112569.

44 Parangusan, H., Ponnamma, D., and Al Ali Almaadeed, M. (2017). Flexible tri-layer piezoelectric nanogenerator based on PVDF-HFP/Ni-doped ZnO nanocomposites. *RSC Adv.* 7 (79): 50156–50165. https://doi.org/10.1039/c7ra10223b.

45 Sukumaran, S., Chatbouri, S., Rouxel, D. et al. (2021). Recent advances in flexible PVDF based piezoelectric polymer devices for energy harvesting applications. *J. Intell. Mater. Syst. Struct.* 32 (7): 746–780. https://doi.org/10.1177/1045389X20966058.

46 Cao, S., Zou, H., Jiang, B. et al. (2022). Incorporation of ZnO encapsulated MoS_2 to fabricate flexible piezoelectric nanogenerator and sensor. *Nano Energy* 102: 107635. https://doi.org/10.1016/j.nanoen.2022.107635.

47 Wu, T., Song, Y., Shi, Z. et al. (2021). High-performance nanogenerators based on flexible cellulose nanofibril/MoS_2 nanosheet composite piezoelectric films for energy harvesting. *Nano Energy* 80: 105541. https://doi.org/10.1016/j.nanoen.2020.105541.

48 Xu, M., Wu, T., Song, Y. et al. (2021). Achieving high-performance energy harvesting and self-powered sensing in a flexible cellulose nanofibril/MoS_2/$BaTiO_3$ composite piezoelectric nanogenerator. *J. Mater. Chem. C* 9 (43): 15552–15565. https://doi.org/10.1039/d1tc03886a.

49 Jiang, L., Xie, H., Hou, Y. et al. (2019). Enhanced piezoelectricity of a PVDF-based nanocomposite utilizing high-yield dispersions of exfoliated few-layer MoS_2. *Ceram. Int.* 45 (9): 11347–11352. https://doi.org/10.1016/j.ceramint.2019.02.213.

50 Kumar Prajapati, G., Katla, R., and Singh, B. (2021). Effect of variation of MoS_2 concentration on the piezoelectric performance of PVDF-MoS_2 based flexible nanogenerator. *Mater. Today Proc.* 47: 4861–4865. https://doi.org/10.1016/j.matpr.2021.06.084.

51 Mitra, R., Priyadarshini, B.S., Ramadoss, A., and Manju, U. (2022). Stretchable polymer-modulated PVDF-HFP/TiO_2 nanoparticles-based piezoelectric nanogenerators for energy harvesting and sensing applications. *Mater. Sci. Eng., B* 286: 116029. https://doi.org/10.1016/j.mseb.2022.116029.

52 Chakhchaoui, N., Farhan, R., Eddiai, A. et al. (2021). Improvement of the electroactive β-phase nucleation and piezoelectric properties of PVDF-HFP thin films influenced by TiO_2 nanoparticles. *Mater. Today Proc.* 39: 1148–1152. https://doi.org/10.1016/j.matpr.2020.05.407.

53 Kulkarni, N.D. and Kumari, P. (2023). Development of highly flexible PVDF-TiO_2 nanocomposites for piezoelectric nanogenerator applications. *Mater. Res. Bull.* 157: 112039. https://doi.org/10.1016/j.materresbull.2022.112039.

54 Pang, S., He, Y., Zhong, R. et al. (2019). Multifunctional ZnO/TiO_2 nanoarray composite coating with antibacterial activity, cytocompatibility and piezoelectricity. *Ceram. Int.* 45 (10): 12663–12671. https://doi.org/10.1016/j.ceramint.2019.03.076.

55 Hu, S., Shi, Z., Zhao, W. et al. (2019). Multifunctional piezoelectric elastomer composites for smart biomedical or wearable electronics. *Compos. Part B: Eng.* 160: 595–604. https://doi.org/10.1016/j.compositesb.2018.12.077.

56 Kwak, J., Kingon, A.I., and Kim, S. (2012). Medical sensor applications. *Mater. Lett.* 82: 130–132. https://doi.org/10.1016/j.matlet.2012.05.079.

57 Bo, L., Youbo, W., and Jianying, F. (2014). Weak signal acquisition and processing system for piezoelectric ceramic transducer actuator. In: *2014 9th International Forum on Strategic Technology (IFOST)*, Cox's Bazar, Bangladesh (21–23 October 2014)., 209–215. IEEE.

58 Park, J., Kim, M., Lee, Y. et al. (2015). Fingertip skin-inspired microstructured ferroelectric skins discriminate static/dynamic pressure and temperature stimuli. *Sci. Adv.* 1 (9): e1500661.

59 Genchi, G.G., Ceseracciu, L., Marino, A. et al. (2016). P(VDF-TrFE)/$BaTiO_3$ nanoparticle composite films mediate piezoelectric stimulation and promote differentiation of SH-SY5Y neuroblastoma cells. *Adv. Healthcare Mater.* 5 (14): 1808–1820. https://doi.org/10.1002/adhm.201600245.

60 Zhang, Y., Kim, H., Wang, Q. et al. (2020). Progress in lead-free piezoelectric nanofiller materials and related composite nanogenerator devices. *Nanoscale Adv.* 2 (8): 3131–3149. https://doi.org/10.1039/c9na00809h.

61 Niu, X., Jia, W., Qian, S. et al. (2019). High-performance PZT-based stretchable piezoelectric nanogenerator. *ACS Sustainable Chem. Eng.* 7 (1): 979–985. https://doi.org/10.1021/acssuschemeng.8b04627.

62 Li, J., Chen, S., Liu, W. et al. (2019). High performance piezoelectric nanogenerators based on electrospun ZnO nanorods/poly(vinylidene fluoride) composite membranes. *J. Phys. Chem. C* 123 (18): 11378–11387. https://doi.org/10.1021/acs.jpcc.8b12410.

63 Ferreira, O.B., Venkat, R.S, Adam, J., and Boller, C. (2016). Development of the fabrication process and characterization of piezoelectric $BaTiO_3$/epoxy composite used for coated ultrasonic transducer patterns in structural health monitoring. *19th World Conference on Non-Destructive Testing* (13–17 June 2016), München. Cambridge: arXiv. pp. 1–9.

64 Kai Dong, Z.L.W. and Peng, X. (2019). Fiber/fabric-based piezoelectric and triboelectric nanogenerators for flexible/stretchable and wearable electronics and artificial intelligence. *Adv. Mater.* 32: 1–43.

65 Song, G., Mo, Y.L., Otero, K., and Gu, H. (2006). Health monitoring and rehabilitation of a concrete structure using intelligent materials. *Smart Mater. Struct.* 15 (2): 309–314. https://doi.org/10.1088/0964-1726/15/2/010.

66 Jiao, P., Egbe, K.J.I., Xie, Y. et al. (2020). Piezoelectric sensing techniques in structural health monitoring: a state-of-the-art review. *Sensors* 20 (13): 1–21. https://doi.org/10.3390/s20133730.

10

Cellulose-Based Hybrid Piezoelectric Materials

Srijan Das[1], Mukta Nitin Mirlekar[1], Sourav Banerjee[2], Anupam Chowdhury[1], and S. Wazed Ali[1,2]

[1] *Department of Textile and Fibre Engineering, Indian Institute of Technology Delhi, Hauz Khas, New Delhi, India*
[2] *School of Interdisciplinary Research (SIRe), Indian Institute of Technology Delhi, Hauz Khas, New Delhi, India*

10.1 Introduction

As the world's appetite for fossil fuels expands, the exploration of renewable energy resources to substitute them has risen to the forefront of global research demand. Renewable energy sources such as wind power, hydropower, ocean energy, and solar power are constantly being improved due to decades of research. This provides a significant portion of the energy required by the population. Total renewable energy consumption rose dramatically from 4098 TWh in 2010 to 7627 TWh in 2020 (shown in Figure 10.1), as reported by the global energy review of the International Energy Agency (IEA) in 2021 [1, 2].

Increasing numbers of countries are turning to renewable energy sources for their power needs. A big reason for this change is that people are becoming more aware of how damaging the traditional way of making electricity is to the environment. Furthermore, India has planned to generate 175 GW of power exclusively from renewable sources by 2022 [3, 4]. Considering the above aspirational goals, countries are putting forth a pressing need for alternative sources of energy that meet the requirements and the financial, adoption, and implementation requirements in real-world scenarios.

It is well known that mechanical strain can be converted into electrical current in piezoelectric materials. The production of renewable energy from this material is also of promising scope [5]. Combining mechanical, wind, and wave energy conversion systems with some conceptual designs that

Hybrid Materials for Piezoelectric Energy Harvesting and Conversion, First Edition.
Edited by S. Wazed Ali, Satyaranjan Bairagi, and Shahid Ul Islam.
© 2024 John Wiley & Sons, Inc. Published 2024 by John Wiley & Sons, Inc.

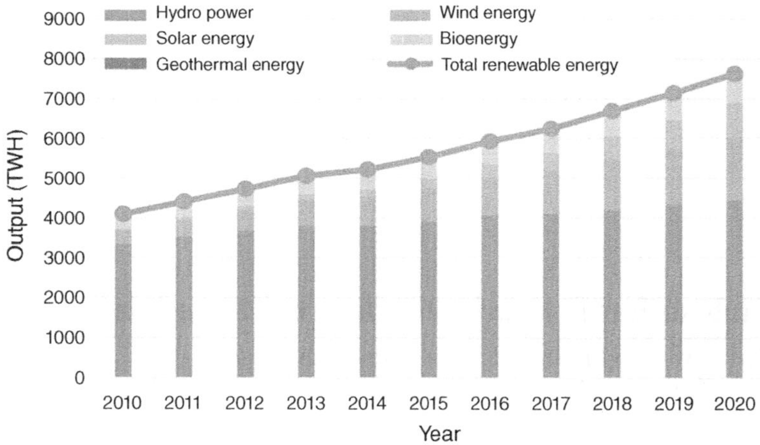

Figure 10.1 The year-wise trend of renewable energy usage. Source: [1]/Elsevier/ CCBY 4.0/Public domain.

utilize piezoelectric transducers for energy production is possible. Using piezoelectric-based charging panels, we can create a power-generating sidewalk that stores energy from the pressure of footsteps [6]. Gyms, offices, vehicle clutches and gears, mobile keypads and keyboards, soldier boots and floor mats, and human-powered clubs are excellent areas for energy harvesting applications. Some reports have suggested that the global market for energy harvesting systems is expected to expand from its current valuation of US$415.5 million to a new valuation of US$746.6 million at a growth rate of 8.7% from 2020 to 2027 [7].

Single crystals, ceramics, polymers, polymer composites, and nanocomposites are the primary piezoelectric materials used for energy harvesting. In the case of a single crystal, aligned dipoles are primarily responsible for piezoelectric responses resulting from ordered positive and negative ions throughout the structure [8]. The most common examples of piezoelectric single crystals are lithium niobate ($LiNbO_3$), lead magnesium niobate-lead titanate solid solution [$Pb(Mg_{1/3}Nb_{2/3})–PbTiO_3$)], lead zinc niobate-lead titanate (PZN-PT), and lead indium niobate-lead magnesium niobate-lead titanate [$Pb(In_{1/2}Nb_{1/2})–Pb(Mg_{1/3}Nb_{2/3})–PbTiO_3$ (PIN–PMN–PT)] [9]. Piezo ceramics have a perovskite crystal structure, which means they are made up of many smaller crystals with the same chemical composition. However, the distance between the ions varies slightly from crystal to crystal due to the orientation. Some examples are lead zirconate-titanate (PZT) [10], barium titanate ($BaTiO_3$) [11], and potassium niobate ($KNbO_3$) [12]. The inherent flexibility of piezoelectric polymers allows them to withstand much higher strain, making them more suitable for applications requiring significant bending or twisting. Some well-known piezoelectric polymers are polyvinylidene fluoride (PVDF), polyvinylidene fluoride-trifluoroethylene (PVDF-TrFE), cellulose and its derivatives, polyamides (PA), polylactic acid (PLA), etc. [13–19]. Comparing the composites to only ceramic and polymer materials, it has been discovered that polymer composites and nanocomposites exhibit several improved properties. Additional fillers (like ceramics, metal-based materials, hybrid materials, and biomaterials) of different sizes (macro, micro, and nano) are also being researched for better performance and compatibility.

Crystalline cellulose, a well-known polymeric piezoelectric material, has a piezoelectric property due to the linear alignment of glucose units via glycosidic linkage along the non-centrosymmetric C2 monoclinic lattice. Even piezoelectricity in natural wood was identified more than 50 years ago, but recently working with cellulose paper or cellulose derivative-based piezoelectric energy harvesting has become a hot topic [20, 21]. However, because of growing concerns about lead leaching, the PZT-based materials with the best performance are losing some of their potential. The lead-free alternatives have a limited range of applications due to their rigidity and difficulty recycling their waste, such as potassium sodium niobate (KNN) or boron tetraoxide (BTO). Hydrogen fluoride, which is dangerous to both the environment and people, could be liberated during the breakdown of PVDF [22]. As a result, the most recent generation of implantable and wearable bioelectronics is heading toward innovative piezoelectric materials with the best biodegradability.

The main benefits of cellulose are its abundance and its high biocompatibility. While there have been various papers on exceptionally high piezoelectricity, their average piezoelectric coefficients must be enhanced before they can be used in any practical application. This chapter's primary goal is to provide a comprehensive overview of the current state of and potential applications for piezoelectric research based on cellulose.

10.2 Origin of Piezoelectricity in Cellulose

The basic piezoelectricity of cellulose was discovered by Fukada in 1955. The shear piezoelectricity ($d_{25} = -d_{14}$) in natural cellulose is reported to be in the order of 10^{-9} magnitude, which confirms the origin of piezoelectricity. The orderly arrangement of polar hydroxyl groups in non-centrosymmetric manner results in a higher net dipole moment, thus showing higher piezoelectricity in wood. In each cellulose unit cell, the strong dipole moments exist between O2-H and O6, O3-H and O5, and O3-H and O6. The dipole moments, as calculated by DFT, were 36.4, 10, and 4.3 pm/V. The molecular chains of cellulose are linked by inter-chain hydrogen bonding between (110)t and (200)m planes. These bonds are arranged in non-centrosymmetric order, resulting in a high dipole moment. But the piezoelectricity of natural cellulose is low. The piezoelectric property is generally determined by the transverse piezoelectric constant (d_{31}), the longitudinal piezoelectric constant (d_{33}), and the shear piezoelectric constant (d_{14}). The longitudinal piezoelectric constant, d_{33}, of natural cellulose is only 0.4 pC/N, which is far lower than piezoelectric PVDF film (21 pC/N) [14].

10.3 Different Crystal Structures and Forms of Cellulose

Cellulose, the primary element of plant cell walls, is the most common biomaterial in the wild. Cellulose-based materials have broad-ranging functions due to their effective chemical structure and properties, as well as their biodegradability, recyclability, and biocompatibility. Cellulosic polymers exhibit piezoelectric properties due to dipolar alignment and confined charges. Biopolymers based on cellulose, like wood, amylase, and starch, have shown shear piezoelectricity due to their structure [23].

There are four distinct crystal structures for cellulose, and the most common two are Cellulose I and Cellulose II. Cellulose I has two polymorphs, triclinic type I_a and monoclinic type I_b, that are distinguished by their respective cellulose sources. The most stable thermodynamic state is cellulose II [25]. It's monoclinic and typically made by dissolving and regenerating Cellulose I or treating it with an alkali. Crystals of all types of cellulose have asymmetric dipoles because the monoclinic and triclinic structures lack symmetrical centers, and the hydroxyl group is highly polar. This provides the required structural conditions for cellulose crystals to function as piezoelectric materials. Researchers have studied the fundamental difference in piezo responses of cellulose I and II [26]. Wood and electroactive paper have been chosen for study, representing cellulose I and cellulose II, respectively. Hydrogen bonding occurs significantly differently in the crystalline conformations of cellulose I and cellulose II, which may have implications for piezoelectricity [27]. Cellulose I only shows intra-sheet intermolecular hydrogen bonding along the a–c plane, whereas cellulose II also shows inter-sheet hydrogen bonds along the b-oriented plane.

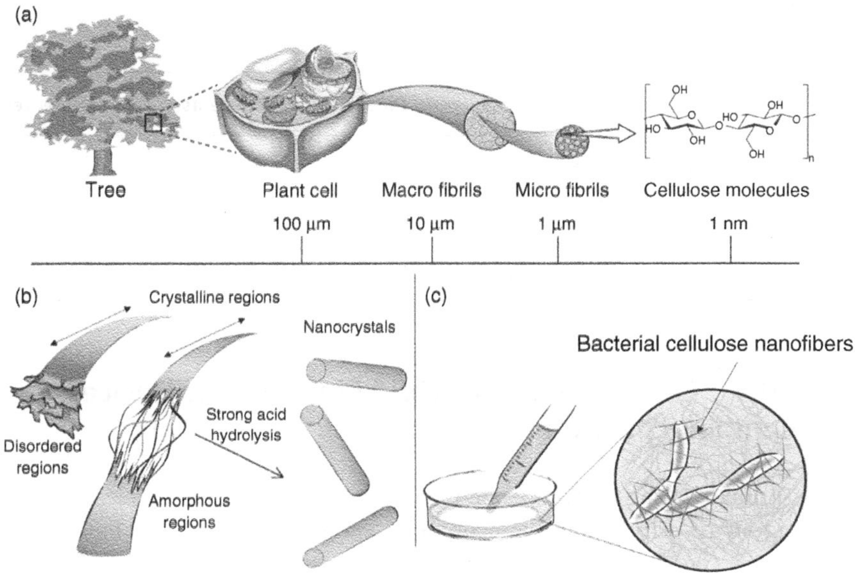

Figure 10.2 (a) Cellulose present in plant cell from meter to the nanometer scale, (b) nano cellulose crystals and its preparation process through acid hydrolysis and (c) bacterial cellulose nanofibers. Source: [24]/Frontiers/CC By 4.0/Public domain.

Conversely, cellulose crystals or nanoforms have been used as fillers for various polymeric substrates to improve overall piezoelectric performance. Different forms of cellulose, nanocrystals and microfibrils have been shown in Figure 10.2. Basically, nanocellulose covers most of the research due to its intrinsic properties. In the case of cellulose nanocrystals (CNC) and cellulose nanofibrils (CNF), nanocellulose is fabricated by cutting the wood cellulose or cellulosic fibers into smaller units until their dimensions become nanoscale, which is actually a top-down approach. CNF can be obtained, for example, from softwood by sulfate digestion and subsequent bleaching, followed by miniaturization processing through a grinder and homogenizer [28–31]. When some bacteria like *Komagataeibacter*, *Enterobacter*, *Rhodococcus*, and *Sarcina* oxidatively ferment polysaccharides, it creates bacterial cellulose (BC) films. This process is known as a "bottom–up approach". Bacterial cellulose has certain intriguing characteristics over wood-based cellulose, including biodegradability, high purity, water-holding ability, and improved mechanical and structural capabilities [32, 33]. The overall piezoelectric responses of different cellulosic forms have been tabulated in Table 10.1.

Table 10.1 Piezoelectric sensitivity of different cellulose-based materials along with PVDF.

Types of material	Piezoelectric coefficients	References
Cellulose nanocrystals (CNC)	7.3 pC/N	[34, 35]
Cellulose nanofibrils (CNF)	4.7 pC/N	[36]
Bacterial cellulose (BC)	5–15 pC/N	[33]
Polyvinylidene fluoride (PVDF)	24–34 pC/N	[37]

10.4 Cellulose-Based Hybrid Piezoelectric Devices Containing Cellulose as Matrix

Throughout the literature, one reliable way of improving the piezoelectric output of cellulose is to incorporate certain fillers into it. Fillers like $BaTiO_3$ and MoS_2 can be used for such purposes. Some of these hybrid cellulose-based piezoelectric nanogenerators are discussed in details in succeeding sections, where cellulose is used in liaison with conventional piezoelectric fillers to augment its piezoelectric property.

10.4.1 Molybdenum Disulphide (MoS_2) as Filler

Molybdenum disulphide (MoS_2) is known to show strong piezoelectric effect when used as a monolayer. The odd numbers of layers in MoS_2 flakes were found to show piezoelectricity [38]. Increasing thickness and using bilayers result in decreased or no piezoelectric output because of the opposite orientation of adjacent layers.

Yang and co-workers developed 2,2,6,6-tetramethylpiperidine-1-oxyl (TEMPO) oxidized cellulose nanofibril (TOCN) and molybdenum disulphide nanosheet-based composite films with enhanced piezoelectric properties whose d_{33} value reached 31 pC/N as compared to 12 pC/N of the pristine TEMPO-oxidized film [39]. The nanogenerator using such composite film showed an output voltage of 4.1 V and 0.21 μA short-circuit current. The performance of the nanogenerator reached a maximum at 4% content of MoS_2 and decreased with further addition of MoS_2. At higher concentrations, the MoS_2 films aggregated and stacked. This resulted in poorer piezoelectric performance. The process used for making the composite films included exfoliation of MoS_2 powder using triethanolamine, formation of MoS_2 nanosheets, then blending with TOCN dispersion, stirring, and subsequent sonicating, followed by vacuum drying.

Song et al. prepared porous aerogel films using TEMPO-oxidized cellulose nanofibrils and MoS_2 nanosheets [40]. An open-circuit voltage of 42 V and short-circuit current of 1.1 μA were obtained using the aerogel-based piezoelectric nanogenerator with 6% weight of MoS_2. The MoS_2 nanosheets and the air present in a large number of mesopores in the aerogel structure were polarized, and this resulted in increased energy generation.

10.4.2 Barium Titanate ($BaTiO_3$) as Filler

Wang et al. prepared a multilayered piezoelectric nanogenerator with maleic anhydride grafted PVDF nanofibers, cellulose nanofibrils, and barium titanate nanoparticles modified with polydopamine [41]. The barium titanate nanoparticle layer would act as a mediator between maleic hydride grafted PVDF nanofibers, which are hydrophobic in nature, and the hydrophilic cellulose nanofibrils. The nanogenerator was also used for detecting a variety of human motions, including fist clenching, finger bending, knee bending, and so on. At 5% $BaTiO_3$ loading, a maximum output of 3.2 V was achieved and a d_{33} value of 27.2 pC/N could be reached.

Choi and Jeong reported a nanocellulose/$BaTiO_3$ composite film for use in piezoelectric nanogenerators [42]. A maximum output voltage of 2.86 V was obtained, and current of 262.4 nA was obtained when the $BaTiO_3$ filler content was about 40%. Thereafter, on further increase in add on, the piezoelectric property of the film decreased. This was attributed to increased mechanical stiffness upon the addition of $BaTiO_3$ beyond 40% weight. The dielectric constants of films with $BaTiO_3$ were significantly greater than the dielectric constants of neat films.

10.4.3 Other Fillers and Blends

Li et al. prepared a piezoelectric sensor consisting of CNC and polyethylene glycol (PEG) blends [43]. Graphene was used as an electrode. The flexible PEG acted as a piezoelectric activator, and maximum voltage was achieved when the compression frequency and relaxation frequency of the PEG molecules became the same. No external polarization was required for proper use of the material as sensor.

Recently, a methyl cellulose scaffold containing Barium titanate nanoparticles impregnated with (PVDF-TrFE) was proposed [44]. The high Young's modulus of the methyl cellulose scaffold made the transfer of stress to the nanoparticles easier. The device showed the best performance, with an optimal cellulose content of 3%. The power density of the device was 42 $\mu W/cm^3$.

Antimony sulfoiodide (SbSI) nanowires, along with flexible cellulose having good elastic properties, were prepared to develop piezoelectricity

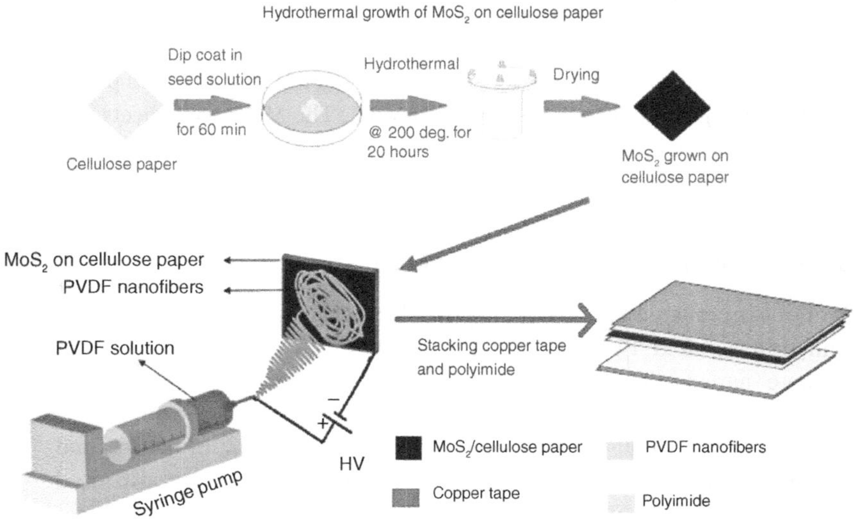

Figure 10.3 Schematic of the preparation of cellulose/PVDF/MoS_2-based nanogenerator. Source: Copyright Elsevier 2018. Reproduced with permission from Sahatiya et al. [46].

nanogenerators [45]. This nanocomposite of SbSI/cellulose provided excellent piezoelectric and electrochemical coefficients due to the presence of SbSI. SbSI nanowires were synthesized using pure elements sonochemically until the red-orange color of nanowires was formed. After this, cellulose fibers were dispersed in SbSI nanowires and sonicated in water till a homogenous mixture was obtained. Prepared sheets thus fabricated were used in nanogenerators, and gold sheets were used as electrodes to which copper wires were connected. The whole device was then coated with silicone rubber to protect it from outside environment. The maximum power output generated for the device was around 0.21 nW/cm^2.

In another study, it was shown that energy might be harvested from the simple human activities like handwriting and human touch using piezo-triboelectric hybrid nanogenerator by growing few layers of molybdenum disulphide on cellulose paper using hydrothermal method and, after that, by depositing the electrospun in situ poled PVDF nanofibers using the same MoS_2-cellulose paper as collector [46]. The composite formed provided 50 V as output voltage. The schematic of the device is shown in Figure 10.3.

A flexible piezoelectric nanogenerator made up of ZnS nanowires via hydrothermal process grown on cellulose matrix has been used to harvest energy. The cellulose paper was sandwiched between carbon tapes, which acted as electrodes. The energy generated from the hybrid nanogenerator

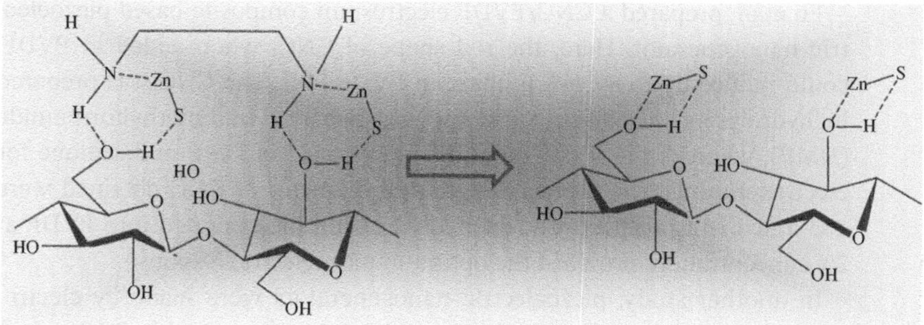

Figure 10.4 The probable mechanism of interaction of ZnS with cellulose. Source: Copyright American Institute of Physics 2018. Reproduced with permission from Sultana et al. [47].

was mainly because, when stress was applied to it, compressive strain was produced in the cellulose matrix, which generated dipole moments in the nanowires. Another possible reason could be that, because of the deformation of the cellulose matrix, the nanowires of ZnS present on its surface experienced rubbing action, which generated the piezoelectric potential energy along the nanodiameter. The device had an output voltage of 4 V [47]. A schematic of the probable mechanism of interaction of ZnS with cellulose is shown in Figure 10.4.

10.5 Cellulose-Based Hybrid Piezoelectric Devices Containing Cellulose as Filler

This section provides a description of piezoelectric nanogenerators that have utilized cellulose in some form as filler. Alam and Mandal reported a cellulose microfibres/polydimethylsiloxane (PDMS)/multi wall carbon nanotube (MWCNT)-based piezoelectric device that was found to show an open-circuit voltage of 30 V and power density of 9 $\mu W/cm^3$ under hand punching [20]. The MWCNT added, worked as both a conductive filler and a dispersing agent. The addition of MWCNT at higher concentrations than 0.5 wt% leads to current leakage. Similarly, addition of cellulose microfibres was found to be optimum at 5% concentration, after which the same agglomerated, leading to lower output voltage. This nanogenerator gives superior output voltage without poling because of the bond-breaking mechanism of the PDMS, producing more of the OH groups, which forms hydrogen bonds with glucose resulting in more piezoelectric charge generation. Cellulose itself is also a source of electric dipoles.

Fu et al. prepared a CNC/PVDF electrospun composite-based piezoelectric nanogenerator. Here, the rod shape of CNC, when added to PVDF, could induce up to 89.96% β phase in PVDF [48]. The CNC was prepared by hydrolysis using sulfuric acid and was dispersed in dimethylformamide (DMF), whereas the PVDF solutions were made in DMF and acetone for electrospinning. The average length and diameter of CNC obtained were 183 and 15 nm, respectively. Such CNC, when incorporated into PVDF at 2% concentration, provided the optimum piezoelectric response.

In another study, piezoelectric nanogenerators were made by electrospinning PVDF-hexafluoropropylene (PVDF-Hp) doped with CNC as one layer and PVDF-Hp with Fe-doped nanoZnO as filler for another layer [49]. Fe-doped ZnO nanoparticles were synthesized via hydrothermal route. The fibers obtained through electrospinning showed evidence of intertwining and better aspect ratio. A maximum output voltage of 12 V and current density of 1.9 $\mu A/cm^2$ could be obtained using the said device. The device, when subjected to finger tapping, ultrasonication, and elbow and shoulder movements, could harvest an appreciable amount of energy.

Mousa et al. prepared membranes of CNC and PVDF for oil–water separation applications [50]. Interestingly, PVDF doped with CNC showed less output than the original PVDF membrane. But the PVDF membrane containing CNC exhibited better water flux as compared to pristine PVDF. The CNC was prepared by hydrolysis of bleached bagasse pulp, and the electrospinning solution consisted of 12.5% PVDF and 1% CNC.

CNF/PDMS aerogel films were used to fabricate the piezoelectric nanogenerators (Figure 10.5) [51]. It consists of five layers, as shown in Figure 10.5b. Around porous CNF/PDMS aerogel film, two pure PDMS sheets were attached in order to make it more mechanically sturdy when the highest stress would be applied. Also, the PDMS attached around the aerogel was in order to make it smooth and soft so that there would not be any leakage of charge, along with more efficient transfer of the compressive stress to aerogel. Aluminum foil was used as an electrode in the structure. Also, additional electric dipoles could be generated under mechanical stress by homolytic cleavage of Si—O—Si bonds. The level of compression was more in the CNF/PDMS aerogel than in the pristine structure under the same set of stresses.

In another study, cellulose microfibrils were used to reinforce the PDMS matrix forming nanocomposite, which gave enhanced piezoelectric response [52]. Gold nanoparticles were incorporated into the structure, which reduced the dielectric loss. The nanoscale addition of gold filler increases the interfacial area between the filler and matrix, which gives the enhanced dipole interaction and high polarization. The preparation process

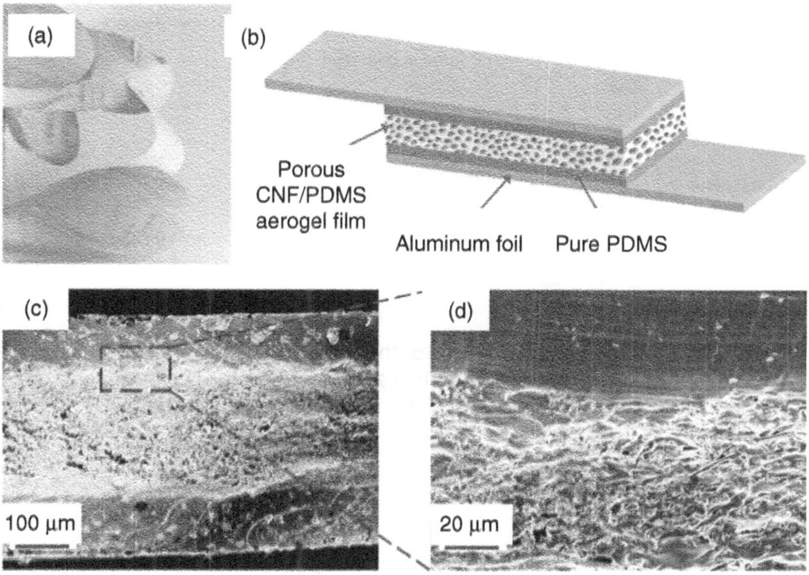

Figure 10.5 (a) Optical image of CNF/PDMS aerogel, (b) schematic showing the entire device structure of CNF/PDMS. Cross-sectional SEM images of the aerogel at (c) a low magnification and (d) a high magnification. Source: Reproduced with permission from Zheng et al. [51], © 2020, Elsevier.

is shown in Figure 10.6. The gold nanoparticles (Au-NP) have high number of electrons, which forms ionic interaction with the hydrogen atoms of the cellulose, and the mechanism of interaction has also been shown in Figure 10.6. After curing PDMS, aluminum tape was attached to the top of the composites, followed by attaching the conductive wires. Finally, the structure was covered with polypropylene tapes to avoid triboelectric effect. To protect the device from mechanical stress and environmental damage, it was finally encapsulated with PDMS. The device had an open-circuit voltage of 6 V and short-circuit current of 700 nA.

Biowaste orange peel has cellulose in its constituents and hence can be used as a piezoelectric energy harvester. The energy harvester hybrid has a fine powder of orange peel as a filler and polyvinylidene fluoride as a matrix. The polymer nanohybrid film was coated with silver paste, along with attaching the copper wires on both the sides of the film and finally, it was wrapped in the polypropylene tape and then protected by PDMS. The rotation of polar atomic group during the application of stress or the formation of new dipoles is responsible for the piezoelectric effect. The generated electric dipole, which gives the induced crystallization of PVDF, is due to the hydroxyl group of cellulose, which is interconnected via hydrogen

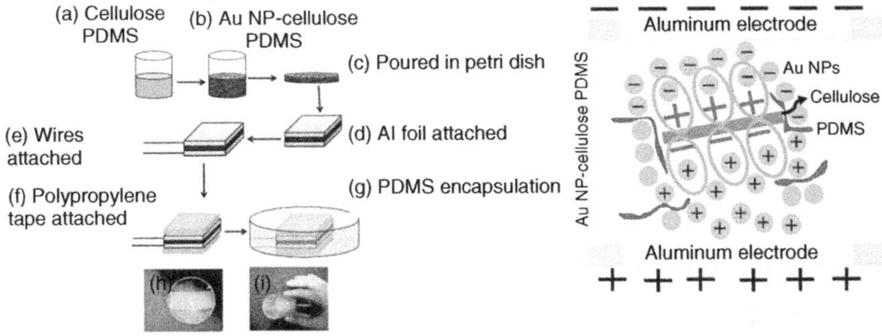

Figure 10.6 The preparation process of the Au-NP/cellulose/PDMS nanogenerator (left) and possible mechanism of piezoelectric charge generation (right). Source: Reproduced with permission from Pusty and Shirage [52], © 2020, Royal Society of Chemistry/CC BY 4.0.

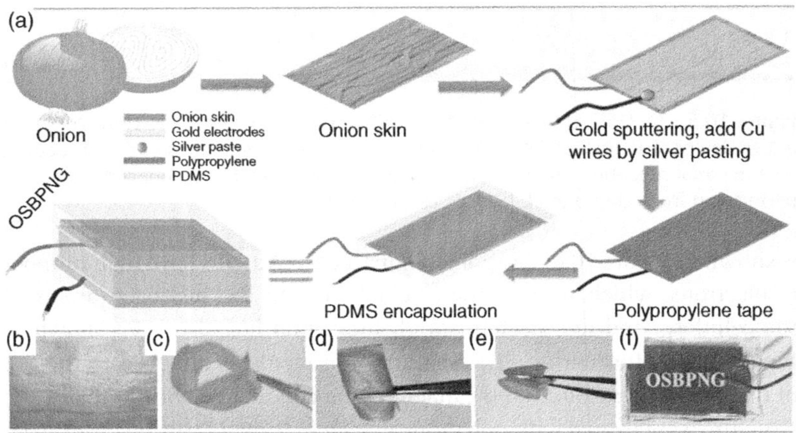

Figure 10.7 Schematic of fabrication of onion skin-based piezoelectric nanogenerator. (a) Optical image of (b) onion skin and images showing the flexibility of skin during (c) bending, (d) rolling, and (e) twisting, respectively. (f) Optical image of the assembled nanogenerator. Source: Reproduced with permission from Maiti et al. [54], © 2017, Elsevier.

bonding in the structure. The resultant structure produced an open-circuit voltage of 90 V by finger tapping [53].

Bio-piezoelectric nanogenerators using the natural and biodegradable cellulose nanofibrous onion skin have been used as a piezoelectric material [54]. Onion skin has α-cellulose, which is mainly responsible to generate the self-powered nanogenerator. The device preparation procedure is shown in Figure 10.7. The various hydroxyl groups of the glucose in the cellulosic structures form hydrogen bonding and keep the chains side by side tightly, which generates the electric dipole inside the crystal lattice of cellulose microfibrils and hence when the stress is applied to the structure the reorientation of these electric dipoles takes place inside the crystal lattice. This generates the maximum output voltage of 106 V when six devices are connected in series.

10.6 Conclusion

In summary, this chapter describes the various cellulose-based hybrid piezoelectric devices as reported in the literature, while also trying to elaborate on the reason for incorporation of cellulose in such devices. Different types of cellulose used in manufacturing piezoelectric devices have been described first. Cellulose is biodegradable and renewable. However, the piezoelectric property of cellulose only is not high enough and is not good as compared to its synthetic counterparts. This demands the use of hybrid cellulose materials. Mostly, the devices reported here have been divided into two groups; the first one where, cellulose is acting as the matrix or as a component of the blend and the second group where cellulose is mainly acting as filler. In future, cellulose may be functionalized and their piezoelectric properties may be looked at. Also, cellulose in different forms like aerogels and foams and their hybrids may be investigated for piezoelectric properties.

References

1 Ang, T.Z., Salem, M., Kamarol, M. et al. (2022). A comprehensive study of renewable energy sources: classifications, challenges and suggestions. *Energ. Strat. Rev.* 43: 100939. https://doi.org/10.1016/J.ESR.2022.100939.

2 International Energy Agency (2021). *Global Energy Review 2021*. Glob Energy Rev 2020, 1–36.

3 Lal, S.R.S., Herbert, G.M.J., Arjunan, P., and Suryan, A. (2022). Advancements in renewable energy transition in India: a review. *Energy Sources, Part A* https://doi.org/10.1080/15567036.2021.2024921.

4 Chaturvedi, P. (2022). Future of renewable energy in India. In: *Sustainable Energy Development and Innovation*, 625–628. Springer International Publishing https://doi.org/10.1007/978-3-030-76221-6_68.

5 Nguyen, Q.H., Hoai Ta, Q.T., and Tran, N. (2022). Review on the transformation of biomechanical energy to green energy using triboelectric

and piezoelectric based smart materials. *J. Clean. Prod.* 371: 133702. https://doi.org/10.1016/J.JCLEPRO.2022.133702.

6 Wang, S., Wang, C., Yuan, H., and Ji, X. (2022). Design and performance of piezoelectric energy output promotion system for road. *Renew. Energy* 197: 443–451. https://doi.org/10.1016/J.RENENE.2022.07.150.

7 Energy Harvesting Systems: Global Strategic Business Report. (2023) https://www.researchandmarkets.com/reports/4804844/energy-harvesting-systems-global-strategic ().

8 Dineva, P., Gross, D., Müller, R., and Rangelov, T. (2014). Piezoelectric materials. In: *Solid Mechanics and Its Applications*, vol. 212, 7–32. Springer https://doi.org/10.1007/978-3-319-03961-9_2/COVER.

9 Luo, L., Zhao, X., and Luo, H. (2010). Single crystal PZN-PT, PMN-PT, PSN-PT, and PIN-PT-based piezoelectric materials. *Adv. Piezoelectric Mater.* 271–318. https://doi.org/10.1016/B978-0-08-102135-4.00007-2.

10 Kang, M.G., Jung, W.S., Kang, C.Y., and Yoon, S.J. (2016). Recent progress on PZT based piezoelectric energy harvesting technologies. *Actuators* 5: 5. https://doi.org/10.3390/ACT5010005.

11 Niu, Y., Yu, K., Bai, Y., and Wang, H. (2015). Enhanced dielectric performance of BaTiO3/PVDF composites prepared by modified process for energy storage applications. *IEEE Trans. Ultrason. Ferroelectr. Freq. Control* 62: 108–115. https://doi.org/10.1109/TUFFC.2014.006666.

12 Safari, A. and Hejazi, M. (2013;9781441995988). Lead-free KNN-based piezoelectric materials. In: *Lead-Free Piezoelectrics*, 139–175. Springer https://doi.org/10.1007/978-1-4419-9598-8_5/COVER.

13 Fukada, E. (1995). Piezoelectricity of biopolymers. *Biorheology* 32: 593–609. https://doi.org/10.1016/0006-355X(95)00039-C.

14 Fukada, E. (1955). Piezoelectricity of wood. *J. Phys. Soc. Jpn.* 10: 149–154. https://doi.org/10.1143/JPSJ.10.149.

15 Lu, L., Ding, W., Liu, J., and Yang, B. (2020). Flexible PVDF based piezoelectric nanogenerators. *Nano Energy* 78: 105251. https://doi.org/10.1016/J.NANOEN.2020.105251.

16 Bairagi, S. and Ali, S.W. (2020). A hybrid piezoelectric nanogenerator comprising of KNN/ZnO nanorods incorporated PVDF electrospun nanocomposite webs. *Int. J. Energy Res.* 44: 5545–5563. https://doi.org/10.1002/ER.5306.

17 Bairagi, S., Banerjee, S., Kalyaniyan, P. et al. (2021). New insights toward casein/polyvinyl alcohol electrospun nanofibrous webs as a piezoelectric-cum-triboelectric energy harvester. *ACS Appl. Electron. Mater.* 3: 4348–4361. https://doi.org/10.1021/acsaelm.1c00532.

18 Bairagi, S., Banerjee, S., Chowdhury, A., and Ali, S.W. (2021). Development of a sustainable and flexible piezoelectric-cum-triboelectric energy harvester comprising a simple commodity cotton fabric. *ACS Sustain. Chem. Eng.* 9: 4004–4013. https://doi.org/10.1021/acssuschemeng.0c07274P4.

19 Bairagi, S., Ghosh, S., and Ali, S.W. (2020). A fully sustainable, self-poled, bio-waste based piezoelectric nanogenerator: electricity generation from pomelo fruit membrane. *Sci. Rep.* 101: 1–13. https://doi.org/10.1038/s41598-020-68751-3.

20 Alam, M.M. and Mandal, D. (2016). Native cellulose microfiber-based hybrid piezoelectric generator for mechanical energy harvesting utility. *ACS Appl. Mater. Interfaces* 8: 1555–1558. https://doi.org/10.1021/acsami.5b08168.

21 Song, Y., Shi, Z., Hu, G.H. et al. (2021). Recent advances in cellulose-based piezoelectric and triboelectric nanogenerators for energy harvesting: a review. *J. Mater. Chem. A* 9: 1910–1937. https://doi.org/10.1039/D0TA08642H.

22 Danz, P., Aryan, V., Möhle, E., and Nowara, N. (2019). Experimental study on fluorine release from photovoltaic backsheet materials containing PVF and PVDF during pyrolysis and incineration in a technical lab-scale reactor at various temperatures. *Toxics* 7: 47. https://doi.org/10.3390/TOXICS7030047.

23 Ghozali, M., Triwulandari, E., and Restu, W.K. (2022). Biopolymers in electronics. In: *Biopolymers*, 289–321. https://doi.org/10.1007/978-3-030-98392-5_14.

24 Trache, D., Tarchoun, A.F., Derradji, M. et al. (2020). Nanocellulose: from fundamentals to advanced applications. *Front. Chem.* 8: 392. https://doi.org/10.3389/FCHEM.2020.00392/BIBTEX.

25 Wang, H., Li, D., Yano, H., and Abe, K. (2014). Preparation of tough cellulose II nanofibers with high thermal stability from wood. *Cellulose* 21: 1505–1515. https://doi.org/10.1007/S10570-014-0222-6/FIGURES/7.

26 Gindl, W., Emsenhuber, G., Plackner, J. et al. (2010). Converse piezoelectric effect in cellulose I revealed by wide-angle X-ray diffraction. *Biomacromolecules* 11: 1281–1285. https://doi.org/10.1021/BM1000668/ASSET/IMAGES/LARGE/BM-2010-000668_0009.JPEG.

27 Wang, X., Yao, C., Wang, F. et al. (2017). Cellulose-based nanomaterials for energy applications. *Small* 13: 1702240. https://doi.org/10.1002/SMLL.201702240.

28 Fashandi, H., Abolhasani, M.M., Sandoghdar, P. et al. (2016). Morphological changes towards enhancing piezoelectric properties of PVDF electrical generators using cellulose nanocrystals. *Cellulose* 23: 3625–3637. https://doi.org/10.1007/S10570-016-1070-3/TABLES/5.

29 Csoka, L., Hoeger, I.C., Rojas, O.J. et al. (2012). Piezoelectric effect of cellulose nanocrystals thin films. *ACS Macro Lett.* 1: 867–870. https://doi.org/10.1021/mz300234a.

30 Kim, J. (2021). Nanocellulose-based paper actuators. In: *Nanocellulose Based Compositses for Electronics*, 163–183. Elsevier https://doi.org/10.1016/B978-0-12-822350-5.00007-2.

31 Tayeb, P. and Tayeb, A.H. (2019). Nanocellulose applications in sustainable electrochemical and piezoelectric systems: a review. *Carbohydr. Polym.* 224: 115149. https://doi.org/10.1016/J.CARBPOL.2019.115149.

32 Klemm, D., Schumann, D., Udhardt, U., and Marsch, S. (2001). Bacterial synthesized cellulose—artificial blood vessels for microsurgery. *Prog. Polym. Sci.* 26: 1561–1603. https://doi.org/10.1016/S0079-6700(01)00021-1.

33 Mangayil, R., Rajala, S., Pammo, A. et al. (2017). Engineering and characterization of bacterial nanocellulose films as low cost and flexible sensor material. *ACS Appl. Mater. Interfaces* 9: 19048–19056. https://doi.org/10.1021/acsami.7b04927.

34 Matsouka, D. and Vassiliadis, S.G. (2018). *Piezoelectricity – Organic and Inorganic Materials and Applications.* IntechOpen.

35 Tuukkanen, S. and Rajala, S. (2015). A survey of printable piezoelectric sensors. In: *2015 IEEE SENSORS – Proceedings 2015.* https://doi.org/10.1109/ICSENS.2015.7370542.

36 Rajala, S., Siponkoski, T., Sarlin, E. et al. (2016). Cellulose nanofibril film as a piezoelectric sensor material. *ACS Appl. Mater. Interfaces* 8: 15607–15614. https://doi.org/10.1021/acsami.6b03597.

37 Gomes, J., Nunes, J.S., Sencadas, V., and Lanceros-Mendez, S. (2010). Influence of the β-phase content and degree of crystallinity on the piezo- and ferroelectric properties ofpoly (vinylidene fluoride). *Smart Mater. Struct.* 19: 065010. https://doi.org/10.1088/0964-1726/19/6/065010.

38 Wu, W., Wang, L., Li, Y. et al. (2014). Piezoelectricity of single-atomic-layer MoS2 for energy conversion and piezotronics. *Nature* 514: 470–474. https://doi.org/10.1038/nature13792.

39 Wu, T., Song, Y., Shi, Z. et al. (2021). High-performance nanogenerators based on flexible cellulose nanofibril/MoS2 nanosheet composite piezoelectric films for energy harvesting. *Nano Energy* 80: 105541. https://doi.org/10.1016/J.NANOEN.2020.105541.

40 Song, Y., Wu, T., Bao, J. et al. (2022). Porous cellulose composite aerogel films with super piezoelectric properties for energy harvesting. *Carbohydr. Polym.* 288: 119407. https://doi.org/10.1016/J.CARBPOL.2022.119407.

41 Wang, L., Cheng, T., Lian, W. et al. (2022). Flexible layered cotton cellulose-based nanofibrous membranes for piezoelectric energy harvesting and self-powered sensing. *Carbohydr. Polym.* 275: 118740. https://doi.org/10.1016/J.CARBPOL.2021.118740.

42 Choi, H.Y. and Jeong, Y.G. (2019). Microstructures and piezoelectric performance of eco-friendly composite films based on nanocellulose and barium titanate nanoparticle. *Compos. Part B Eng.* 168: 58–65. https://doi.org/10.1016/J.COMPOSITESB.2018.12.072.

43 Li, S., He, Y., Ye, X. et al. (2022). Improved piezoelectricity of porous cellulose material via flexible polarization-initiate bridge for self-powered sensor. *Carbohydr. Polym.* 298: 120099. https://doi.org/10.1016/J.CARBPOL.2022.120099.

44 Zhang, H., Jeong, C.K., Shen, Z. et al. (2022). Ultrahigh augmentation of flexible composite-based piezoelectric energy harvesting efficiency via polymer-impregnated nanoparticles network within 3D cellulose scaffold. *Compos. Part B Eng.* 236: 109813. https://doi.org/10.1016/J.COMPOSITESB.2022.109813.

45 Toroń, B., Szperlich, P., Nowak, M. et al. (2018). Novel piezoelectric paper based on SbSI nanowires. *Cellulose* 25: 7–15. https://doi.org/10.1007/S10570-017-1597-Y/TABLES/1.

46 Sahatiya, P., Kannan, S., and Badhulika, S. (2018). Few layer MoS2 and in situ poled PVDF nanofibers on low cost paper substrate as high performance piezo-triboelectric hybrid nanogenerator: energy harvesting from handwriting and human touch. *Appl. Mater. Today* 13: 91–99. https://doi.org/10.1016/J.APMT.2018.08.009.

47 Sultana, A., Middya, T.R., and Mandal, D. (2018). ZnS-paper based flexible piezoelectric nanogenerator. *AIP Conf. Proc.* 1942: 120018. https://doi.org/10.1063/1.5029058.

48 Fu, R., Chen, S., Lin, Y. et al. (2017). Improved piezoelectric properties of electrospun poly(vinylidene fluoride) fibers blended with cellulose nanocrystals. *Mater. Lett.* 187: 86–88. https://doi.org/10.1016/J.MATLET.2016.10.068.

49 Ponnamma, D., Parangusan, H., Tanvir, A., and AlMa'adeed, M.A.A. (2019). Smart and robust electrospun fabrics of piezoelectric polymer nanocomposite for self-powering electronic textiles. *Mater. Des.* 184: 108176. https://doi.org/10.1016/J.MATDES.2019.108176.

50 Mousa, H.M., Fahmy, H.S., Abouzeid, R. et al. (2022). Polyvinylidene fluoride-cellulose nanocrystals hybrid nanofiber membrane for energy harvesting and oil-water separation applications. *Mater. Lett.* 306: 130965. https://doi.org/10.1016/J.MATLET.2021.130965.

51 Zheng, Q., Zhang, H., Mi, H. et al. (2016). High-performance flexible piezoelectric nanogenerators consisting of porous cellulose nanofibril (CNF)/poly(dimethylsiloxane) (PDMS) aerogel films. *Nano Energy* 26: 504–512. https://doi.org/10.1016/J.NANOEN.2016.06.009.

52 Pusty, M. and Shirage, P.M. (2020). Gold nanoparticle–cellulose/PDMS nanocomposite: a flexible dielectric material for harvesting mechanical energy. *RSC Adv.* 10: 10097–10112. https://doi.org/10.1039/C9RA10811D.

53 Gaur, A., Tiwari, S., Kumar, C., and Maiti, P. (2020). Bio-waste orange peel and polymer hybrid for efficient energy harvesting. *Energy Rep.* 6: 490–496. https://doi.org/10.1016/J.EGYR.2020.02.020.

54 Maiti, S., Kumar Karan, S., Lee, J. et al. (2017). Bio-waste onion skin as an innovative nature-driven piezoelectric material with high energy conversion efficiency. *Nano Energy* 42: 282–293. https://doi.org/10.1016/J.NANOEN.2017.10.041.

11

Collagen-Based Hybrid Piezoelectric Material

Adrija Ghosh[1], Suprakas Sinha Ray[3,4], Jonathan Tersur Orasugh[3,4], and Dipankar Chattopadhyay[1,2]

[1] *Department of Polymer Science and Technology, University of Calcutta, Kolkata, West Bengal, India*
[2] *Center for Research in Nanoscience and Nanotechnology, Acharya Prafulla Chandra Roy Sikhsha Prangan, University of Calcutta, Kolkata, India*
[3] *Department of Chemical Sciences, University of Johannesburg, Johannesburg, South Africa*
[4] *DST-CSIR National Centre for Nanostructured Materials, Council for Scientific and Industrial Research, Pretoria, South Africa*

11.1 Introduction

It is crucial to use natural resources to harness sustainable as well as ecologically friendly energy in order to reduce future global energy concerns.

The usage of hazardous and non-biocompatible materials created in laboratories/industries for energy generation and storage devices has a serious negative influence on the global environment and the health of users. Consequently, bio-organic natural materials are required for a civilization with low carbon emissions and no pollution. Researchers are leaning toward natural materials in the recent decades for the creation of various energy tools, self-powered sensors, such as roll-up display equipment, as well as wearable electronics [1–3]. These non-traditional harvesting techniques can protect the environment and improve user safety. Numerous studies have been done to increase the efficiency, power generation, and biocompatibility of energy harvesting as tribo-electric nanogenerators (TENGs) in the wake of the evolutionary transition made by the Song group in this sector [3]. The piezo-electric nanogenerators (PENGs) are also garnering an increasing amount of interest these days as a more robust, extremely

Hybrid Materials for Piezoelectric Energy Harvesting and Conversion, First Edition.
Edited by S. Wazed Ali, Satyaranjan Bairagi, and Shahid Ul Islam.
© 2024 John Wiley & Sons, Inc. Published 2024 by John Wiley & Sons, Inc.

stable, efficient, and superbly sensitive gadget [3]. The mechanical motions caused by wind, ocean currents, and human movement can easily generate electrical energy from TENG/PENG [3]. These technologies have so far been able to provide voltage, current, and power that are remarkably high while also being simple to make, lightweight, and exceptionally durable [4]. For the PENGs and TENGs, the researchers have previously reported on a variety of materials that have been produced in the lab (ZnO, $BaTiO_3$, $SrTiO_3$, $ZnSnO_3$, etc.) [5]. However, due to their hazardous characteristics and labor-intensive manufacturing procedures, these synthetic materials seriously confuse the environment.

Piezoelectric behavior, which is a kind of linear electromechanical coupling, forms the basis of mechanisms behind maximum biological processes. Piezoelectric materials produce electric potential, or polarization, in response to applied mechanical stress. Also, on the contrary, piezoelectric materials undergo mechanical strain on application of electric field. Fukada and Yasuda were the pioneers in accounting for the piezoelectric behavior of bone. It was further discovered that piezoelectricity is present in tendon and tooth. Currently, it is evident that piezoelectricity is omnipresent in biological tissues. Electrostatic interactions and surface charge play an imperative role in regulating various cell functions, which include contraction of muscles, communication between charged biomolecules, folding of proteins, etc. In a similar way, electric charge produced due to piezoelectric effect plays a significant role in interaction between cells and extracellular matrix, restoration of connective tissues during rehabilitation, and remodeling of tissues. For instance, bone can be defined as a composite consisting of seven stages of hierarchy, where collagen forms the base level. In addition to that, collagen itself is made up of numerous levels of hierarchy, such as triple helix, microfibril, amino acids, etc. Collagen in bone is further infused with hydroxyapatite nanocrystals. It can be inferred from various research experiments that collagen fibrils possess piezoelectric behavior. It is being further approved that the dipole orientations cannot be swapped in case of collagen, as it is not ferroelectric. While biomaterials like elastin are found to be ferroelectric in nature. Hydroxyapatite nanocrystals are also piezoelectric in nature. Conclusively, the piezoelectric nature of bone can be attributed to the presence of collagen and hydroxyapatite components.

However, due to the presence of various complex biological processes and contending mechanisms in our body, the in vivo significance of piezoelectricity is quite questionable. In spite of this, collagen is widely used as a piezoelectric biomaterial for various applications, including sensors, nanogenerators, memory devices, etc. We can thus expect to witness increasing applications of collagen in upcoming days [6].

11.2 Origin of Piezoelectricity in Collagen

Though it is quite evident that presence of collagen makes the bones piezoelectric in nature but the mechanism behind electromechanical coupling in collagen is unresolved. However, there are various hypotheses in the literature which explain collagen's piezoelectric nature. Few hypotheses are based on either non-centrosymmetric protein structure or concurrent existence of optical property and polar bonds that impart piezoelectric nature to them. Other hypotheses include the rotation of amide bond in α-helix structure of collagen and polarized hydrogen bonds in the polypeptide chain. Conversely, none of these hypotheses have been established by testing on real collagen models.

Ravi et al. [7] used molecular dynamics simulations to examine the influence of mechanical stress or heat on electrostatic nature of collagen. Collagen helices selected for this study were glycine, proline, and hydroxyproline. It was concluded that mechanical stress stimulates considerable amount of change in the polarization of collagen fibrils. This phenomenon can be explained on the basis of re-organization of the backbone polar groups in response to mechanical strain. Such effect is further smoothly progressed due to the unusual folding of collagen helix, that is distinguished due to the lack of hydrogen-bonding in the collagen network. The stretch and compress action of the helix is connected with the wind/unwind movement of the globular structure, making the shear constituents of the stress tensor induce change in structure due to piezoelectric stimulation. In addition to that, calculations also display that thermal stimulation causes pyroelectric behavior. However, this pyroelectric nature is superseded by secondary components responsible for piezoelectric tensor.

On the other hand, Zhou et al. [6] presented an atomistic simulation study to find out the origination of piezoelectricity in collagen, making use of an experimentally developed "super-twisted" collagen microfibril model. The conclusion drawn from the experiment reveals that piezoelectricity in collagen stems from each collagen molecule and is retained in super-twisted microfibril form that constitutes collagen fibrils and fibers. It further proves that polar and charged groups cause the cumulative polarization effect in collagen. On application of mechanical strain, the dipole moments of the residues undergo reorientation to arrange along the long axis of the collagen molecule. Also, their magnitude alters in response to mechanical strain. These consequences cumulatively result in piezoelectricity in collagen. Moreover, the outcomes indicate that polarization is uni-directional along the axis of collagen molecule and microfibril. These results could be further applied to other biopolymers exhibiting piezoelectric nature. Table 11.1 represents the piezoelectricity effect of collagen extracted from various sources.

Table 11.1 Presentation of collagen-based hybrid piezoelectric (CBHP) materials with respect to literature.

Formulated CBHP	Fabrication technique	Collagen source	Output voltage	Output current	Electrode	Operation mode	piezoelectric effect	References
Eggshell membrane/PVDF	Spin coating	Eggshell membrane	15 V	150 nA	Ag	Pushing	Direct	[8]
Collagen nanofibril coated cotton fabric	Deposition	Purchased	5 V	250 nA	Al	Compression	Direct	[9]
Barium titanate/collagen/ hydroxyapatite	Sol–gel	Purchased	—	—	—	—	Direct	[10]
Collagen-CTMA3 (Cationic surfactant hexadecyltrimethy-lammonium chloride (CTMA))	Spin-coated	Purchased	13.6 V	—	Au & ITO-glass	Bending	Direct	[11]
Polythene-collagen	Lamination	Fish scale	4–14 V	1.5 μA	Au	Compression	Direct	[12]
Fish swim bladder (FSB)		FSB	10 V	51 nA	Au	Compression	Direct	[13]
Collagen-graphene	Chicken feet	Coating	100 mV/s	up to 10 A/g	Graphene	Compression	Direct	[14]
N-carbon-sodium carboxymethyl cellulose/collagen	Sol–gel	Purchased	—	1 A/g	—	Compression	Direct	[15]

Ag-collagen-Ti	Spin coating	Pig skin	−1.3 V	0.1 mA	Ag and Ti	Electrochemically	Direct	[16]
Anionic collagen	Solution intercalation and film casting	Bovine serosa	—	—	Ag	Pressure	Direct	[17]
Collagen-hydroxyapatite (HA)	Solution intercalation and film casting	Bovine serosa	—	—	Ag	Pressure	Direct	[17]
Collagen nanofibrils-cotton fabric	Layer-by layer coating	Purchased	45 V	250 nA	Ag	Pressure	Direct	[9]
Collagen-PVDF	Solution intercalation and compression moulding	Fish scales	22 V	—	Ag	Pressure, bending, and twisting	Direct	[18]

11.3 Application of Collagen-based Hybrid Piezoelectric Systems

We know that researchers globally have turned their attention to bioderived sustainable materials for the fabrication of advanced materials, of which piezoelectric materials, as presented in Table 11.1, are not left out. Collagen-based piezoelectric systems are one of the bioderived advanced materials of interest within the recent decades. Their applications have spanned tissue regeneration [17], enhanced wound healing [18, 19], battery-free humidity sensors [9, 20], supercapacitors [14], temperature and pressure sensors [3], self-powered cathodic protection [21], bio-organic field effect transistors with improved memory characteristics [11], etc.

11.4 Collagen-Based Piezoelectric Nanogenerator

Nanogenerators are used for production of electricity from mechanical or thermal energy induced by small-scale physical alterations. A nanogenerator can be either piezoelectric, triboelectric, or pyroelectric in nature. Collagen has been used to make both piezoelectric and triboelectric nanogenerators [18, 21, 22].

Ghosh and Mandal [13] synthesized a nanogenerator from collagen nano-fibrils extracted from the swim bladder of fishes. Near edge X-ray absorption spectroscopic analysis confirmed the highly ordered structure of the procured collagen. The nanogenerator could be driven by creating normal stress by finger contact leading to generation of 10 V voltage and 51 nA current. The electricity generated has the potential to turn on about 50 blue LEDs. It exhibited an energy conversion efficiency of 0.3% and instantaneous power generation of 4.15 $\mu W/cm^2$. Thus, it could be used to fabricate self-powered biosensors.

In another report, Vivekananthan et al. fabricated a sustainable, eco-friendly energy harvesting as well as battery-free humidity sensor made by coating a cotton fabric with nanocollagen fibrils [9]. This system presented by the authors possessed a dual function as a sensor and an energy harvesting device. Their bioderived collagen-based hybrid piezoelectric system could generate 45 V vs. 250 nA of voltage and current upon the application of five newton force even as presented in Table 11.1 [9].

Another report by Silver and his group presented their study of the performance of anionic-based collagen as well as collagen@hydroxyapatite (HA) composites' dielectric, piezoelectric, as well as their physicochemical performance [17]. Their report revealed that the C=O, C–N, as well as N–H functionalities in collagen, along with the $(PO_4)^{3-}$, $(CO_3)^{2-}$ interfacially interacted amongst themselves, resulting in better performance of the piezoelectric material.

In another instance, Kumar et al. have shown in their work that fish scales (FS) derived self-aligned collagen could be utilized as an efficient energy harvesting system where the FS induces the electroactive phase on the structure of the PVDF-collagen piezo-generator via its synergy [18]. The collagen reinforcing filler induced the formation of the γ-phased PVDF in the composite system in situ during the composite formation process leading to a hybrid piezo-generator exhibiting a very high voltage of 22 V as well as an energy density of 28.5 μW cm^{-2} with the aid of pressure applied by human finger as well as foot tapping, bending, walking upon, twisting, etc. of the device [18].

Though researchers are nowadays exploring collagen-based systems as piezo-generators, there still exist a large vacuum to be filled within this niche.

Researchers also extracted collagen from fish scale, which could function as a piezoelectric material having strength of −5.0 pC/N. X-ray absorption spectroscopy was used to confirm the dipole orientation of the fish scale collagen. The nanogenerator could track movements associated with machines, wind, or bodies. It produced 4 V of output voltage, 1.14 μW/cm^2 of output power density, and 1.5 μA of circuit current under compressive stress of about 0.17 MPa. Serially arranged, four piezoelectric nanogenerators could generate 14 V of voltage, which had the power to turn on around fifty light-emitting diodes [12].

11.5 Collagen-based Supercapacitors

Supercapacitors store energy via repeated cycles of charging and discharging at large amount of current and short amount of time.

Qin et al. [23], synthesized a supercapacitor made from teeth of animals. The collagen matrix of the demineralized tooth served as the gel electrolyte, while the porous carbon of calcinated demineralized tooth served as the electrode of the supercapacitor. The highly aligned hierarchal tube-like structure facilitated ionic mobility in the gel electrolyte. On the other hand, the porosity of the carbon electrode promoted adsorption of ions. The supercapacitor displayed specific resistance of 131.7 mF/cm^2. It further exhibited capacitance retention of 84.6 and 90.1% following 6000 charging–discharging cycles and 600 times bending, respectively.

Yu et al. [15] fabricated carbon aerogels and further doped them with nitrogen. Here, collagen was used as a source of nitrogen, and ferric

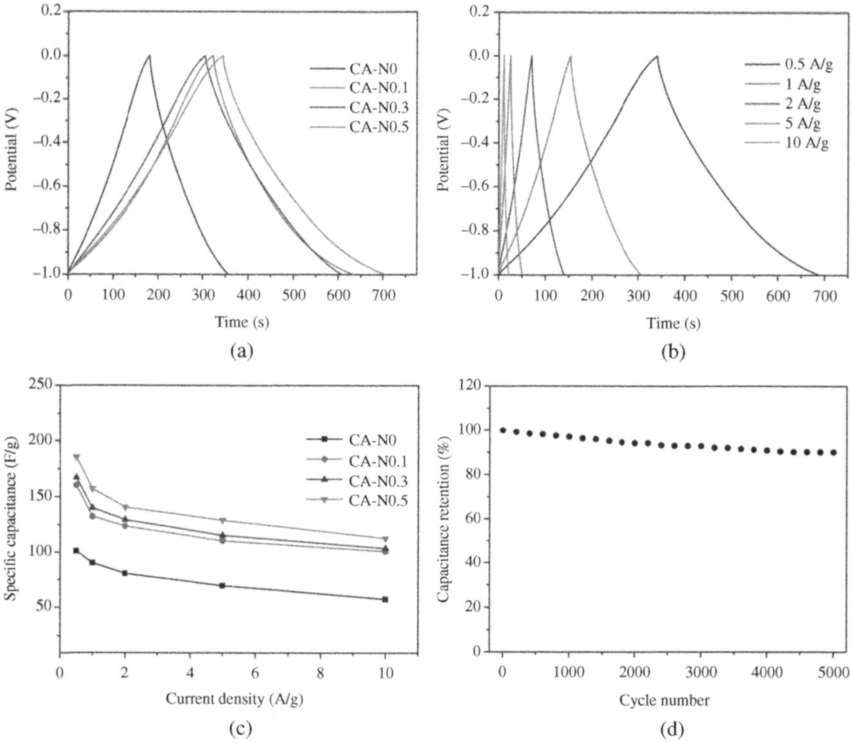

Figure 11.1 (a) Charge/discharge curves of CA-Nx electrodes measured by galvanostatic charge/discharge at current densities of 0.5 A/g; (b) charge/discharge curves of CA-N0.5 electrodes measured at various current densities; (c) specific capacitances of CA-Nx electrodes calculated from discharge curves at various current densities; and (d) cycling performance of CA-N0.5 electrodes measured at current density of 10 A/g. Source: Reproduced with copyright permission from Yu et al. [15], Elsevier 2018.

trichloride was used as a cross-linker. The aerogels thus produced had high porosity, large surface area, and exceptional magnetic characteristics. The maximum specific capacitance exhibited by the supercapacitor (CA-N0.5) was 185.3 F/g. The specific capacitance retention measured following 5000 charge–discharge cycles was found to be 90.2% (Figure 11.1). In addition to that, the nitrogen-doped carbon aerogels also had the potential of adsorbing dyes. It had maximum adsorption capability of 238.2 mg/g for malachite green and 230.4 mg/g for methylene blue.

Collagen-based hybrid systems, though presently focused on the preparation of piezo-capacitors, could also be largely utilized as hybrid materials for UV as well as electromagnetic interference shielding, which is another area of interest within the recent decades.

Subhani et al. [14], fabricated collagen sponges using collagen extracted from chicken feet. The sponges were further incorporated with graphene. The sponge, possessing a surface area of 1087 m^2/g exhibited cyclic stability following 10,000 cycles and capacitance of 365 F/g at 1 mV/s. Thus, the porous carbon sponges could efficiently function as a supercapacitor.

11.6 Collagen-based Sensors

Collagen is a structural protein which is widely distributed in all mammals, as is known. Collagen is the most common component of the extracellular matrix in humans, making up one-third of the total protein and three-quarters of the dry weight of skin (ECM). In vertebrates, there are 28 different forms of collagen that consists of minimum 46 different polypeptide chains, and numerous additional proteins have collagenous domains. For instance, collagen VI makes up a substantial portion of the sciatic nerve, which serves as a sensory organ by enabling us to sense feelings in our legs. Additionally, the dermis's loose collagen VI network tightens up around the dermal-epidermal, nerve, and endothelial basement membranes. This gives skin special abilities, maintaining critical blood vessels and nerves while also allowing for suppleness: a perfect example from nature showing us that collagen will serve well for the fabrication of bioderived sensors.

In this vein, Zhang et al. [19], fabricated soft hydrogels consisting of the polymers poly(N-isopropylacrylamide) and collagen, along with guar gum, graphene oxide, and borax. COL–GG–PNIPAMs finally prepared were easily injectable and processible in nature. It was appreciably stretchable, having the capability to stretch up to 50 times its original length before rupture. Also, it possessed self-healing qualities. Moreover, the hydrogels were thermally sensitive and responsive to NIR, making them capable of strain sensing. Conclusively, the hydrogel could sense human movements and also healing dermal wounds of mouse skin, making it multifunctional in nature.

Aronoff et al. [24], fabricated a collagen-based sensor that could sense diseased tissue. Lysyl oxidase (LOX), which generally cross-links collagen fiber during its maturation, can cause damaged tissue function, leading to cancer-like diseases in cases of abnormal activity. The sensor can monitor LOX activity via chemoselective targeting of endogenous aldehydes produced by LOX. The synergistic probe fabricated anchors covalently to light up in reaction to LOX at places where cross-linking takes place, as made obvious from straining of normal and cancer cells. Thus, the sensor had the potential to diagnose fibrotic and cancer-like diseases.

Collagen-based materials have been used for humid sensing. A humidity sensor measures the relative humidity (RH) of the air or senses the quantity of water vapor in the air. Shapardanis et al. [20] used collagen as a cheap substitute for dielectric materials for developing humidity sensors. Collagen layers were transformed into strips and placed in between copper electrodes to achieve structures like capacitors. The shift in capacitance was measured using the sensing capacitors at relative humidity of 50 and 75% at varying temperatures. The results showed that biocompatible collagen has the potential to act as a humidity sensor with an adequate temperature compensation scheme.

Vivekananthan [25], fabricated a humidity sensor driven by collagen piezoelectric nanogenerator. The collagen nanogenerator could generate 45 V/250 nA at 5 N, along with functioning as a humidity sensor. It exhibited a sensitivity of 0.1287 μA/% RH (relative humidity) when the RH percentage was 50–90%.

Wang et al. [26], synthesized a sensor which had a number of layers and a core 3D network morphology which mimics the hierarchy of skin. It consisted of a substrate made of collagen aggregates extracted from waste leather. On the other hand, multiwalled carbon nanotubes acidified with polyaniline were used as the conductive material. The sensor exhibited rapid and wide dynamic range of response. It acted as an excellent humid sensor and was compatible for wearing. Thus, the multi-functional sensor has the potential to function as an artificial skin which can detect weak pressure and variations in humidity, exhibiting further ability to be used for health monitoring and intelligent robots.

11.7 Collagen-based Memory Devices

Nature presents examples as well as the clue toward solvent diverse problems with respect to science and technology. For instance, the human or animal brain has no external physical wires or power connection to it from external source(s) but generates its power within through a complex mechanism amongst the body tissues. In this way, scientists globally are turning to naturally occurring materials such as collagen, nanocellulose, HA, etc. to solve scientific and technological challenges without polluting mother nature.

Zeng et al. [16], fabricated a memory device consisting of an insulated mid-layer of collagen extracted from pig skin. Two kinds of biomemoristive memory devices were constructed having silver electrode as top layer in both cases. While the first and second devices had indium tin oxide (ITO) and titanium as their bottom electrodes, respectively. It could be corroborated from the experimental results that the proportion of high-resistance states to low-resistance states increased 10 times when the bottom electrode was changed from ITO to titanium. The mechanism behind the switching of the device could be explained on the basis of ions' redox and defect state filling. Thus, collagen-based materials have the potential to serve as green bioelectronic devices.

Benas et al. [11], fabricated a bio-thin film transistor which consisted of a dielectric layer of hexadecyltrimethylammonium chloride (CTMA) blended collagen hydrolysate. The mobility and on/off proportion displayed by a 3-weight percentage blend of collagen-CTMA was equal to 3.36×10^{-2} $cm^2/V/s$ and 2.7×10^6 respectively. Further 1-aminopyrene was incorporated into the mixture to construct BioFET (field-effect transistor) memory device. The embedment of 1-aminopyrene further enhanced its properties by changing the on/off proportion and average mobility to 5.22×10^5 and 1.23×10^{-2} $cm^2/V/s$ respectively. The amount of added 1-aminopyrene was found to be directly proportional to the charge-trapping potential of collagen. Also, when tested under water, the BioFET device performed better because of hydroxyl groups, even though the authors concluded that the environment in which the BioFET was tested had great influence on the output result. In order to track the development of the dielectric characteristics in a pentacene-based BioTFT shown in Fig. 11.2, these scientists used a thin film of random coil collagen hydrolysate and

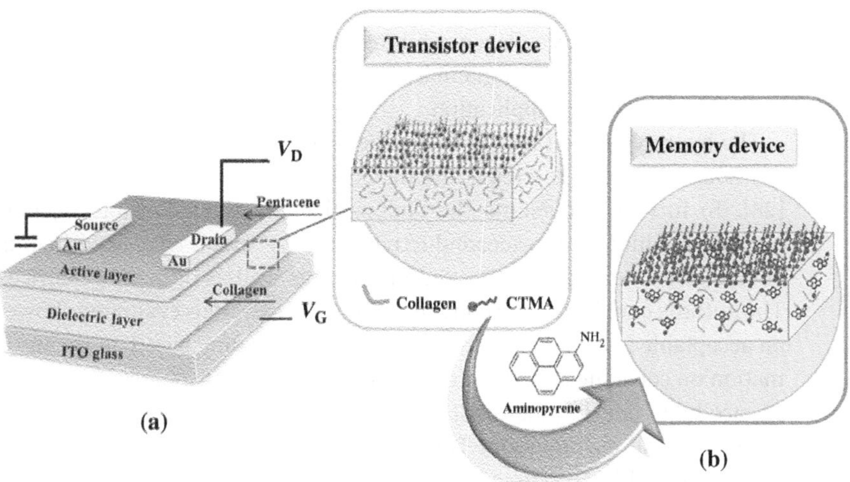

Figure 11.2 Pentacene-based BioFET memory device schematic diagram. (a) Col-CTMA combination BioTFT device (b) a Col-CTMA3-AP mixture memory device. Source: Reproduced with copyright permission from Benas et al. [11], Elsevier, 2020.

cationic surfactant producing Col-CTMA mixes to generate the appropriate dielectric layers [11]. Their conclusion was that the inclusion of collagen in CTMA3 resulted in enhanced material performance such as average mobility, mechanical properties, piezoelectric character, etc.

All we can deduce is that nature is the best answer to challenges faced by humans in all niches of science and technology.

11.8 Collagen-based Tissue Engineering Scaffolds

For many tissue engineering applications, it is crucial to assemble artificial collagenous tissues with structural, functional, and mechanical characteristics that resemble genuine tissues. Although it is believed that collagen's electro-mechanical characteristics contribute to, for instance, bone formation and remodeling, synthetic tissues have not sufficiently taken into account this functional attribute. Using piezoresponse force microscopy, the electro-mechanical characteristics of rat tail tendon were contrasted with those of dried, isoelectrically focused collagen hydrogels by Denning et al. [27]. D-periodic type I collagen fibrils were seen in both the natural tissue and the synthetic hydrogel and displayed shear piezoelectricity. Both tissues have fibrils that are parallel in orientation, but according to Fourier transform analysis, the tendon's fibrils are three times more parallel than those in the dried hydrogel. The findings showed that isoelectrically focused collagen exhibited structural and electro-mechanical characteristics that are comparable to those of tendon, which is important for tissue engineering applications.

The hydrolysis of either the bovine pericardium for 24 (BP 24) or 48 (BP 48) hours was used to create anionic collagen scaffolds by Moreira et al. [28]. To further understand the biological interactions causing excellent biocompatibility and osteoconductivity shown in *in vivo* experiments, bovine osteoblasts were cultivated on them as well as on native matrices. The authors could draw the conclusion that the materials evaluated showed an acceptable level of biocompatibility for bone healing based on the information on cell adhesion, the production of alkaline phosphatase (ALP) and extracellular matrix macromolecules, and cellular proliferation acquired by histological examination. The most encouraging findings were shown with BP 48, which promoted ALP production, in vitro mineralization, and potential osteoblast differentiation.

Kakudo et al. [29], conducted a study which sought to determine whether adipose-derived stem cells (ASCs) might be cultured on a honeycomb collagen scaffold for bone tissue engineering. Histology and cbfa-1 expression testing proved the osteogenic potential of ASCs in vitro. ASCs were then cultivated in the honeycomb scaffold for up to 14 days to provide a high-density, 3D culture. Data from scanning electron microscopy (SEM) revealed that mature ASCs filled up the scaffold, and von Kossa's black staining verified calcification. Additionally, the ASC-loaded honeycomb collagen scaffolds were subcutaneously implanted into mice and removed after 8 weeks of culture. HE, von Kossa, and osteocalcin immunostain were used to assess bone development in vivo. HE, von Kossa, and osteocalcin immunostain were used to assess bone development in vivo. These histological images revealed notable positive stains in the three types of stains in the osteogenic media samples. These findings imply that this carrier is an effective scaffold for ASCs and that it may be used both in vitro and in vivo as a 3D bone tissue engineering scaffold.

Through the use of a combination phase-separation technology and an expedited biomimetic coating process, novel scaffolds made of a bone-like apatite or apatite/collagen composite over a poly(l-lactic acid) (PLLA) skeleton were created by Chen et al. [30]. In order to assess the cellular behavior of these biomimetic coatings, Saos-2 osteoblast-like cells were utilized. SEM was used to examine the morphologies of the cells on the surfaces of PLLA scaffolds, apatite coated PLLA, and an apatite/collagen coated PLLA scaffolds. The MTT test was used to gauge cell viability. Alkaline phosphatase activity was measured to track differentiated osteoblastic activity. These findings showed that the coatings made of apatite and an apatite/collagen composite produced by accelerated biomimetic procedures could enhance the interactions between osteoblasts and polymeric scaffolds. Such interactions were improved more by the apatite/collagen composite coating than by the apatite coating alone. When used as a possible 3D substrate for bone tissue synthesis, apatite/collagen composite covered PLLA scaffold showed promise.

11.9 Conclusion and Future Prospects

Collagen is an integral part of extracellular matrix like bone, teeth, and tendon, where it provides mechanical and structural strength. This piezoelectric biopolymer possesses a piezoelectric coefficient in the range of 0.2–2.0 pC/N. As it is biocompatible, hydrophilic, and less antigenic in nature, it is apposite for tissue engineering purposes too. Collagen scaffolds promote bone healing. Collagen-hydroxyapatite composites also support cell proliferation and bone healing. Also, collagen-calcium phosphate composites could be used for cartilage tissue engineering. It can be further used as sensors, nanogenerators, and memory devices. However, it has less mechanical strength and show fast degradation.

Though the presence of piezoelectricity in collagen is quite evident yet its role has not been fully explored. Thus, a lot more research needs to be dedicated toward exploring the origin of piezoelectricity in collagen.

We know that the human or animal body generates its electricity via the movement of diverse tissues, cells, or organs, of which collagen fibers are the major structural components. For instance, the wound healing mechanism or inherent tissue regeneration largely relies upon the coordination and interfacial interaction between various biological as well as immunological systems, which could be aided by advanced materials technology such as hybrid collagen-based piezoelectric as well as triboelectric nanogenerator systems. This is because the wounds' epithelial damaged layer generates its endogenous bio-electric field in situ for the regulation of the wound healing process. However, researchers have neglected this profound research area, which is of great importance to the biomedical materials niche. We also propose that researchers withing the field of collagen-based piezoelectric materials should focus on scaling-up the fabrication process toward commercialization. Another suggestion we have herewith is that there exist diverse materials which can be utilized by researchers as raw materials for the synthesis of collagen for the fabrication of their collagen-based piezoelectric hybrid systems. These raw material sources include waste fish scales [31, 32], waste fish swim bladder, pig skin, chicken feet's, waste fruits, vegetables, etc. in order to also create a circular and sustainable economy. It is widely known that the intermolecular hydrogen bonds in collagen fibers produce a uniaxial orientation of the molecular dipoles that give organic molecules their piezoelectricity. Also, the utilization of simulation software's such as density functional theory (DFT) [33], response surface methodology [34], etc. aimed at effective optimization of the process parameters for fast scaling up is necessary. Researchers in this niche are therefore encouraged to move their research study beyond in vitro or ex vivo to in vivo so as to enable fast and effective scaling up of the engineering/fabricated collagen-based hybrid generators. Comparative research into collagen-based piezoelectric material fabrication approaches with the aim of identifying the best technique should also be a major point of interest for researchers within thin niche. Again, the synergistic utilization of biocompatible carbon-based materials such as graphene oxide nanoparticles as well as its quantum dots along with collagen aimed at fabricating effective/efficient hybrid piezoelectric materials should also be given attention.

References

1 Liu, Y., Wang, Z., He, W. et al. (2020). Quantifying contact status and the air-breakdown model of charge-excitation triboelectric nanogenerators to maximize charge density. *Nat. Commun.* 11 (1): 1–8.

2 Zou, Y., Liao, J., Ouyang, H. et al. (2020). A flexible self-arched biosensor based on combination of piezoelectric and triboelectric effects. *Appl. Mater. Today* 20: 100699.

3 Song, K., Zhao, R., Wang, Z.L., and Yang, Y. (2019). Conjuncted pyro-piezoelectric effect for self-powered simultaneous temperature and pressure sensing. *Adv. Mater.* 31 (36): 1902831.

4 Karan, S.K., Maiti, S., Lee, J.H. et al. (2020). Recent advances in self-powered tribo-/piezoelectric energy harvesters: all-in-one package for future smart technologies. *Adv. Funct. Mater.* 30 (48): 2004446.

5 Saqib, Q.M., Khan, M.U., Song, H. et al. (2021). Natural hierarchically structured highly porous tomato peel based tribo-and piezo-electric nanogenerator for efficient energy harvesting. *Adv. Sustain. Syst.* 5 (7): 2100066.

6 Zhou, Z., Qian, D., and Minary-Jolandan, M. (2016). Molecular mechanism of polarization and piezoelectric effect in super-twisted collagen. *ACS Biomater. Sci. Eng.* 2 (6): 929–936.

7 Ravi, H.K., Simona, F., Hulliger, J. et al. (2012). Molecular origin of piezo-and pyroelectric properties in collagen investigated by molecular dynamics simulations. *J. Phys. Chem. B* 116 (6): 1901–1907.

8 Badatya, S., Bharti, D.K., Srivastava, A.K., and Gupta, M.K. (2021). Solution processed high performance piezoelectric eggshell membrane – PVDF layer composite nanogenerator via tuning the interfacial polarization. *J. Alloys Compd.* 863: 158406.

9 Vivekananthan, V., Alluri, N.R., Purusothaman, Y. et al. (2018). Biocompatible collagen nanofibrils: an approach for sustainable energy harvesting and battery-free humidity sensor applications. *ACS Appl. Mater. Interfaces* 10 (22): 18650–18656.

10 Busuioc, C., Voicu, G., Jing, S.-I. et al. (2019). The influence of barium titanate on the biological properties of collagen-hydroxiapatite composite scaffolds. *Mater. Lett.* 253: 317–322.

11 Benas, J.-S., Veeramuthu, L., Chuang, Y.-Y. et al. (2020). Eco-friendly collagen-based bio-organic field effect transistor with improved memory characteristics. *Org. Electron.* 86: 105925.

12 Ghosh, S.K. and Mandal, D. (2016). High-performance bio-piezoelectric nanogenerator made with fish scale. *Appl. Phys. Lett.* 109 (10): 103701.

13 Ghosh, S.K. and Mandal, D. (2016). Efficient natural piezoelectric nanogenerator: electricity generation from fish swim bladder. *Nano Energy* 28: 356–365.

14 Subhani, K., Jin, X., Hameed, N. et al. (2022). Porous carbon sponges from collagen-rich biomass waste for high-performance supercapacitors. *Mater. Today Sustain.* 18: 100152.

15 Yu, M., Han, Y., Li, J., and Wang, L. (2018). Magnetic N-doped carbon aerogel from sodium carboxymethyl cellulose/collagen composite aerogel for dye adsorption and electrochemical supercapacitor. *Int. J. Biol. Macromol.* 115: 185–193.

16 Zeng, Y., Sun, B., Yu, H.-Y. et al. (2019). A sustainable biomemristive memory device based on natural collagen. *Mater. Today Chem.* 13: 18–24.

17 Silva, C., Pinheiro, A.G., Figueiró, S.D. et al. (2002). Piezoelectric properties of collagen-nanocrystalline hydroxyapatite composites. *J. Mater. Sci.* 37 (10): 2061–2070.

18 Kumar, C., Gaur, A., Tiwari, S. et al. (2019). Bio-waste polymer hybrid as induced piezoelectric material with high energy harvesting efficiency. *Compos. Commun.* 11: 56–61.

19 Zhang, M., Deng, F., Tang, L. et al. (2021). Super-ductile, injectable, fast self-healing collagen-based hydrogels with multi-responsive and accelerated wound-repair properties. *Chem. Eng. J.* 405: 126756.

20 Shapardanis, S., Hudspeth, M., and Kaya, T. (2013). Design and implementation of collagen-based capacitive relative humidity sensors. In: *Sensors, 2013 IEEE*. IEEE.

21 Cai, T., Liu, X., Ju, J. et al. (2022). Flexible cellulose/collagen/graphene oxide based triboelectric nanogenerator for self-powered cathodic protection. *Mater. Lett.* 306: 130904.

22 Li, L., Zhang, J., Wang, M. et al. (2022). Electrospun hydrolyzed collagen from tanned leather shavings for bio-triboelectric nanogenerators. *Mater. Adv.* 3 (12): 5080–5086.

23 Qin, C., Wu, X., Huang, C. et al. (2022). Tooth-derived flexible supercapacitor. *J. Energy Storage* 52: 104728.

24 Aronoff, M.R., Hiebert, P., Hentzen, N.B. et al. (2021). Imaging and targeting LOX-mediated tissue remodeling with a reactive collagen peptide. *Nat. Chem. Biol.* 17 (8): 865–871.

25 Vivekananthan, V. (2020). *Development of Flexible Composite Film Based Hybrid Energy Harvesters for Self-Powered Sensors*. Jeju National University Graduate School.

26 Wang, X., Yue, O., Liu, X. et al. (2020). A novel bio-inspired multi-functional collagen aggregate based flexible sensor with multi-layer and internal 3D network structure. *Chem. Eng. J.* 392: 123672.

27 Denning, D., Abu-Rub, M.T., Zeugolis, D.I. et al. (2012). Electromechanical properties of dried tendon and isoelectrically focused collagen hydrogels. *Acta Biomater.* 8 (8): 3073–3079.

28 Moreira, P.L., An, Y.H., Santos, A.R. Jr., and Genari, S.C. (2004). In vitro analysis of anionic collagen scaffolds for bone repair. *J. Biomed. Mater. Res. B Appl. Biomater.* 71B (2): 229–237.

29 Kakudo, N., Shimotsuma, A., Miyake, S. et al. (2008). Bone tissue engineering using human adipose-derived stem cells and honeycomb collagen scaffold. *J. Biomed. Mater. Res. A* 84A (1): 191–197.

30 Chen, Y., Mak, A.F.T., Wang, M. et al. (2006). PLLA scaffolds with biomimetic apatite coating and biomimetic apatite/collagen composite coating to enhance osteoblast-like cells attachment and activity. *Surf. Coat. Technol.* 201 (3): 575–580.

31 Orasugh, J.T., Dutta, S., Das, D. et al. (2019). Sustained release of ketorolac tromethamine from poloxamer 407/cellulose nanofibrils graft nanocollagen based ophthalmic formulations. *Int. J. Biol. Macromol.* 140: 441–453.

32 Samanta, A.P., Ali, M.S., Orasugh, J.T. et al. (2022). Crosslinked nanocollagen-cellulose nanofibrils reinforced electrospun polyvinyl alcohol/methylcellulose/polyethylene glycol bionanocomposites: study of material properties and sustained release of ketorolac tromethamine. *Carbohydr. Polym. Technol. Appl.* 3: 100195.

33 Orasugh, J.T. and Ray, S.S. (2022). Prospect of DFT utilization in polymer-graphene composites for electromagnetic interference shielding application: a review. *Polymers* 14 (4): 704.

34 Zaman, A., Orasugh, J.T., Banerjee, P. et al. (2020). Facile one-pot in-situ synthesis of novel graphene oxide-cellulose nanocomposite for enhanced azo dye adsorption at optimized conditions. *Carbohydr. Polym.* 246: 116661.

12

Chitin and Chitosan—Foremost Hybrid Piezoelectric Materials for Energy Harvesting Applications

Sourav Banerjee[1], *S. Wazed Ali*[1,2], *and Satya Narayan Naik*[1,3]

[1] *School of Interdisciplinary Research (SIRe), Indian Institute of Technology Delhi, Hauz Khas, New Delhi, India*
[2] *Department of Textile and Fibre Engineering, Indian Institute of Technology Delhi, Hauz Khas, New Delhi, India*
[3] *Centre for Rural Development and Technology, Indian Institute of Technology Delhi, Hauz Khas, New Delhi, India*

12.1 Introduction

Curie brothers were the first to blaze a trail for researchers to explore piezoelectricity and its effects. They started it in 1880 with quartz and Rochelle salt crystals. One of the reasons behind this electrical response was the absence of center of symmetry in the crystalline structure. Gradually, the research trend in inorganic materials started to lean toward polymeric substrates like polyvinylidene fluoride (PVDF). Molecular dipole orientation in polymeric chains is one of the critical aspects that generate piezoelectric performance. Young's modulus of piezoelectric polymers is relatively lower than that of ceramic materials; flexibility is another potential factor that helps widen the application in biomedical segments. Researchers are currently confronted with the persistent task of improving piezoelectric material qualities through material design and engineering to fulfill various demands of medical sciences and the living universe. In-depth studies have been conducted on cellulose [1–4], wood [5, 6], bone [7, 8], skin [9, 10], tendon [11, 12], and many other asymmetric biocomponents to reveal their piezoelectric responses after finding the polarization effect in wool and keratin [13]. Chitin piezoelectricity has been studied since the '70s. More research into piezoelectric materials has increased in recent years due to increased interest in this class of materials [14].

Hybrid Materials for Piezoelectric Energy Harvesting and Conversion, First Edition.
Edited by S. Wazed Ali, Satyaranjan Bairagi, and Shahid Ul Islam.
© 2024 John Wiley & Sons, Inc. Published 2024 by John Wiley & Sons, Inc.

Past research has shown that chitin- or chitosan-based films have significantly greater piezoelectric responses than proteins [15]. Natural systems are known to have the improper ordering of large molecules due to the presence of amorphous or distinct molecules. As a result, the piezoelectric effect in such scenarios is projected to be less than that in fully ordered piezoelectric biocrystals [16]. Some performance loss may nevertheless be tolerable in situations where energy conversion efficiency is not of primary concern, given the ease of access and processing of these natural resources [14]. No exact information was given about how much mechanical energy was put in, but the voltage output was high. There are 127 publications on chitin and chitosan related to piezoelectricity in Scopus (Figure 12.1). The literature's rising need for piezoelectric sensors and devices based on biomaterials is reflected in the field's rising popularity graph.

In this chapter, we have provided a complete overview of current developments in the fabrication of chitin/chitosan-based piezoelectric biomaterials, their properties, and the biological domains in which they are applied. Research on various methods to enhance their piezoelectric properties has been covered. In conclusion, a perspective has been given on the prospects and difficulties associated with further research into piezoelectric chitin or chitosan for different applications. Additionally, it has been suggested that this chapter will set an effective and progressive reference and accelerate the development of new piezoelectric biomaterials.

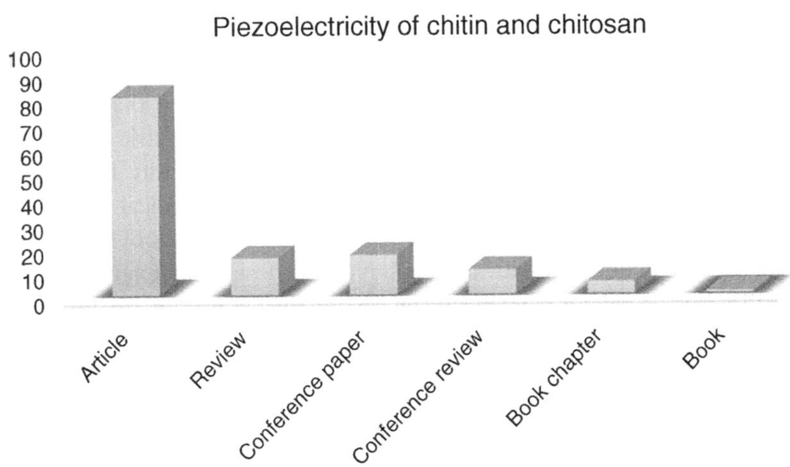

Figure 12.1 Number of literatures in different document types of chitin and chitosan-related piezoelectricity from 1972 to 2023. Source: https://www.scopus.com.

12.2 Chitin and its Application as Piezoelectric Material

A biopolymer called chitin can be found in certain fungi's cell walls as well as the exoskeletons of arthropods, including crabs, shrimp, and insects. It is a fibrous, strong, and insoluble substance made of linear chains of N-acetylglucosamine (GlcNAc) monomers attached by β-(1→4) glycosidic linkages (Figure 12.2a). Chitin is a polysaccharide that is second only to cellulose in terms of abundance. The traditional chemical approach is commonly used for chitin extraction. Three main processes are involved in the extraction of chitin: demineralization, deproteination, deacetylation, and decolorization (as shown in Figure 12.3). In industrial processing, calcium carbonate is first dissolved by an acid treatment, and then proteins are hydrolyzed in an alkaline solution. Both of these steps are needed to get to the chitin. Moreover, decolorization is often used to eliminate colors and achieve a pure, colorless chitin [17]. Chitin occurs in nature as three distinct crystalline polymorphs. Different forms of chitin have microfibrils that are oriented in different ways, including antiparallel chains in α-chitin, parallel chains in β-chitin, and a hybrid of parallel and anti-parallel chains in γ-chitin (Figure 12.4) [18]. Fukuda was the first to discover a piezoelectric effect in α-chitin, which he examined across a broad temperature range of 130–140 °C [19]. Chitin has three distinct crystallographic forms (α, β, and γ) and a structure that is very similar to cellulose. The piezoelectricity

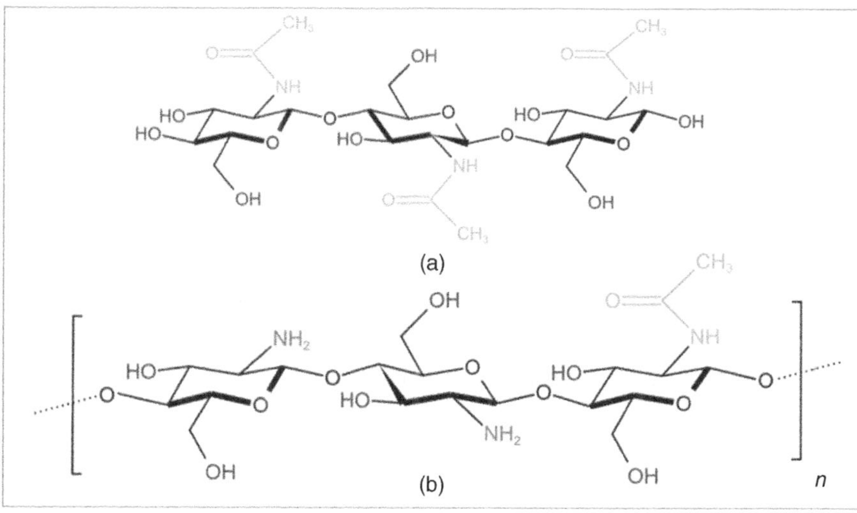

Figure 12.2 Structure of (a) chitin and (b) chitosan (self-drawn).

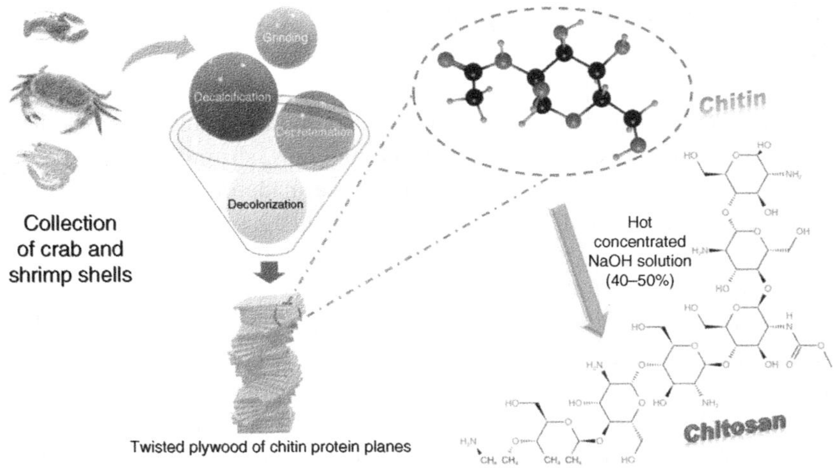

Figure 12.3 Preparation of chitin and chitosan from crab and shrimp shells (self-drawn).

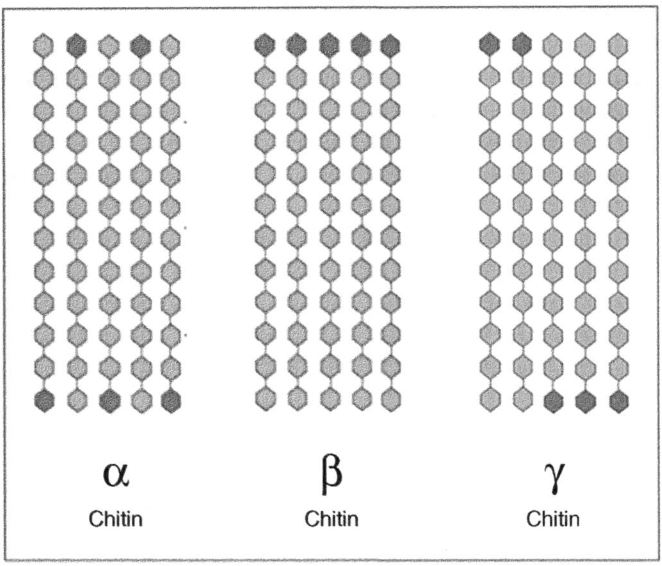

Figure 12.4 Different forms of chitin (self-drawn).

observed by Ghosh and Mandal in chitin-based prawn shell nanofibers has been attributed to the influence of nonlinear electrostriction driven by spontaneous polarization, and has a piezoelectric charge constant ($d33$) of 2 pC/N. Despite its antiparallel molecular orientation, α-chitin also exhibited intramolecular or intermolecular H-bonding inside its crystalline structure [20, 21].

Due to its special characteristics, chitin finds several uses in many different industries. Wound healing, drug delivery, and tissue engineering are just a few of the biological uses for chitin. For these applications, it is the perfect material due to its biodegradability, biocompatibility, and antimicrobial qualities. As it may inhibit the development of specific bacteria and fungi, it is used in the food industry as a food preservative. It is also used as a dietary fiber supplement and a fat replacement in low-fat diets. Because it can promote resistance to pests and plant diseases, chitin is employed as a natural insecticide. It has been demonstrated to improve crop growth and yields, and it also holds the potential to cut down on the usage of chemical pesticides [22–24]. It has been used for the elimination of heavy metals and dyes from wastewater due to its strong adsorption capacity. Furthermore, it is capable of being chemically altered to create the biodegradable and biocompatible polymer chitosan. Chitosan may be used to create bioplastics, which are environmentally friendly substitutes for traditional plastics [25, 26]. Research on chitin-based piezoelectric materials is still in its early stages, but initial results are promising. One study found that chitin films showed piezoelectric properties when stretched, indicating that chitin has the potential to be used in flexible, wearable sensors. Another study explored the use of chitin in energy harvesting, showing that chitin-based nanogenerators could convert mechanical energy into electricity [27].

A chitin-based electrospun [28] has been explored for its natural piezoelectricity. Chitin nanofibers have shown around 300% mechanical strength improvement and 400% piezoelectricity enhancement when collected using the standard air-gap method. The air-gap electrospinning procedure has resulted in a greater crystallinity of the chitin fibers, which led to a greater piezoelectric response. The studies have also shown that the nanofibers' perimeter is more crystalline and, hence, more active than the center, which is less crystalline and less piezoelectric. The morphology of electrospun nanofibers may have been caused by forces of elongation or fast solvent evaporation at the surfaces of nanofibers [29]. In the presence of external mechanical pressure, the conveniently controllable ferroelectric chitin layer has been provided with exceptional piezoelectricity and an equivalent performance compared with that of traditional fluorine-based piezoelectric polymers [30]. The excellent piezoelectric properties of the chitin film have provided a decent 177 nA/cm^2 output current density and dependable endurance performance (5000 cycles of a 100 kPa vertical pressure). Another group of scientists [31] looked into the possibility of making piezoelectric nanogenerators that are biocompatible, biodegradable, flexible, and efficient out of chitin nanofiber (CNF) recovered from biowaste crab shells and poly(vinylidene fluoride) (PVDF). Electroactive β-phase nucleation is improved by almost 81% when CNF is included in a PVDF matrix, with a better $d33$ of 35.56 pC/N. Nanofibers made of chitin have a high aspect ratio, strong binding energy, and powerful hydrogen bonding energy because of their anti-parallel arrangement. Within the nanofiber, electric dipoles are generated by these groups. The nanofiber dipoles exhibit a robust piezoelectric response to an external strain. After being compressed mechanically, highly crystalline CNFs develop stress-induced polarization, leading to piezoelectricity.

12.3 Chitosan and its Applications as Piezoelectric Materials

Chitosan comes from chitin, a biopolymer that can be found in the shells of shrimp, crab, and lobster. Using diluted aqueous sodium hydroxide to remove the protein from the rest of the material is a way to keep chitin products from having protein in them. The resultant chitin undergoes a 1–3-hour deacetylation process in a solution of 40–45% sodium hydroxide at 160 °C, as shown in Figure 12.3. Chitosan, on the other hand, has primary and secondary hydroxyl groups on each repeat unit, as well as an amine group on each deacetylated unit, making it chemically more active than chitin [32] (as shown in Figure 12.2b). Due to its special characteristics, chitosan has a wide range of potential uses in the culinary, cosmetic, biomedical, and agricultural industries [33–35]. The agriculture sector uses chitosan in one of its most important ways. It has been demonstrated to have several advantageous impacts on plants, including promoting plant growth, raising agricultural yields, and strengthening pest and disease resistance. Chitosan is a substance that may be placed on seeds, soil, or leaves. It has been found to be particularly effective in reducing the damage caused by plant pathogens such as fungi and bacteria. The plant's immune system is also stimulated by chitosan, which increases its tolerance to environmental stresses, including dehydration and excessive salt [36, 37]. In the food industry, chitosan is used as a natural preservative due to its antimicrobial properties, which can extend the shelf life of food products. The possibility of using it for biomedical purposes has also been researched. Its ability to fight off microbes has been demonstrated, and it may be applied to wounds as a dressing to shield them from infection and speed up recovery. Furthermore, the ability of chitosan to encapsulate pharmaceuticals and release them gradually over time makes it a useful drug delivery mechanism. Since medications may now be supplied specifically to particular tissues or organs, it is very helpful for targeted drug delivery [38, 39].

The term "piezoelectricity" refers to the phenomenon in which some materials produce an electric charge when subjected to a mechanical stress. When stretched or squeezed, chitosan has piezoelectric capabilities due to its semi-crystalline structure. Chitosan's unique combination of characteristics makes it an attractive candidate for use in new piezoelectric devices and sensors. The molecular structure of chitosan gives it piezoelectric characteristics. The piezoelectric characteristics of chitosan come from the organized arrangement of polar and nonpolar areas in its molecules. Chitosan undergoes a structural rearrangement in response to mechanical stress, producing an electrical charge that may be collected and put to use in a number of ways. In addition to its piezoelectric properties, chitosan's biocompatibility, biodegradability, and low toxicity make it a desirable material for use in biomedical applications. Electrospinning, film casting, and filler reinforcement are only a few of the methods that have been successfully used to fabricate chitosan-based piezoelectric materials, as reported by many research studies [40]. For instance, solution-cast chitosan/PVDF composite films have been demonstrated to have exceptional piezoelectric characteristics. Electrospun chitosan nanofibers have also been utilized to increase the piezoelectric characteristics of PVDF-based piezoelectric composites [41, 42]. Pressure sensors, vibration monitors, and energy harvesters are just a few of the many sensor types that have benefited from the usage of chitosan-based piezoelectric materials. For instance, piezoelectric nanogenerators based on chitosan have been created to harness mechanical energy for the purpose of providing electricity for miniature electronic devices [43–45].

The crystalline structure of chitosan contains non-centrosymmetrical elements that cause piezoelectricity. In order to display piezoelectricity, the ions in the crystalline structure of piezoelectric crystals are compressed to occupy locations in a certain repeating relationship to one another. Advanced piezoelectric materials like PZT and PVDF have shown excellent mechanical quality factors due to their capacity to optimize dipole moment orientation, which enhances the effect of piezoelectricity inside the crystal unit of thin film. During their study, Zamli et al. measured the mechanical quality factor (Qm) [46]. They found that it was 3.22 and that the lowest dissipation factor was 0.327. It has been shown that the Qm value increased as the thickness of the film increased, and the crystallinity also improved. Dipole–dipole movement can be facilitated by the increase in crystallinity of the thin film. As a result, it may contribute to improved piezoelectric responsiveness [47]. Their research has also shown that the increase in localized charge carriers inside the crystalline chitosan structure was responsible for the inverse relationship between the dielectric dissipation factor and the thickness of the material. Pure chitosan and chitosan derived from fungi have been used to assess the dissipation factor. The rate of energy dissipation during electric polarity reversal is denoted by the dissipation factor. The study has also revealed that the dissipation factor, or tan value, for pure chitosan (0.327) was higher than chitosan derived from fungi (0.248). Pure chitosan probably has a greater dielectric loss tangent because of its crystalline structure, which allows for a greater dipole interaction. Also, the loss tangent of fungal chitosan has increased because electron–photon scattering went down because photons did not interact with the lattice structure as much because of the surface nonuniformity [48]. Finally, Qm values have also been investigated for fungal chitosan, which was 3.68 (higher than pure chitosan). This observation has suggested that the chitosan film produced by fungi may be somewhat more efficient in transforming mechanical energy into electrical energy [48, 49].

A substrate composed of PVDF-chitosan-gelatin nanofibers and mesoporous silica nanoparticles (known as MCM41) was developed by researchers and encased in gellan-PVA hydrogel [41]. Nanoparticles made of MCM41 are utilized to enhance output voltage and eradicate microorganisms. Although none of these skin patches have been tried on humans, they do provide a new method of healing wounds [50]. Researchers have also investigated the use of polydopamine hydrogel-coated chitosan films, which are based on the shells of mussels. Using the photothermal action of polydopamine (which makes heat when exposed to NIR or near-infrared light) and the piezoelectric effect of chitosan, these researchers described the initial approach to speed up wound healing by combining electrical stimulation and heat. Although experts have admitted that the particular mechanism underlying this improvement in chitosan piezoelectricity is uncertain [51, 52]. Chitosan can be used as a versatile energy harvester in piezoelectric applications when it is mixed with other polymers. The amount of polar crystalline area in the hydroxyl and amine groups of chitosan determines its piezoelectric properties. In the crystalline phase, hydrogen bonds between and within molecules can also have an effect on piezoelectric responses [53]. In another study, foams made from chitosan and konjac glucomannan were created using the directed freezing method (as shown in Figure 12.5); these foams have shown good strain

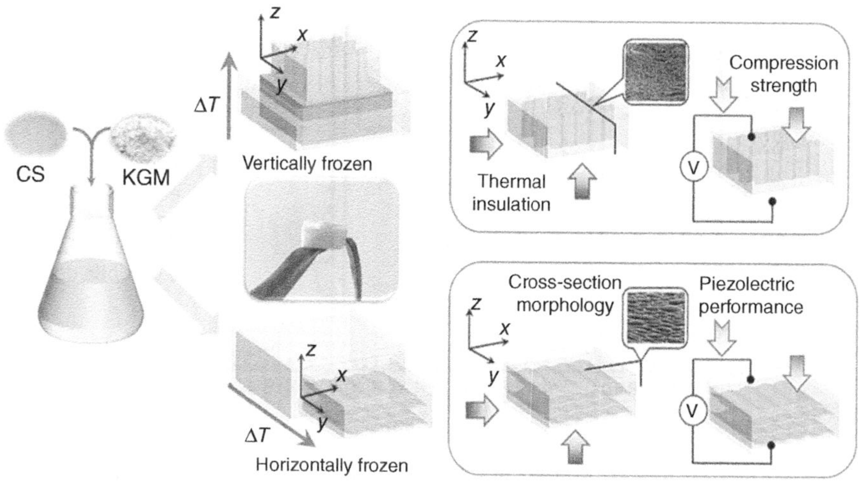

Figure 12.5 Two techniques of ice templating Chitosan-konjac glucomannan foams and their characterization. Source: Song et al. [54]/Elsevier.

recovery, excellent piezoelectric production and sensing characteristics, and a thermally insulating nature [54]. They showed that the chitosan's piezoelectricity contributed to a stable piezoelectric generating ability, which measured between 0.80 and 2.460 V during the compression cycle procedure. The foam has been compressed along the z-axis, which is perpendicular to its horizontal freezing direction. The piezoelectric performance of a bridge is affected by its structural design and its material composition.

Prokhorov and his co-workers [55] have reported in their study about the chitosan-zinc oxide nanoparticle films and analyzed piezoelectric performances. Chitosan-ZnO nanocomposite films with 15 wt% of nanoparticles had a greater piezoelectric coefficient $d33$ ($d33 = 65.9$ pC/N) relative to other polymer-ZnO nanocomposites due to the combined effects of the chitosan and nanoparticles. The piezoelectricity of polymer nanocomposites may be traced back to a fundamental hypothesis based on the concentration of nanoparticles with a significant dielectric constant (such as lead zirconate titanate, Barium titanate, and Barium zirconate titanate), which helps to raise $d33$ with an increase in the dielectric constant with nanoparticle content. A separate study [56] demonstrated that the piezoelectric sensitivity of plain chitosan films is 4–6 pC/N, which is more than twice as high as that of quartz crystals (2 pC/N). This demonstrates the feasibility of using microcrystalline chitosan as a biodegradable, inexpensive piezoelectric energy harvesting or sensor material. Researchers have looked at how tweaking factors like film thickness, surface roughness, and molecular alignment might affect a film's piezoelectric sensitivity.

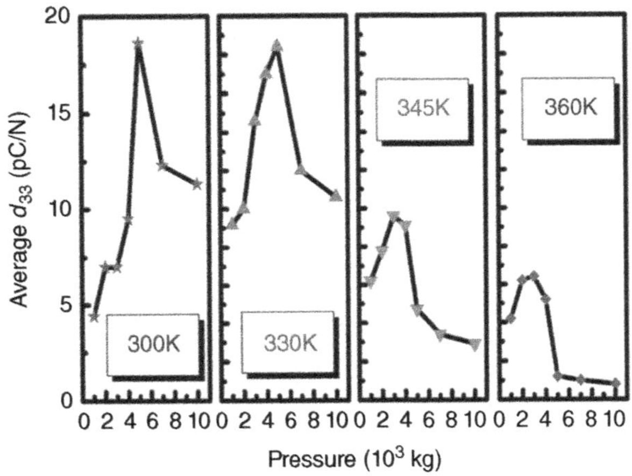

Figure 12.6 *d*33 of chitosan formate at different pressures and sintering temperatures. Source: Praveen et al. [57]/Royal Society of Chemistry/CC By 3.0/Public domain.

Some researchers [57] have investigated the impact of the sintering temperature on the piezoelectric strain coefficient in chitosan formate, and the results are shown in Figure 12.6. The *d*33 coefficient measurement has shown that the sample is durable up to 57 °C, with an utmost *d*33 coefficient of 18.4 pC/N under 5 tons of load. Perhaps as a result of chitosan losing its non-centrosymmetry at higher temperatures, the *d*33 coefficient drops as temperature is increased. Silva and his group [58] have researched the dielectric and piezoelectric properties of collagen–chitosan films. Collagen has been measured to have a piezoelectric strain constant (*d*14) of 0.096 pC/N. The values discovered in the literature and by employing other approaches are quite consistent with our results. Furthermore, it was found that the piezoelectricity was raised to 0.212 pC/N after 15% chitosan was added. Incorporation of glycine crystals in a chitosan matrix to generate a flexible piezoelectric composite has been described in a separate study on glycine-chitosan-based biosensors as a means of fabricating homogenous and very stable β-glycine films [59]. They have synthesized crystals of β-glycine that are both crystallographically aligned and stable inside a chitosan matrix. Charge separation occurs in a composite piezoelectric film when external pressure is applied. A high voltage is produced between the electrodes. Although each spherulite exhibits radial crystallization, it is the sum of these individual responses that accounts for the overall piezoelectric response. The sensor based on glycine and chitosan has a sensitivity of 2.82 ± 0.2 mV/kPa and an output voltage of 190 mV at 60 kPa pressure (Figure 12.7).

The electroactive nanofibrous scaffold is an important tool in bioelectronics, regenerative medicine, and other areas of biological research [60–62]. The biocompatibility and piezoelectricity of chitosan make it particularly

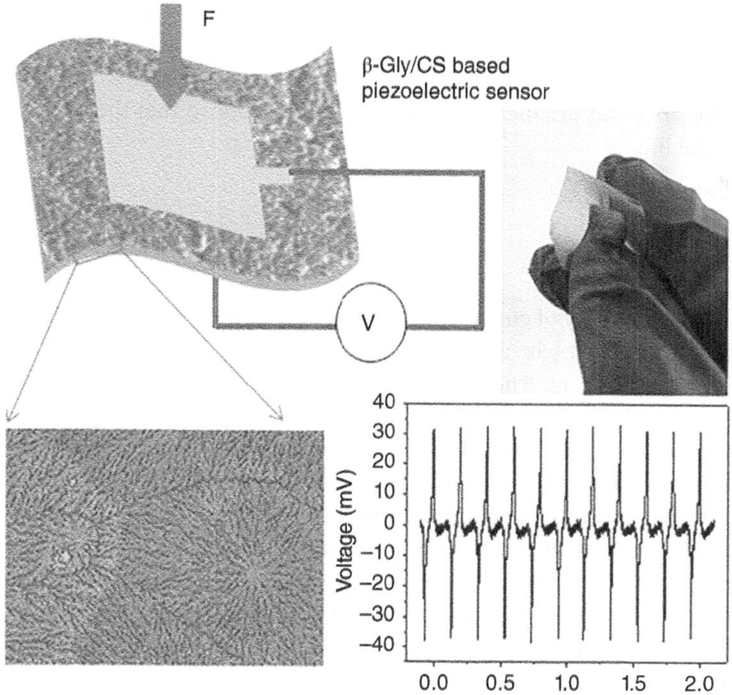

Figure 12.7 Glycine-chitosan based piezoelectric sensor. Source: Hosseini et al. [59]/American Chemical Society/CC BY 4.0.

useful in this context. By depositing a negligible amount of PEDOT:PSS onto the surface of chitosan nanofibers using a non-in situ permeation approach, researchers have designed novel biodegradable and biocompatible electroactive polymers [63]. Because of their non-centrosymmetric molecular structure, materials made from chitosan have shown unique piezoelectric properties. It has been possible to stimulate cell behavior using chitosan-based polymers; however, this approach presents a new obstacle in the form of the materials' exceptional sensitivity required to detect extremely minute pressures resulting from either the external environment or the cell itself. A biosensor is an integrated, miniature device used for signal recognition that has a transducer linked to it and employs a biological component as a detecting element. Various literatures have exhibited chitosan-based physical biosensors. Carboxymethyl chitosan has been used with a crosslinker to utilize as biosensors, which can inhibit lysozyme and bovine serum albumin from binding. The main advantages of these biosensors were that they provided an inexpensive and quick immunoglobulin [64–66].

Overall, the biodegradability, biocompatibility, and piezoelectric properties of chitosan-based piezoelectric materials have shown that they have a lot of potential for many different uses. But more research is needed to improve how these materials are made and how well they work so that they can be used in real life.

12.4 Problems

There has been plenty of curiosity in natural biopolymers recently because of their potential uses in different fields like energy harvesting, sensor, and tissue engineering. The utilization of chitin and chitosan as piezoelectric materials face several difficulties, even though their piezoelectric characteristics have been demonstrated. The low piezoelectric coefficient of chitin and chitosan is a significant barrier. The piezoelectric coefficient quantifies the capacity of a material to produce an electric charge as a result of a certain amount of mechanical pressure or deformation. Compared to other piezoelectric materials, chitin and chitosan have a relatively low piezoelectric coefficient, which limits their potential use in equipment that requires great sensitivity and accuracy [67]. The average piezoelectric coefficient for this class of biobased materials is less than 10 pC/N. Their piezoelectric coefficients are equivalent to certain synthetic polymers at the high end, but they are far away from some commonly used ceramic piezoelectric substances. Furthermore, because of their structural composition, they often generate excellent piezoelectric effects in a shear mode. On a macroscopic level, their crystal polarization domains' frequently random orientation typically causes the net polarization to be reduced or canceled. Another challenge is processing chitin and chitosan into piezoelectric devices' desired shapes and forms [68]. Extensive processing is required to eliminate impurities and generate the proper form and structure after obtaining chitin and chitosan from natural sources like shrimp and crab shells. This processing can be time-consuming and expensive, which may limit the scalability of chitin- and chitosan-based piezoelectric devices. Furthermore, chitin and chitosan can degrade over time, especially if they are exposed to extreme heat or humidity. This may affect how stable and dependable they are over time as piezoelectric materials. The majority of studies with piezoelectric biomaterials are currently conducted on moderately sized, often for elementary characteristics research or proof-of-concept experiments [69].

12.5 Future Scope

The intrinsic properties of chitin and chitosan put them in a position of exceptional biocompatibility, biosafety, and environmental sustainability, making them desirable, especially for application in the biomedical field. There is a wide range of intriguing possibilities for piezoelectric devices based on chitin or chitosan. The following are just a few instances of how this may be put to use. Biomedical devices, including biosensors, implantable devices, and drug delivery systems, can all make use of chitin- and chitosan-based films, electrospun fibers, and composites. These devices can monitor physiological factors, including blood pressure, glucose levels, and heart rate. In response to changes in these parameters, the gadgets may either deliver drugs or stimulate nerves. These may be used to capture energy from environmental mechanical vibrations and motions, including wind, river flow, and human footsteps. Small electrical devices like wireless transmitters, actuators, and sensors can be powered by this energy. These may also be used to keep an eye on environmental variables, including air quality, humidity, and temperature. These sensors might be used in intelligent homes and buildings as filtration or antibacterial solutions to save energy and improve indoor air quality. If these biopolymer-based devices are used to monitor, regulate, and control industrial processes, they could make production more efficient. The difficulties we discussed above are not anticipated to limit their potential. The creation of fresh and cutting-edge technologies in this field will have a positive impact on society in a variety of ways. We predict that the class of naturally occurring piezoelectric biomaterials will soon emerge as a potential multidisciplinary topic at the intersection of material science and biomedical engineering.

References

1 Bairagi, S., Banerjee, S., Chowdhury, A., and Ali, S.W. (2021). Development of a sustainable and flexible piezoelectric-cum-triboelectric energy harvester comprising a simple commodity cotton fabric. *ACS Sustain. Chem. Eng.* 9 (11): 4004–4013. https://doi.org/10.1021/acssuschemeng.0c07274.

2 Mangayil, R., Rajala, S., Pammo, A. et al. (2017). Engineering and characterization of bacterial nanocellulose films as low cost and flexible

sensor material. *ACS Appl. Mater. Interfaces* https://doi.org/10.1021/acsami.7b04927.

3 Fashandi, H., Abolhasani, M.M., Sandoghdar, P. et al. (2016). Morphological changes towards enhancing piezoelectric properties of PVDF electrical generators using cellulose nanocrystals. *Cellulose* 23 (6): 3625–3637. https://doi.org/10.1007/S10570-016-1070-3/TABLES/5.

4 Rajala, S., Siponkoski, T., Sarlin, E. et al. (2016). Cellulose nanofibril film as a piezoelectric sensor material. *ACS Appl. Mater. Interfaces* https://doi.org/10.1021/acsami.6b03597.

5 Fukada, E. (1955). Piezoelectricity of Wood. *J. Phys. Soc. Jpn.* 10 (2): 149–154. https://doi.org/10.1143/JPSJ.10.149.

6 Placner, J. (2009). *The Converse Piezoelectric Effect in Wood and Cellulose Materials*. No. September, 64.

7 Shamos, M.; Lavine, L.; Nature, M. S.; 1963. Piezoelectric Effect in Bone. Springer.

8 Bassett, C. and Beckker, R.O. (1962). Generation of electric potentials by bone in response to mechanical stress. *Science* 137 (3535): 1063–1064. https://doi.org/10.1126/science.137.3535.1063.

9 De Rossi, D., Domenici, C., and Pastacaldi, P. (1986). Piezoelectric properties of dry human skin. *IEEE Trans. Electr. Insul.* EI-21 (3): 511–517. https://doi.org/10.1109/TEI.1986.349102.

10 Wu, Y., Ma, Y., Zheng, H., and Design, S.R.-M. (2021). Piezoelectric materials for flexible and wearable electronics: a review. *Mater. Des.* 211: 110164.

11 Williams, W. (1975). Piezoelectricity in tendon and bone. *J. Biomech.* 8: 407–413.

12 Kim, D., Han, S.A., Kim, J.H. et al. (2020). Biomolecular piezoelectric materials: from amino acids to living tissues. *Adv. Mater.* 32 (14): 1906989. https://doi.org/10.1002/ADMA.201906989.

13 Martin, A. (1926). Tribo-electricity in wool and hair. *Proc. Phys. Soc.* 53: 186.

14 Pohanka, M. (2018). Overview of piezoelectric biosensors, immunosensors and dna sensors and their applications. *Materials* 11: 448.

15 Jain, N., Hussain, A., Ramadan, K.S. et al. (2014). A review of piezoelectric polymers as functional materials for electromechanical transducers. *Smart Mater. Struct.* 23: 33001–33027. https://doi.org/10.1088/0964-1726/23/3/033001.

16 Cao, L., Qiu, X., Jiao, Q. et al. (2021). Polysaccharides and proteins-based nanogenerator for energy harvesting and sensing: a review. *Int. J. Biol. Macromol.* 173: 225–243.

17 Percot, A., Viton, C., and Domard, A. (2003). Optimization of chitin extraction from shrimp shells. *Biomacromolecules* 4 (1): 12–18. https://doi.org/10.1021/BM025602K.

18 Kaya, M., Lelešius, E., Nagrockaite, R. et al. (2015). Differentiations of chitin content and surface morphologies of chitins extracted from male and female grasshopper species. *PLoS One* 10 (1): https://doi.org/10.1371/JOURNAL.PONE.0115531.

19 Fukada, E. (1995). Piezoelectricity of biopolymers. *Biorheology* 32 (6): 593–609. https://doi.org/10.1016/0006-355X(95)00039-C.

20 Maiti, S., Karan, S.K., Kim, J.K., and Khatua, B.B. (2019). Nature driven bio-piezoelectric/triboelectric nanogenerator as next-generation green energy harvester for smart and pollution free society. *Adv. Energy Mater.* 9 (9): 1803027. https://doi.org/10.1002/AENM.201803027.

21 Ghosh, S.K. and Mandal, D. (2017). Bio-assembled, piezoelectric prawn shell made self-powered wearable sensor for non-invasive physiological signal monitoring. *Appl. Phys. Lett.* 110 (12): 123701. https://doi.org/10.1063/1.4979081.

22 Fan, Z., Wang, L., and Qin, Y. (2023). Activity of chitin/chitosan/chitosan oligosaccharide against plant pathogenic nematodes and potential modes of application in agriculture: a review. *Carbohydr. Polym.* 306: 120592.

23 Lee, S., Hao, L.T., Park, J. et al. (2022). Nanochitin and nanochitosan: chitin nanostructure engineering with multiscale properties for biomedical and environmental applications. *Adv. Mater.* 35 (4): https://doi.org/10.1002/adma.202203325.

24 Wang, F., Xie, C., Tang, H. et al. (2023). Development, characterization and application of intelligent/active packaging of chitosan/chitin nanofibers films containing eggplant anthocyanins. *Food Hydrocoll.* 139: 108496.

25 Yu, Z., Ji, Y., Bourg, V. et al. (2020). Chitin- and cellulose-based sustainable barrier materials: a review. *Emergent Mater.* 3 (6): 919–936. https://doi.org/10.1007/S42247-020-00147-5.

26 Shamshina, J.L., Berton, P., and Rogers, R.D. (2019). Advances in functional chitin materials: a review. *ACS Sustain. Chem. Eng.* 7 (7): 6444–6457. https://doi.org/10.1021/ACSSUSCHEMENG.8B06372.

27 Xu, Q., Gao, X., Zhao, S. et al. (2021). Construction of bio-piezoelectric platforms: from structures and synthesis to applications. *Adv. Mater.* 33 (27): 2008452. https://doi.org/10.1002/ADMA.202008452.

28 Street, R.M., Huseynova, T., Xu, X. et al. (2018). Variable piezoelectricity of electrospun chitin. *Carbohydr. Polym.* 195: 218–224. https://doi.org/10.1016/J.CARBPOL.2018.04.086.

29 Richard-Lacroix, M. and Pellerin, C. (2013). Molecular orientation in electrospun fibers: from mats to single fibers. *Macromolecules* 46 (24): 9473–9493. https://doi.org/10.1021/MA401681M.

30 Kim, K., Ha, M., Choi, B. et al. (2018). Biodegradable, electro-active chitin nanofiber films for flexible piezoelectric transducers. *Nano Energy* 48: 275–283. https://doi.org/10.1016/J.NANOEN.2018.03.056.

31 Hoque, N.A., Thakur, P., Biswas, P. et al. (2018). Biowaste crab shell-extracted chitin nanofiber-based superior piezoelectric nanogenerator†. *J. Mater. Chem.* https://doi.org/10.1039/c8ta04074e.

32 Islam, S., Bhuiyan, M.A.R., and Islam, M.N. (2016). Chitin and chitosan: structure, properties and applications in biomedical engineering. *J. Polym. Environ.* 25 (3): 854–866. https://doi.org/10.1007/S10924-016-0865-5.

33 Riseh, R., Vatankhah, M., Hassanisaadi, M., and Kennedy, J.F. (2023). Chitosan-based nanocomposites as coatings and packaging materials for the postharvest improvement of agricultural product: a review. *Carbohydr. Polym.* 309: 120666.

34 Gal, M., Rahmaninia, M., and Polymers, M.H.-C. (2023). A comprehensive review of chitosan applications in paper science and technologies. *Carbohydr. Polym.* 309: 120665.

35 Chen, Y., Liu, Y., Dong, Q. et al. (2023). Application of functionalized chitosan in food: a review. *Int. J. Biol. Macromol.* 235: 123716. https://doi.org/10.1016/J.IJBIOMAC.2023.123716.

36 Morin-Crini, N., Lichtfouse, E., Torri, G., and Crini, G. (2019). Applications of chitosan in food, pharmaceuticals, medicine, cosmetics, agriculture, textiles, pulp and paper, biotechnology, and environmental chemistry. *Environ. Chem. Lett.* 17 (4): 1667–1692. https://doi.org/10.1007/S10311-019-00904-X.

37 Katiyar, D., Hemantaranjan, A., and Singh, B. (2015). Chitosan as a promising natural compound to enhance potential physiological responses in plant: a review. *Indian J. Plant Physiol.* 20 (1): https://doi.org/10.1007/s40502-015-0139-6.

38 Sonia, T.A. and Sharma, C.P. (2011). Chitosan and its derivatives for drug delivery perspective. *Adv. Polym. Sci.* 243 (1): 23–54. https://doi.org/10.1007/12_2011_117.

39 Gadkari, R.R., Garg, H., Chowdhury, A., and Ali, W. (2022). Chitosan-based bionanocomposites for food packaging applications. In: *Bionanocomposites for Food Packaging Applications*, 181–200. Elsevier https://doi.org/10.1016/B978-0-323-88528-7.00006-X.

40 De Marzo, G., Mastronardi, V.M., Algieri, L. et al. (2022). Sustainable, flexible, and biocompatible enhanced piezoelectric chitosan thin film for compliant piezosensors for human health. *Adv. Electron. Mater.* https://doi.org/10.1002/aelm.202200069.

41 Mohseni, M., Delavar, F., and Rezaei, H. (2021). The piezoelectric gel-fiber-particle substrate containing short PVDF-chitosan-gelatin nanofibers and mesoporous silica nanoparticles with enhanced antibacterial activity as a potential of wound dressing applications. *J. Macromol. Sci. Part A Pure Appl. Chem.* 58 (10): 694–708. https://doi.org/10.1080/10601325.2021.1927754.

42 Mohseni, M. and Ramazani Saadatabadi, A. (2021). Highly conductive self-electrical stimuli core-shell conduit based on PVDF-chitosan–gelatin filled with in-situ gellan gum as a possible candidate for nerve regeneration: a rheological, electrical, and structural study. *Appl. Nanosci.* 11 (8): 2199–2213. https://doi.org/10.1007/S13204-021-02012-1.

43 Karrat, A. and Amine, A. (2020). Recent advances in chitosan-based electrochemical sensors and biosensors. *Arabian J. Chem. Environ. Res.* 7: 66–93.

44 Nasution, T., Nainggolan, I., and Hutagalung, S.D. (2013). The sensing mechanism and detection of low concentration acetone using chitosan-based sensors. *Sensors Actuators B Chem.* 177: 522–528.

45 Annu and Raja, A.N. (2020). Recent development in chitosan-based electrochemical sensors and its sensing application. *Int. J. Biol. Macromol.* 164: 4231–4244.

46 Zamli, M.I., Ahmad, F.B., and Akmal, M.H. (2021). Extraction of microbial chitosan for piezoelectric application. *IOP Conf. Ser.: Mater. Sci. Eng.* https://doi.org/10.1088/1757-899X/1045/1/012037.

47 Singh, D., Choudhary, A., and Garg, A. (2018). Flexible and robust piezoelectric polymer nanocomposites based energy harvesters. *ACS Appl. Mater. Interfaces* 10 (3): 2793–2800. https://doi.org/10.1021/acsami.7b16973.

48 Zamli, M., Akmal, M., and Niazi, F. (2022). Piezoelectric performance of microbial chitosan thin film derived from *Aspergillus oryzae*. *Int. J. Renewable Energy Dev.* 11: 1157–1164.

49 Syauqi, N., Mohammed, S., Akmal, M. et al. (2022). Fabrication of biopolymer-based piezoelectric thin film from chitosan using solvent casting method. *J. Adv. Res. Appl. Sci. Eng. Technol.* 28: 80–89. https://doi.org/10.37934/araset.28.2.8089.

50 Du, Y., Du, W., Lin, D. et al. (2023). Recent progress on hydrogel-based piezoelectric devices for biomedical applications. *Micromachines* 14 (1): 167. https://doi.org/10.3390/MI14010167.

51 Chen, Y., Ye, M., Song, L. et al. (2020). Piezoelectric and photothermal dual functional film for enhanced dermal wound regeneration via upregulation of Hsp90 and HIF-1α. *Appl. Mater. Today* 20: 100756.

52 Ji, J., Yang, C., Shan, Y. et al. (2022). Research trends of piezoelectric nanomaterials in biomedical engineering. *Adv. Nanobiomed Res.* 3 (1): 2200088. https://doi.org/10.1002/anbr.202200088.

53 Bairagi, S., Banerjee, S., Banerjee, S., and Ali, S.W. (2023). Biomaterials toward flexible nanogenerators. *Adv. Appl. Biobased Mater.* 655–672. https://doi.org/10.1016/B978-0-323-91677-6.00021-0.

54 Song, Y., Zhao, Q., Qu, M. et al. (2022). Chitosan-based thermal insulation compressible foam enhanced with high performance of piezoelectric generation and sensing. *Carbohydr. Polym.* 294: 119775. https://doi.org/10.1016/J.CARBPOL.2022.119775.

55 Prokhorov, E., Luna-Bárcenas, G., Limón, J.M.Y. et al. (1991). Chitosan-ZnO nanocomposites assessed by dielectric, mechanical, and piezoelectric properties. *Polymers* 12 (9): https://doi.org/10.3390/POLYM12091991.

56 Hänninen, A., Sarlin, E., Lyyra, I. et al. (2018). Nanocellulose and chitosan based films as low cost, green piezoelectric materials. *Carbohydr. Polym.* 202: 418–424. https://doi.org/10.1016/J.CARBPOL.2018.09.001.

57 Praveen, E., Murugan, S., and Jayakumar, K. (2017). Investigations on the existence of piezoelectric property of a bio-polymer – chitosan and its application in vibration sensors. *RSC Adv.* 7 (56): 35490–35495. https://doi.org/10.1039/C7RA04752E.

58 Silva, C.C., Lima, C.G.A., Pinheiro, A.G. et al. (2001). On the piezoelectricity of collagen–chitosan films. *Phys. Chem. Chem. Phys.* 3 (18): 4154–4157. https://doi.org/10.1039/B100189M.

59 Hosseini, E.S., Manjakkal, L., Shakthivel, D., and Dahiya, R. (2020). Glycine-chitosan-based flexible biodegradable piezoelectric pressure sensor. *ACS Appl. Mater. Interfaces* 12 (8): 9008–9016. https://doi.org/10.1021/acsami.9b21052.

60 Elsabee, M.Z. (2015). Chitosan-based edible films. In: *Polysaccharides Bioactivity and Biotechnology*, 829–870. Springer https://doi.org/10.1007/978-3-319-16298-0_7.

61 Elsabee, M.Z. and Abdou, E.S. (2013). Chitosan based edible films and coatings: a review. *Mater. Sci. Eng. C* 1: 1819–1841. https://doi.org/10.1016/j.msec.2013.01.010.

62 Feng, Z.Q., Leach, M.K., Chu, X.H. et al. (2010). Electrospun chitosan nanofibers for hepatocyte culture. *J. Biomed. Nanotechnol.* 6 (6): 658–666. https://doi.org/10.1166/JBN.2010.1159.

63 Du, L., Li, T., Jin, F. et al. (2020). Design of high conductive and piezoelectric poly (3,4-ethylenedioxythiophene)/chitosan nanofibers for enhancing cellular electrical stimulation. *J. Colloid Interface Sci.* 559: 65–75. https://doi.org/10.1016/J.JCIS.2019.10.003.

64 Chauhan, S. and Thakur, A. (2023). Chitosan-based biosensors-a comprehensive review. *Mater. Today Proc.* https://doi.org/10.1016/j.matpr.2023.01.123.

65 Wu, S., Wu, S., Zhang, X. et al. (2023). Chitosan-based hydrogels for bioelectronic sensing: recent advances and applications in biomedicine and food safety. *Biosensors* 13: 93. https://doi.org/10.3390/bios13010093.

66 Taokaew, S., Kaewkong, W., and Gels, W.K. (2023). Recent development of functional chitosan-based hydrogels for pharmaceutical and biomedical applications. *Gels* https://doi.org/10.3390/gels9040277.

67 ElMessiery, M.A. (1981). Physical basis for piezoelectricity of bone matrix. *IEE Proc. A Phys. Sci. Meas. Instrument. Manag. Educ. Rev.* 128 (5): 336–346. https://doi.org/10.1049/IP-A-1.1981.0049.

68 Kapat, K., Shubhra, Q.T.H., Zhou, M., and Leeuwenburgh, S. (2020). Piezoelectric nano-biomaterials for biomedicine and tissue regeneration. *Adv. Funct. Mater.* 30 (44): https://doi.org/10.1002/ADFM.201909045.

69 Wang, R., Sui, J., and Wang, X. (2022). Natural piezoelectric biomaterials: a biocompatible and sustainable building block for biomedical devices. *ACS Nano* 16 (11): 17708–17728. https://doi.org/10.1021/ACSNANO.2C08164.

Index

Hybrid Materials for Piezoelectric Energy Harvesting and Conversion, First Edition.
Edited by S. Wazed Ali, Satyaranjan Bairagi, and Shahid Ul Islam.
© 2024 John Wiley & Sons, Inc. Published 2024 by John Wiley & Sons, Inc.

C

d

e

t

Printed and bound by CPI Group (UK) Ltd, Croydon, CR0 4YY
17/09/2026
14964256-0001